AF356150

Nano/Micro Biosensors for Biomedical Applications

Nano/Micro Biosensors for Biomedical Applications

Editors

Jin-Ho Lee
Ki Su Kim

Basel • Beijing • Wuhan • Barcelona • Belgrade • Novi Sad • Cluj • Manchester

Editors

Jin-Ho Lee
School of Biomedical
Convergence Engineering
Pusan National University
Yangsan
Republic of Korea

Ki Su Kim
School of Chemical
Engineering/Department of
Organic Materials Science
and Engineering
Pusan National University
Busan
Republic of Korea

Editorial Office
MDPI
St. Alban-Anlage 66
4052 Basel, Switzerland

This is a reprint of articles from the Special Issue published online in the open access journal *Biosensors* (ISSN 2079-6374) (available at: https://www.mdpi.com/si/biosensors/nano_micro_biosens).

For citation purposes, cite each article independently as indicated on the article page online and as indicated below:

Lastname, A.A.; Lastname, B.B. Article Title. *Journal Name* **Year**, *Volume Number*, Page Range.

ISBN 978-3-7258-0273-9 (Hbk)
ISBN 978-3-7258-0274-6 (PDF)
doi.org/10.3390/books978-3-7258-0274-6

Contents

About the Editors

Jin-Ho Lee

Dr. Jin-Ho Lee is an Associate professor at Pusan National University in the School of Biomedical Convergence Engineering. Dr. Lee's group has a primary interest in developing and integrating nanotechnologies for bio-inspired platforms for biomedical applications, including the development of biosensors and the modulation of stem cell differentiation. Through their efforts, the group has successfully developed innovative technology platforms, which effectively overcome substantial challenges in biosensing and realizing the complete therapeutic potential of stem cells.

Ki Su Kim

Dr. Ki Su Kim is an Associate Professor in the School of Chemical Engineering at PNU. He received his Ph.D. degree in 2012 from the Department of Materials Science and Engineering, POSTECH. In 2012, he moved to Harvard Medical School and worked as a research fellow in the Wellman Center for Photomedicine at Massachusetts General Hospital (2012-2015), and then as a postdoctoral research fellow in the Center for Nanomedicine, Brigham and Women's Hospital (2016-2017). His Smart NanoBiomaterials Laboratory conducts cutting-edge research that focuses on making breakthroughs in the development of nano-enabled smart biomaterials and technologies for various biomedical applications, such as photomedicine, drug delivery, tissue engineering, bioimaging, and medical devices.

Editorial

Nano/Micro Biosensors for Biomedical Applications

Jin-Ho Lee [1,2,3,*] and Ki Su Kim [4,5,6,*]

1 School of Biomedical Convergence Engineering, Pusan National University, Yangsan 50612, Republic of Korea
2 Department of Information Convergence Engineering, Pusan National University, Yangsan 50612, Republic of Korea
3 Department of Convergence Medical Sciences, School of Medicine, Pusan National University, Yangsan 50612, Republic of Korea
4 School of Chemical Engineering, Pusan National University, Busan 46241, Republic of Korea
5 Department of Organic Material Science & Engineering, Pusan National University, Busan 46241, Republic of Korea
6 Institute of Advanced Organic Materials, Pusan National University, Busan 46241, Republic of Korea
* Correspondence: leejh@pusan.ac.kr (J.-H.L.); kisukim@pusan.ac.kr (K.S.K.)

Citation: Lee, J.-H.; Kim, K.S. Nano/Micro Biosensors for Biomedical Applications. *Biosensors* **2024**, *14*, 58. https://doi.org/10.3390/bios14020058

Received: 11 January 2024
Accepted: 17 January 2024
Published: 23 January 2024

Introduction

Advances in nano/micro technologies in recent years have significantly improved biosensors in terms of their viability for biomedical purposes, from diagnostic to therapeutic applications, allowing for effective early detection and personalized treatment modalities. Specifically, the introduction of a variety of nano/micro technologies has offered new opportunities to improve the sensitivity, selectivity, response time, and biocompatibility of biosensors through outstanding physical, chemical, electrical, and electrochemical properties. This Special Issue aims to highlight the most recent and promising nano/micro technologies utilized in the development of biosensors for biomedical applications. We thus collated 10 original research papers and review articles aligned with these themes, to lead the charge towards new approaches to and solutions for a next-generation biosensor for biomedical applications.

Anara Molkenova et al. (Contribution 1) presents the research trend of plasmon modulated upconversion biosensors. Luminescent behavior UCNPs have been widely utilized for background-free biorecognition and biosensing. Currently, a paramount challenge exists in how to maximize NIR light harvesting and upconversion efficiencies for achieving a faster response and better sensitivity without damaging the biological tissue upon laser-assisted photoactivation. In this article, they offer the reader an overview of the recent updates, exciting achievements, and challenges in the development of plasmon-modulated upconversion nanoformulations for biosensing applications.

Hanbin Park et al. (Contribution 2) reports on the recent advancements in nanomaterial-based microcystin (MC) biosensors. The study addresses the environmental issue of eutrophication in lakes and rivers, where harmful toxins are produced by cyanobacterial algae. The consumption of these contaminants from the water, particularly microcystins, poses serious health risks, increasing the risk, for example, of liver failure and hepatocirrhosis. Recognizing the importance of precise MC detection in water samples, this review highlights the promising developments in nanomaterial-based MC biosensors, emphasizing their potential to overcome the limitations of traditional detection methods.

Hye Kyu Choi et al. (Contribution 3) presents an enzymatic electrochemical/fluorescent nanobiosensor for the detection of small chemicals. As techniques used to implement the sensing function of such enzymatic biosensors, electrochemical and fluorescence techniques have mostly been used for the detection of small molecules because of their advantages. In addition, through the incorporation of nanotechnologies, the detection property of each technique-based enzymatic nanobiosensor can be improved to measure harmful or important small molecules accurately. This review provides interdisciplinary information related to developing enzymatic nanobiosensors for small molecule detection, such as widely used enzymes, target small molecules, and electrochemical/fluorescence techniques.

Jin-Ha Choi (Contribution 4) presents the utilization of functional nanomaterials in the development of proteolytic biosensors. A proteolytic enzyme can be selectively quantified by changing the detectable signals causing the degradation of the peptide chain. In addition, by combining polypeptides with various functional nanomaterials, proteolytic enzymes can be measured more sensitively and rapidly. In this paper, proteolytic enzymes that can be measured using a polypeptide degradation method are reviewed, and recently studied functional nanomaterials-based proteolytic biosensors are discussed.

Rowoon Park et al. (Contribution 5) emphasizes the potential use of molecularly imprinted polymer (MIP) in the development of molecularly imprinted polymer (MIP)-based in vitro diagnostic medical devices for point-of-care testing (POCT). The article highlights the potential impact of an artificial bioreceptor, molecularly imprinted polymer (MIP), in the development of POCT devices. Taking advantage of their physico-chemical stability, MIP could improve their analytical performance in physiological conditions along with their stability. As such, the article discusses new point-of-care testing (POCT) devices designed to detect biomarkers in clinical biofluids like sweat, tears, saliva, or urine.

Yoo-Kyum Shin et al. (Contribution 6) focus on the micro/nano-structured biodegradable pressure sensors designed for biomedical applications. The growing interest in these sensors arises from their temporal nature in wearable and implantable devices while eliminating biocompatibility concerns. Recent advancements in micro/nano-technologies, including device structures and materials, have significantly enhanced the performance and functionality of biodegradable pressure sensors. The review emphasizes the improved device-level capabilities achieved through adjustments in geometrical design parameters at the micro and nanometer scale. The discussion covers material choices and sensing mechanisms, historical developments in biodegradable pressure sensors, fabrication methods, device performance, and biocompatibility.

Kwang-Ho Lee et al. (Contribution 7) reports the recent advances in multicellular tumor spheroid generation and its applications in drug screening. Multicellular tumor spheroids (MCTs) have been employed in biomedical fields owing to their advantage in designing a three-dimensional (3D) solid tumor model. In drug cytotoxicity assessments, MCTs provide better mimicry of conventional solid tumors that can precisely represent anticancer drug candidates' effects. To generate and incubate multicellular spheroids, researchers have developed several 3D multicellular spheroid culture technologies to establish a research background and a platform using tumor modeling advanced materials science and biosensing techniques for drug screening. In application, drug screening is performed in both an invasive and non-invasive manner, according to their impact on the spheroids.

Chuntae Kim et al. (Contribution 8) reports the recent trends in exhaled breath diagnosis using an artificial olfactory system. Artificial olfactory systems are needed in various fields that require real-time monitoring, such as healthcare. This review introduces cases of detection of specific volatile organic compounds (VOCs) in a patient's exhaled breath and discusses trends in disease diagnosis technology development using artificial olfactory technology that analyzes exhaled human breath. By regularly monitoring health status, diseases can be prevented or treated at an early stage, thus increasing the human survival rate and reducing overall treatment costs. This review introduces several promising technical fields to develop technologies that can monitor health conditions and diagnose diseases early by analyzing exhaled human breath in real time.

Svetlana Mescheryakova et al. (Contribution 9) presents an interesting study that developed a fluorescent-based nanosensor for the detection of doxorubicin by utilizing alloyed semiconductor QDs. Owing to their photostability, uniformity, and colloidal stability, CdZnSeS/ZnS core/shell nanocrystals (QD) were successfully employed to develop a nanosensor. The analytical approach exploits changes in emission intensity through the quenching effect, while an interaction occurs between the nanosensors and modulating molecules. To maximize the quenching efficiency, QDs, alloyed CdZnSeS/ZnS core/shell nanocrystals, were carefully modified with thioglycolic acid (TGA) and 3-mercaptopropionic acid (MPA). The developed fluorescence-based turn-off nanosensors were able to determine DOX from undiluted human plasma. The calculated limit of detection values was 0.08 and 0.03 ug/mL, while QDs were stabilized with thioglycolic and 3-mercaptopropionic acids, respectively. As such, recognizing the critical importance and challenges of clinical sample detection, and continuously improving and developing cutting-edge technologies in clinical sample analysis, will help ensure accurate diagnostics for clinical and medical approaches.

Samantha Nunez et al. (Contribution 10) introduces a fluorescent inducible system that can quantify the free fucose in Escherichia coli. As fucose metabolism generates short-chain fatty acids and plays a key role in facilitating cross-feeding microbial interactions, the bacterial consumption of fucose plays a pivotal role in the assembly of the gut microbiome in infants. In this regard, they designed a molecular quantification method for free fucose using fluorescent Escherichia coli. They assessed low- and high-copy plasmids with and without the transcription factor fucR, along with the corresponding fucose-inducible promoter controlling the reporter gene sfGFP. The proposed system was successfully validated through the supernatant of Bifidobacterium bifidum JCM 1254 supplemented with 2-fucosyl lactose.

Overall, though this Special Issue contains several research papers and review articles that highlight different aspects of biosensors and their applications in various fields, we believe each article provides valuable insights into the utilization of nano/micro technologies in the development of biosensors for biomedical applications.

Funding: This work was implemented within the National Research Foundation of Korea (NRF) grant supported by the Ministry of Science and ICT (No. 2020R1C1C1011876, 2021M3H4A4079509, 2022R1C1C1009533 and 2022R1A5A2027161).

Acknowledgments: The authors are grateful for the valuable opportunity to serve as Guest Editors for this Special Issue, "Nano/Micro Biosensor for Biomedical Applications". We are thankful to all authors for their contributions. We also thank the editorial and publishing staff at *Biosensors* for their support.

Conflicts of Interest: The authors declare no conflicts of interest.

List of Contributions:

1. Molkenova, A.; Choi, H.E.; Park, J.M.; Lee, J.H.; Kim, K.S. Plasmon Modulated Upconversion Biosensors. *Biosensors* **2023**, *13*, 306.
2. Park, H.; Kim, G.; Seo, Y.; Yoon, Y.; Min, J.; Park, C.; Lee, T. Improving Biosensors by the Use of Different Nanomaterials: Case Study with Microcystins as Target Analytes. *Biosensors* **2021**, *11*, 525.
3. Choi, H.K.; Yoon, J. Enzymatic Electrochemical/Fluorescent Nanobiosensor for Detection of Small Chemicals. *Biosensors* **2023**, *13*, 492.
4. Choi, J.H. Proteolytic Biosensors with Functional Nanomaterials: Current Approaches and Future Challenges. *Biosensors* **2023**, *13*, 171.
5. Park, R.; Jeon, S.; Jeong, J.; Park, S.Y.; Han, D.W.; Hong, S.W. Recent Advances of Point-of-Care Devices Integrated with Molecularly Imprinted Polymers-Based Biosensors: From Biomolecule Sensing Design to Intraoral Fluid Testing. *Biosensors* **2022**, *12*, 136.
6. Shin, Y.K.; Shin, Y.; Lee, J.W.; Seo, M.H. Micro-/Nano-Structured Biodegradable Pressure Sensors for Biomedical Applications. *Biosensors* **2022**, *12*, 952.
7. Lee, K.H.; Kim, T.H. Recent Advances in Multicellular Tumor Spheroid Generation for Drug Screening. *Biosensors* **2021**, *11*, 445.
8. Kim, C.; Raja, I.S.; Lee, J.M.; Lee, J.H.; Kang, M.S.; Lee, S.H.; Oh, J.W.; Han, D.W. Recent Trends in Exhaled Breath Diagnosis Using an Artificial Olfactory System. *Biosensors* **2021**, *11*, 337.

9. Mescheryakova, S.A.; Matlakhov, I.S.; Strokin, P.D.; Drozd, D.D.; Goryacheva, I.Y.; Goryacheva, O.A. Fluorescent Alloyed CdZnSeS/ZnS Nanosensor for Doxorubicin Detection. *Biosensors* **2023**, *13*, 596.

10. Nunez, S.; Barra, M.; Garrido, D. Developing a Fluorescent Inducible System for Free Fucose Quantification in Escherichia coli. *Biosensors* **2023**, *13*, 388.

biosensors

Review

Plasmon Modulated Upconversion Biosensors

Anara Molkenova [1], Hye Eun Choi [2], Jeong Min Park [2], Jin-Ho Lee [3,*] and Ki Su Kim [1,2,4,*]

[1] Institute of Advanced Organic Materials, Pusan National University, 2 Busandaehak-ro 63 beon-gil, Geumjeong-gu, Busan 46241, Republic of Korea
[2] School of Chemical Engineering, College of Engineering, Pusan National University, 2 Busandaehak-ro 63 beon-gil, Geumjeong-gu, Busan 46241, Republic of Korea
[3] School of Biomedical Convergence Engineering, Pusan National University, 49 Busandaehak-ro, Yangsan 50612, Republic of Korea
[4] Department of Organic Material Science & Engineering, Pusan National University, 2 Busandaehak-ro 63 beon-gil, Geumjeong-gu, Busan 46241, Republic of Korea
* Correspondence: leejh@pusan.ac.kr (J.-H.L.); kisukim@pusan.ac.kr (K.S.K.)

Abstract: Over the past two decades, lanthanide-based upconversion nanoparticles (UCNPs) have been fascinating scientists due to their ability to offer unprecedented prospects to upconvert tissue-penetrating near-infrared light into color-tailorable optical illumination inside biological matter. In particular, luminescent behavior UCNPs have been widely utilized for background-free biorecognition and biosensing. Currently, a paramount challenge exists on how to maximize NIR light harvesting and upconversion efficiencies for achieving faster response and better sensitivity without damaging the biological tissue upon laser assisted photoactivation. In this review, we offer the reader an overview of the recent updates about exciting achievements and challenges in the development of plasmon-modulated upconversion nanoformulations for biosensing application.

Keywords: upconversion nanoparticles (UCNPs); biosensing; plasmon-enhanced upconversion; plasmon modulated upconversion; upconversion quenching; plasmonic nanoparticles (PNPs); gold nanoparticles (GNPs); surface plasmon resonance (SPR); fluorescence resonance energy transfer (FRET)

Citation: Molkenova, A.; Choi, H.E.; Park, J.M.; Lee, J.-H.; Kim, K.S. Plasmon Modulated Upconversion Biosensors. *Biosensors* **2023**, *13*, 306. https://doi.org/10.3390/bios13030306

Received: 26 January 2023
Revised: 17 February 2023
Accepted: 21 February 2023
Published: 22 February 2023

1. Introduction

Lanthanide-based upconversion nanoparticles (UCNPs) have become the engine of numerous ground-breaking inventions in a wide variety of research areas. By disobeying the Stokes Law, UCNPs are capable of producing higher energy output photons out of multiple (>2) lower energy input photons. For example, one can obtain ultraviolet (UV) or visible emission by exposing UCNPs to near infrared (NIR) laser. Contrary to other light up-converting analogues, such as organic luminophores and quantum dots, lanthanide-based UCNPs indeed possess overwhelming advantages, which include a diversity of emission colors, long lifetime luminescence, large anti-Stokes spectral shifts, weak background autofluorescence, narrow emission bands, nonphotobleaching nature, blinking-free continuous emission capability, and relatively low toxicity [1–3]. So far, UCNPs with tailored multicolor emissions have underpinned a vast array of applications in energy conversion photovoltaics [4], fingerprint detection and anticounterfeiting barcoding [5,6], biosensing [7], super resolution nanoscopy [8], photodetectors [9], drug and gene delivery [10], photodynamic therapy [11], light-triggered on–off tattoo systems [12], photochemical tissue bonding [13] and so on.

In particular, UCNPs have garnered increased interest in the quantitative detection of various biologically relevant targets, such as biomolecules, pH, ions, viruses, bacteria, reactive oxygen species and temperature [14]. However, the development of upconversion biosensing has been seriously hampered by poor NIR harvesting ability and a long-standing issue of quenching, which has multiple origin sources causing a great deal for scientists in finding the straightforward solution. At the edge of the UCNP surface, the excited state electrons undergo parasitic interaction with the surface defects (e.g., stacking faults, vacancies, dislocation, dangling bonds) which deplete the excitation energy leading to a partial decrease or complete "turn off" of the UCNP luminescence.

Composition tailoring strategies by architecting the core with single or multiple shell structures are at the forefront to combat quenching issue. However, luminescence producing dopants are still kept at low doping levels to support quenching stability at the expense of upconversion brightness. In fact, quenching issue and composition limits can be addressed by extremely high NIR laser excitation powers ($\sim 10^6$ W/cm^2 [15,16]), of which the main concern is associated with harmfulness and inapplicability for biomedical applications [17]. NIR laser power densities should be compatible with the tolerances of the skin tissue and kept below the maximum permissible exposure (MPE) values calculated according to ANSI Z136.1-2007 American National Standard for the Safe Use of Lasers [18]. The calculated MPE values for the corresponding wavelength of the NIR lasers are presented in Table 1.

Table 1. Calculated safe power densities of common NIR lasers [19] (MPE—maximum permissible exposure, NIR—near-infrared).

NIR laser, λ	808 nm	915 nm	980 nm	1064 nm
MPE value	~0.329 W/cm^2	~0.538 W/cm^2	~0.726 W/cm^2	~1.0 W/cm^2

The Förster (or fluorescence) resonance energy transfer (FRET) process has vast implications in quantitative and sensitive biosensing applications. In principle, an efficient FRET scenario occurs under specified conditions. First, there should a large degree of overlap in the emission-excitation profiles of the donor (excited state) and acceptor (ground state). Second, the long-range nonradiative dipole-dipole interaction between the donor-acceptor pair should be within the effective range of 10 nm [20]. The FRET process can be utilized for tuning optical properties of UCNPs by sizeable enhancement or depletion of their excitation or emission energy. Therefore, the FRET-based upconversion mechanism has been extensively exploited for bioanalytical sensors and assays, which enlivened modern bioimaging and biosensor research [21–23]. To date, extensive efforts have been focused on developing various energy acceptors, such as plasmonic nanostructures and organic dyes, which can aid in modulation photoluminescence efficiency of energy donating UCNPs via the FRET process. Compared to organic dyes which are prone to photodegradation, plasmon-enhanced upconversion has been considered as a breakthrough engineering strategy to minimize quenching issue, maximize upconversion efficiency and stimulate NIR harvesting offering unprecedented biosensing precision. In particular, the plasmon resonances could serve to achieve control over the luminescence properties of UCNPs by intentional amplification or quenching of the specific emission band [24], which could be beneficial for designing sensing platforms with specificity, low detection limits and linear response to the presence of the target bioanalyte. A recent literature survey has shown that several reviews related to the biosensing applications of UCNPs were published recently [7,14,25,26]. On the other hand, a focused overview of plasmon-modulated upconversion biosensing is still not available in the literature. Therefore, the objective of this review is to offer the reader a brief overview of the up-to-date strategies to incorporate plasmonic nanoparticles to lanthanide UCNPs for potential biosensing applications.

2. Characteristics of UCNPs

2.1. Lanthanide Upconversion Basics: Mechanism and Origin

The upconversion is a nonlinear optical effect which gained widespread scientific interest with the advent of lasers. It is generally acknowledged that the concept of upconversion was proposed in the 1950s by Bloembergen and further developed by Auzel in the 1960s who introduced groundbreaking theoretical studies on the energy transfer upconversion process [27]. Since the 1990s, the emergence of nanoscience has brought a huge diversification of nanofabrication technologies leading to efficient downscaling upconversion phosphors to nanometer regime making them tremendously attractive across a range of biomedical applications [28].

Even now, the field of upconversion keeps evolving. Over time, it is believed to practically benefit health care and different industrial sectors. Ongoing research in the upconversion field continues to uncover and explain the complicated mechanisms behind photon upconversion. Meanwhile, there are several known mechanisms responsible for upconversion origins, such as excited state absorption, energy transfer upconversion, cooperative upconversion, photon avalanche, and energy migration upconversion [2]. This review deals with energy transfer based upconversion, therefore another deep-in depth description of upconversion mechanisms is beyond the scope of this review.

Upconversion in lanthanides originates from their ladder-like arrangement of energy levels which impart effective absorption and subsequent retaining of incident photons from external laser until a sufficient quantity is reached to proceed with upconverted emission. In principle, the upconversion process relies on electronic transitions within partially filled 4f orbitals. Notably, 4f shells in lanthanides are well protected by 5s and 5p shells (secondary electrons) rendering significant advantages, such as high resistance to (environmental influences) photobleaching, photochemical stability and sharp emission bands from 4f-4f transitions. However, the electron shield impedes 4f transitions and imposes limitations on light harvesting properties leading to poor absorption cross sections of lanthanides [29]. For a single and isolated lanthanide ion, these transitions are forbidden by selection rules of quantum mechanics, but this situation drastically changes when lanthanides are embedded as dopants inside the host matrix enhancing the probability of 4f-4f transitions. Selection rules on the spin are relaxed by a larger spin-orbit coupling. In particular, doping distorts site symmetry, which forms a stronger crystal field around lanthanide ions. Such absence of inversion symmetry (noncentrosymmetric system) distorts the electron cloud leading to intermixing with f states. As a result, Coulombic interactions and spin-orbit coupling of 4f subshell electrons lead to abundant energy sublevels and a large number of emitting levels [30].

2.2. Lanthanide Upconversion Composition

The most common configuration of lanthanide based UCNPs is composed of lanthanide ion dopants, such as sensitizers, emissive activators, and sometimes energy migration ions, that reside in the crystal host lattice. The concentration and distance between the sensitizer and emitter ions are critical parameters for tailoring the optical properties of UCNPs. In particular, higher concentration and greater proximity are essential to increase luminosity, but also invoke non-radiative cross-relaxations, known as the "concentration quenching effect". Presently, as the main strategy for overcoming the concentration quenching hurdle, the doping levels of sensitizer ($\sim$20%) and emitter ions (<2%) are kept at low percentages to optimize the luminescence brightness [31].

Host and dopants interaction occur in the form of energy exchange. Thus, some energy is absorbed by the host through vibrational coupling, which explains the choice of host lattices with the lowest phonon energy. Decades of research in the upconversion field have spawned an extensive array of host nanomaterials, such as various metal oxides, oxysulfides, vanadates, halides, phosphates and so on [32]. Among them, the most popular host is hexagonal phase β-NaYF$_4$, which meets the requirements for ideal host materials, such as the lowest phonon frequency, high optical transparency, less hygroscopic and brighter emission than lanthanide fluoride (LnF$_3$) host [33–35]. Sodium ion (Na$^+$) has been widely introduced into the lanthanide fluoride lattice because it has nearly the same ionic radius as lanthanide ions [36]. For instance, Ayadi et al. reported that introducing Na$^+$ ions into GdF$_3$ led to lattice expansion and imparted stabilization of its hexagonal crystal structure [37]. For simplicity, this review specifically concentrates on plasmon modulated upconversion systems constructed using metal lanthanide fluoride hosts.

Activator ions, as their name suggests, receive the excitation energy from the sensitizer or energy migration ion to activate emission. The choice of activator ion plays a decisive role in finely tuning emission profiles of UCNPs, for instance, erbium (Er^{3+}) ion is recruited to produce red/green emissions, while thulium (Tm^{3+}) is broadly exploited for obtaining UV/blue emissions from UCNPs (Figure 1A) [5]. Figure 1A exemplifies a typical transmission electron micrograph of $NaYF_4$:Yb,Tm UCNPs. Digital inset illustrates visible to naked-eye blue emission of $NaYF_4$:Yb,Tm UCNPs under NIR laser illumination. Popular activators, such as erbium (Er^{3+}), thulium (Tm^{3+}) and holmium (Ho^{3+}) ions, possess narrow energy gap of 2000 cm^{-1}. Thereby, they are more efficient for the energy transfer process compared to other luminescent ions with a wide energy gap of 7000 cm^{-1}, e.g., terbium (Tb^{3+}), europium (Eu^{3+}) and dysprosium (Dy^{3+}) which are prone to detrimental nonradiative energy losses. Activators, in general, have a complicated electronic structure and extremely small absorption cross-sections of ~10^{-21} cm^2 responsible for a low quantum yield (<1%) and necessary in photoactivation using high laser power densities [38].

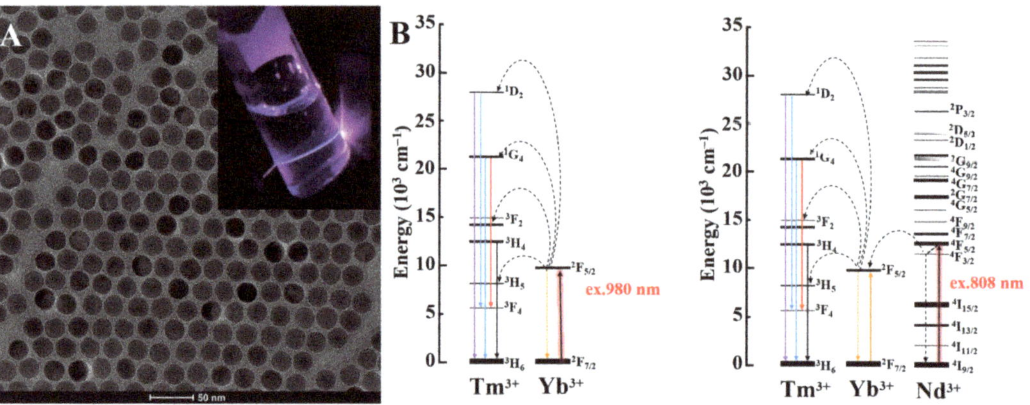

Figure 1. (**A**) Representative TEM image of $NaYF_4$:Yb,Tm UCNPs (scale 50 nm) with digital inset of their UC emission under NIR laser. (**B**) Upconversion process of Tm^{3+} activator under 980 nm and 808 nm excitation.

Sensitizers serve to accumulate incoming NIR photons and further transfer these photons to the activator ions directly or indirectly through energy migration ions. For example, ytterbium (Yb) is utilized as a ~980 nm NIR sensitizer with a single transition $^2F_{7/2}$–$^2F_{5/2}$ levels, which showcases efficient energy transfer when paired with Tm^{3+} activator. Moreover, co-doping Yb ion with neodymium ion (Nd^{3+}, <1 mol.%) allows tailoring 808 nm NIR harvesting upconversion system properties in UCNPs, where Yb^{3+} act as a migration ion facilitating the energy extraction from Nd^{3+} ions and its further transfer to activator ions [39]. Figure 1B displays the schematic energy diagrams of Tm^{3+} ion upconverted emission photoactivation under 980 nm and 808 nm NIR stimulation. Table 2 shows common lanthanide sensitizers and their corresponding absorption cross sections.

Table 2. Common lanthanide sensitizers and their absorption cross-sections [40–42].

Sensitizer	Absorption Wavelength, λ	Absorption Cross-Section, cm^2
Ytterbium (Yb^{3+})	980 nm	~10^{-20}
	740 nm	
Neodymium (Nd^{3+})	800 nm	~10^{-19}
	860 nm	
Erbium (Er^{3+})	1500 nm	1.1×10^{-20} cm^2

Importantly, 808 nm NIR harvesting UCNPs have arisen as a feasible platform for biomedical application to overcome the safety shortcomings of those excitable under conventional 980 nm lasers, since water molecules absorption of 808 nm photons is 20 times weaker than 980 nm ones [31].

Over the past few decades, explosive growth in the development of nanotechnology offered numerous synthetic protocols and optimized designs of UCNPs. By now, several routes to synthesize UCNPs have been proposed and widely utilized, such as co-precipitation [28], thermal decomposition [43], hydrothermal method [44], microwave-assisted synthesis [45], microemulsion method [46] and the liquid–solid solution (LSS) process [47]. Among these synthesis methods, co-precipitation and thermal decomposition methods have found remarkably broad utility in the scientific community for the fabrication of highly monodisperse and bright upconversion nanocrystals. However, excitement over developments in the controlled synthesis of small UCNPs is moderated by luminescence intensity decrease as the size shrinks the quenching issue becomes more prominent. Especially, the enhanced surface-to-volume ratio leads to greater exposure of the UCNPs surface to external quenchers, such as OH impurities, and organic ligands C-H and C-C bonds with a high vibrational energy that present in the dispersion solvent or capping ligands [48,49]. The main strategy so far to remove surface quenching relies on adopting surface passivation with single or multiple outermost shell layers, which mitigate surface defects and preclude the interaction with the external quenchers. However, the core-shell structure fabrication process usually requires prolonged and multi-step synthetic procedures. Therefore, researchers in the upconversion field are faced with the need to accelerate (e.g., by continuous process [50]) or even automate the synthesis process [51].

3. Principles and Applications of Photon Modulation in Upconversion-Based Biosensors

Surface plasmon resonance (SPR) is a light-matter interaction phenomenon inherent to materials with a negative real and small positive imaginary dielectric constant. When incident light interacts with the electron cloud of these materials, electrons start collectively oscillating inducing resonant effect (Figure 2A). Accordingly, surface plasmons could be localized or propagating. In the case of the localized surface plasmons, the excitation wavelength is larger than the nanoparticle causing its electron oscillation in a localized manner. Usually, the evanescent field forms the propagating surface plasmon (also known as surface plasmon polariton), in particular, when the incident light is polarized leading to surface charge oscillation along the metal/dielectric interface in the longitudinal direction and subsequently decay [52].

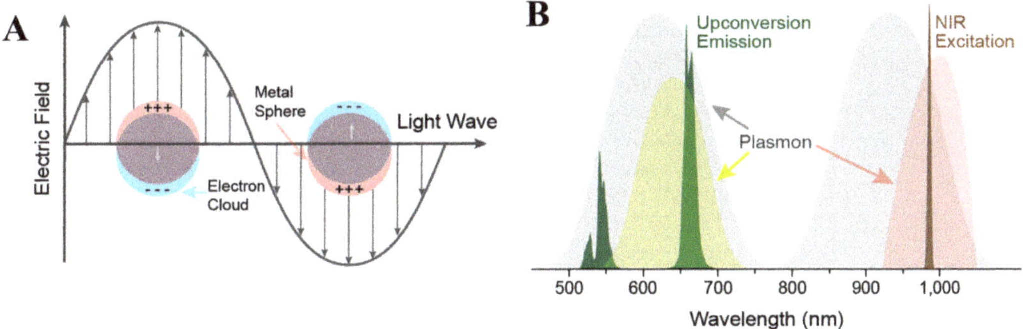

Figure 2. (**A**) Schematics of a metal nanoparticle's electron cloud oscillation. (**B**) Spectral overlap between metal nanoparticle's SPR and upconversion nanoparticles emission or excitation profile. Reprinted with permission from reference [53].

3.1. A Brief Theoretical Principles of Plasmon Modulated Upconversion

SPR extension in UCNPs using plasmonic nanoparticles (PNPs) can serve to augment the functionality of upconversion sensing platforms by resonating its excitation or emission (Figure 2B) [53]. For example, upon plasmon-light coupling energy confinement leads to the amplification of the electromagnetic field benefiting NIR light harvesting abilities and radiative decay rates of UCNPs. Theoretically, plasmonic modulation can afford an enhancement of the upconversion emission intensity E^{2n} time without energy transfer alteration. So far, 100-fold enhancements were verified experimentally [54].

Thus, PNPs are exploited as optical nanoantennae that could also provide modulation over the radiation features of UCNPs through excitation energy redistribution leading to luminescence enhancement (Purcell effect) or intentional quenching [24,55].

The design of plasmon-mode upconversion nanoprobes is commonly based on the matched SPR band of plasmonic nanoparticles and emission/excitation profiles of UCNPs. For example, the aspect ratio of gold nanorods governs the SPR band location and hence could be easily adjusted to modulate upconversion process [56]. Noteworthy, the proximity and isolation between UCNPs and PNPs are critically important for efficient energy transfer and sustaining undesired quenching [57]. For example, the use of a silica shell as a spacer between UCNPs and PNPs enables to prevent of their direct contact and unintentional quenching [58]. Meanwhile, the plasmon-upconversion interaction considerably depends on the geometrical morphology of PNPs. For instance, a variety of plasmonic nanostructures, such as nanoshells, nanofilms, nanowires, nanorods and nanoparticle arrays/cavity, were coupled to UCNPs to achieve plasmon modulated upconversion [59].

Gonzalez et al. investigated the upconversion quenching ability of different sizes of Au NPs ranging from 3.9 nm to 66 nm. Their findings suggest that size range Au NPs between 15 and 20 nm is optimal to initiate intentional quenching in silica coated UCNPs. Conversely, larger Au NPs with the size >50 nm tend to amplify the upconversion emission. In addition, their study also elucidated that an optimal shell thickness for upconversion enhancement is ca. 12 nm, while efficient quenching demands the thinnest possible shell [60].

By exploiting plasmon modulated upconversion strategy, a range of nanoprobes was constructed and proposed for examining a broad variety of analytes in biological environments. We analyzed recent literature published on the practical implications of plasmonic modulation of upconversion biosensors to detect diverse analytes and compiled our review in Table 3, which serves to provide insights into key parameters for designing plasmon modulated upconversion nanoprobes, such as (1) the size and corresponding absorption (or SPR) band of PNPs; (2) the physicochemical and optical characteristic of UCNPs, such as the size, surface coating (spacer), NIR activation laser wavelength, emission profiles; (3) target analytes and achieved the limit of detection (LOD). In the following sections, we will highlight trending topics through discussion of some representative nanoprobes that are expected to provide a brief overview of the current research status in the plasmon-modulated upconversion biosensing field.

Table 3. Photon-modulated upconversion nanoparticles (PNPs—plasmonic nanoparticles, — upconversion nanoparticles, λ abs.—emission wavelength, NIR—near-infrared, λ em.—emission wavelength, LOD—limit of detection).

PNPs	Size	λ abs.	UCNPs	Coating	Size	NIR Laser	λ em.	Analyte	LOD	Ref.
Au NPs	13 nm	522 nm	$NaYF_4$:Yb20%,Er2%	lysine	70 nm	980 nm	540 nm	Cr^{3+}	0.8 nM	[61]
Au NRs Au NBs	9–10 nm 15 nm	515/692 nm 525 nm	$NaYF_4$:Yb20%,Ho2%, Mn	PAA	20–30 nm	980 nm	542 nm, 660 nm	Pb^{2+}, Hg^{2+}	50 pM 150 pM	[64]
Au NPs	20 nm	528 nm	$NaYF_4$:Yb20%,Er5%	CTAB	20 nm	980 nm	-	Cd^{2+} Ache	0.2 μM 0.015 U/mL	[62]
Au NPs	15 nm	521 nm	$NaYF_4$:Yb20%,Er2% @ $NaYF_4$	SiO_2 (2 nm)	37 nm $\times$ 28 nm	980 nm	522/545/654 nm	Cd^{2+} GSH	0.059 μM 0.016 μM	[63]
Ag NPs	12 nm	398 nm	NaY/GdF_4:30%Yb20%,Er2%	NH_2	32 nm	980 nm	545/660 nm	Cr^{3+}	34 nM	[65]
Au NPs	1.7 nm	535 nm	$NaYF_4$:Yb20%,Er2%@ $NaYF_4$:Yb20%	PEI (8 nm)	43 nm	980 nm	655 nm	CN^-	1.53 μM	[66]
Au NRs	80 nm $\times$ 25 nm	-	$NaYF_4$@$NaYF_4$:Yb20%,Er2% @ $NaYF_4$	H1	10 nm	980 nm	543 nm	microRNA	0.036 fM	[67]
Au NRs	~45 nm	520 nm	$NaYF_4$:Yb20%,Tm2%	PAMAM (~2.5 nm)	~90 nm	980 nm	450/470/805 nm	uric acid	1 pM	[68]

Table 3. *Cont.*

PNPs	Size	λ abs.	UCNPs	Coating	Size	NIR Laser	λ em.	Analyte	LOD	Ref.
Ag NCs	1.9 nm	500/620 nm	$NaYF_4$:Yb20%,Tm2%	PEI	30 nm	980 nm	480 nm	biothiols	-	[69]
Au NRs	-	-	$NaYF_4$:Yb20%,Tm2%	PEI	27.7 nm	980 nm	656 nm	DNA methylation	7 pM	[56]
Au arrays	-	-	$NaYF_4$:Yb25%,Tm0.3%	PAA	26 nm	980 nm	345/450/475/800 nm	Vitamin B12	3.0 nM	[70]
Au NPs	13 nm	521 nm	$NaYF_4$:Yb20%,Ho2%	SiO_2 (12 nm)	115 nm	980 nm	483/543/640 nm	ABA aptamer	3.2 nM	[71]
Au NPs	~50 nm	-	$NaYF_4$:Yb20%,Er2%	PAA	~43 nm	980 nm	-	Aflatoxin B1	0.17 ng/mL	[72]
Au NPs	-	544 nm	$NaYF_4$:Yb27%,Tm0.5%	SiO_2/PSA	20 nm	975 nm	-	ssDNA	1 pM	[73]
Au NRs	27 nm × 54 nm	630 nm	$NaYF_4$:Yb,Er	PEI	25 nm	980 nm	545/660 nm	Exosome	1.1×10^3 part./μL	[74]
Au NPs	5 nm	~543 nm	$NaYF_4$:Yb20%,Er2%@$NaYF_4$	LDNA	21 nm	980 nm	543 nm	miR-21	0.54 fM	[75]
Au NPs	~50 nm	~530 nm	$NaYF_4$:Yb20%,Er2%	PSA	~42 nm	980 nm	~550 nm	antibodies (Ab1)	2.3 pM	[76]
Au NPs	30 nm	520 nm	$NaYF_4$:Yb20%,Er2%	Con-A	40–55 nm	980 nm	545/675 nm	glucose	0.02 μM	[77]
Ag NPs	7.8 nm	434 nm	$NaYF_4$:Yb30%,Tm0.5%@$NaYF_4$	bared	24 nm	980 nm	345/360/450/474 nm	glucose H_2O_2	1.41 μM 1.08 μM	[78]
Au NPs	11.9 nm	540 nm	$NaYF_4$:Yb18%,Er2%	PEI	48 nm	980 nm	543/656 nm	Hepatitis B HBV DNA	250 pM	[79]
Au NPs	-	523 nm	$BaGdF_5$:Yb20%,Er2%	NAAO	14 nm	980 nm	523/546/654 nm	Ebola	500 fM	[80]
Au NRs	~78 nm × 15.5 nm	965 nm	$NaYF_4$:Yb11.9%,Tm0.1%	PEI	~30 nm	980 nm	480/800 nm	COVID S protein	1.06 fg/mL	[81]
Au NPs	20 nm	400–700 nm	$NaYF_4$:Yb,20%Er2%	Apt2	35 nm		525/545/650 nm	Shigella	30 CFU/mL	[82]
Au nanofilm	18 nm	980 nm	$NaYF_4$:Yb20%,Er2%	microfiber	38 nm	980 nm	523/545/655 nm	T(K)	325 K–811K	[83]
$W_{18}O_{49}$	100–800 nm × 5–30 nm	600–1400 nm	$NaYF_4$:Yb2%@$NaYF_4$:Er20%	PLA fiber	35 nm	980 nm	520/540/654 nm	T(K)	298 K–358 K	[84]

3.2. Plasmon Modulated Upconversion Sensing of Ions and Small Biomolecules

Anion and cation recognition plays a crucial role to evaluate the operation of vital biological processes and garner information about overall cell health. There is a wealth of literature reporting the ion sensing platforms that utilize plasmon modulated upconversion, while capable of simultaneous measuring changes in the concentration of different small biomolecules. Therefore, we decided to combine the discussion on ions and small biomolecules sensing in one section. The detection principle is based on the ability of PNPs to modulate the radiation properties of UCNPs by disabling them in closer proximity and retrieving "turn on" them upon segregation. For example, Chen et al. constructed an assay for chromium (Cr^{3+}) ion detection, which was composed of electrostatically coupled lysine-capped UCNPs and dimercaptosuccinic acid-capped gold nanoparticles (Au NPs). The presence of Cr^{3+} ions induces a drift away of Au NPs from UCNPs surface recovering the emission profiles in a linear response [61].

Fang et al. developed a bifunctional biosensing platform using plasmon modulated upconversion effect of quantitative detection of acetylcholinesterase (AChE) and cadmium (Cd^{2+}) ions with the help of glutathione (GSH). Two mechanisms account for the tendency of Au NPs to aggregate and drift away from UCNPs to recover the emission profile (Figure 3A,B). The first is that the presence of AChE promotes the hydrolysis of acetylthiocholine (ATC) which tend to destabilize the surface chemistry of Au NPs and lead to their aggregation (Figure 3C–F). The second mechanism relies on the ability of Cd^{2+} ions to detach GSH from the surface of Au NPs to form spherical shaped complex, which disrupts the stability of Au NPs and effectively triggers their gradual isolation from UCNPs. The designed biosensor exhibited LODs of 0.015 mU/mL and 0.2 μM for AchE and Cd^{2+} ions, respectively [62].

Hu et al. reported a bifunctional Au-UCNPs nanoprobe which exhibits a dose dependent response to Cd^{2+} ions and GSH with LODs 0.059 μM and 0.016 μM, respectively. In this study, the presence of GSH restrained the Au NPs aggregation, while co-existence Cd^{2+} ions impaired their stability leading to the gradual weakening of UCNPs red emission [63].

Sun and Gradzielski employed plasmon-modulated upconversion for the detection of poisonous cyanide ions. The sensor operation was based on the redox consumption of Au NPs by CN^- ions which enabled the emancipation of UCNPs to regain their luminescence and produce detection signal. Moreover, the nanoprobe demonstrated excellent selectivity and sensitivity by distinguishing cyanide ions from different interfering ions with a detection limit of 1.53 μM [66].

Remarkably, gold nanorods have gained great momentum from researchers around the world owing to their versatility in SPR band control via aspect ratio adjustment. For example, Kim et al. demonstrated 27-fold upconversion enhancement by altering the aspect ratio of Au NRs to match their SPR band with the emission of UCNPs at 805 nm. The

UCNPs have been preliminarily encapsulated in polyamidoamine generation 1 (PAMAM G1) dendrimer, which served as a spacer to prevent undesired quenching upon coupling with Au NRs. The obtained nanoprobe was post-functionalized with 2-thiouracil for sensitive and selective detection of uric acid, which is an important small biomolecule. Because its elevated content may indicate renal or cardiovascular disorders in the human body [68]. Interestingly, Zhu et al. obtained 50-fold SPR based enhancement of the UCNPs brightness, which is associated with a ~24 nm thick silica layer on the surface Au NRs. Thereby proposing a nanoprobe for potential microRNA-21 detection in human serum samples and human breast cancer cell (MCF-7) lysates [67].

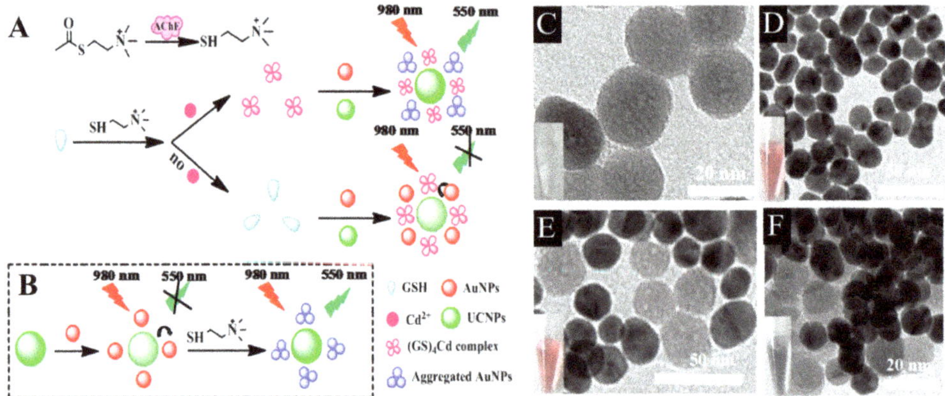

Figure 3. Sensing principle of the bifunctional UCNPs/AuNPs based nanoprobe for detection of (**A**) AchE and (**B**) Cd^{2+} ions with GSH regulation. Representative TEM images of (**C**) UCNPs, (**D**) AuNPs, (**E**) UCNPs/AuNPs, and (**F**) aggregation in UCNPs/AuNPs caused by post-addition of AChE and ATC. Reprinted with permission from reference [62].

Noteworthy, the use of silver nanoparticles for the modulation upconversion biosensors has been relatively limited compared to gold nanoparticles, even though theoretically Ag NPs exhibit stronger SPR [85]. Among recent reports, Liu et al. employed ultrasmall (1.9 nm) silver nanoclusters as acceptors for the construction of plasmon modulated NIR upconversion nanoprobe for intracellular biothiols detection. The authors demonstrated the biosensor performance in the liver tissue of mice to highlight its potential for in vivo sensing [69].

3.3. Plasmon Modulated Upconversion Sensing of Biomacromolecules

Over the past decade, our understanding of how to implement recognition of biomacromolecules, such as RNA or DNA nucleic acids, using plasmon modulated upconversion biosensing systems has significantly advanced. For example, recently, Zhu et al. reported a universal pathway for tumor related noncoding RNA (ncRNA) recognition (Figure 4A). The upconversion recovery principle in the designed biosensor is based on the uncoupling of DNA encapsulated UCNPs from Au NPs bearing a single molecule hairpin DNA (Hp) molecule via exonuclease III (Exo III)-assisted cycling amplification strategy. Notably, the biosensor exhibited a great sensitivity towards the expression level of miR-21 in human breast cancer cell (MCF-7) lysate with LOD of 0.54 fM [75].

Zourob et al. proposed an ssDNA target sequence sensing pathway containing blue-emitting UCNPs for incorporation into silica coated polystyrene-co-acrylic acid nanoparticles (PSA/SiO$_2$) (donor) and Au NPs with immobilized Ir(III) complex (quencher), as shown in Figure 4B. In response to the sequential addition of the target DNA, the quencher could be effectively separated from the donor reactivating the upconversion signal under 975 nm NIR irradiation in a linear manner vs. DNA concentration. The developed sensor sensitivity was indicated at LOD as low as 1 pM. Furthermore, the selectivity of the DNA sensor was confirmed through the titration of the nanoprobe using the DNA conjugated nanohybrids with single base mismatch [73].

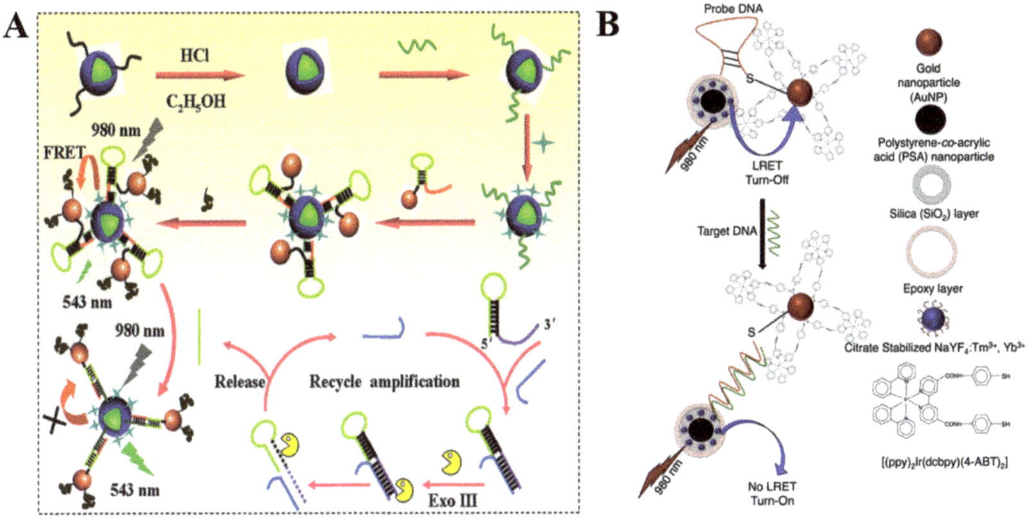

Figure 4. (**A**) Schematics of the UCNPs–AuNPs plasmon-modulated biosensing platform for highly sensitive detection of tumor-related ncRNA via the Exo III-assisted cycling amplification strategy. Reprinted with permission from reference [75]. (**B**) Schematics of the composition and energy transfer mechanism of ssDNA optical sensor composed of PSA/SiO$_2$ coated UCNP and cyclometalated Ir(III)-AuNPs. Reprinted with permission from reference [73].

Conversely, some researchers employ the fluorescence quenching as a signal to recognize the analyte. For instance, Chen et al. employed a simple paper-supported aptasensor to construct Au NR/UCNPs based plasmon modulated nanoprobe for cancer biomarker exosome. The designed nanoprobes operation is activated in the presence of exosome. Upon conjugation with CD63 protein, Au NRs and UCNPs parts are brought to proximity enough to initiate linear quenching which is correlated with exosome concentration. The LOD of exosomes was estimated to be 1.1×10^3 particles/μL [74].

3.4. Plasmon Modulated Upconversion Sensing of Viruses

Despite encouraging progress in the development of safe disease diagnostic tools, such as reverse transcription-polymerase chain reaction (RT-PCR) and enzyme-linked immunosorbent assay (ELISA), rapid and ultrasensitive bioassays for pathogenic viruses detection are still very demanding. So far UCNPs were utilized for the detection of H5N1 Influenza [86], oligonucleotide markers of the SARS-CoV-2 virus [87,88], hepatitis B Virus surface antigen (HBsAg) [89], anti-human immunodeficiency virus (HIV) antibodies [90], thrombocytopenia syndrome virus (SFTSV) total antibodies [91].

There are several reports on deploying plasmon modulated upconversion for construction of novel rapid biosensors for virus recognition. For instance, in 2016, Hao et al. employed plasmon modulated upconversion for ultrasensitive detection of Ebola virus, which outbreak threatened the world significantly between 2014 and 2016 (Figure 5A,B). In their research, they proposed a bioassay composed of BaGdF$_5$:Yb,Er UCNPs and Au NPs that were immobilized on the 3D structured nanoporous alumina (NAAO) substrate. The spectral overlap between nanoparticles allowed to observe effective quenching of naked-eye observable green UC emission upon the combination of probe oligoconjugated with UCNPs and Ebola virus oligoconjugated with AuNPs. Moreover, the authors demonstrated that their sensor could effectively detect Ebola viral RNA in clinical samples with LOD of 500 fM, which is comparable with other conventional detection methods [80].

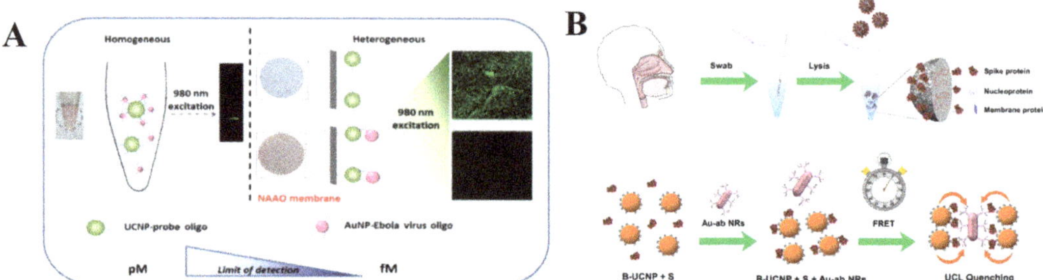

Figure 5. (**A**) Comparison illustration of homogenous and heterogeneous sensor for Ebola virus detection. Reprinted with permission from reference [80]. (**B**) Schematics of COVID S protein detection using plasmon modulated upconversion biosensing system. Reprinted with permission from reference [81].

In another investigation by Hao et al., they introduced plasmon modulated biosensing of SARS-CoV-2 spike protein for COVID-19 point-of-care diagnostics (Figure 5C,D). In this study, the $NaYF_4$:Yb/Tm UCNPs nanoprobe was complemented with Au NRs, which effectively captured and detected S protein endowing LOD of 1.06 fg mL^{-1} [81].

Hepatitis B (commonly abbreviated as HBV) is a DNA virus that causes a serious damage to the human liver. It has been estimated that more than 300 million people around the globe are infected with HBV [92]. Researchers have devoted substantial efforts to the development of sensitive biosensing tools for the prevention of HBV infection spread. For example, Zhu et al. proposed to employ Au-UCNPs-based nanoprobes for HBV DNA detection. In this study, Au NPs were bound to UCNPs via DNA hybridization keeping upconversion emission quenched. The situation drastically changed upon introducing target DNA, which initiated the departure of Au NPs from the UCNPs surface with subsequent restoration of the emission intensity. The proposed nanoprobe exhibited a LOD of 250 pM [79].

3.5. Plasmon Modulated Upconversion Sensing of Temperature

Temperature-dependent properties of UCNPs arising from complex thermally coupled energy levels are suitable for plasmon modulated thermometric biosensing. For example, Li et al. fabricated an optical microfiber (~3 µm) coated with an 18 nm thin gold film and decorated with UCNPs for temperature sensing (Figure 6A). The work principle of the constructed temperature sensor is based on the plasmon enhanced upconversion luminescence, which facilitated temperature dependent upconversion emission enhancement. In particular, plasmonic properties of Au nanofilm were activated by 980 nm laser which selectively increased the green emission intensity of UCNPs at 523 nm by 36 times, which is ascribed to the Au nanofilm-assisted local field enhancement of the incident light. Upon precise control of the UCNPs' amount and laser dosage, the designed sensor can respond to the temperature range of 325–811 K with a resolution of 0.034–0.046 K [83].

Recently, Li et al. developed a flexible temperature sensor, consisting of polyacrylic acid (PLA) fiber and UCNPs/$W_{18}O_{49}$ upconversion/plasmonic semiconductor hybrid optical system for potential application in wearable health monitoring (Figure 6B). Co-doping $W_{18}O_{49}$ nanowires with UCNPs into PLA fiber with a high refractive index of 1.46 facilitated the plasmon enhanced upconversion green emission at 540 nm, which is suitable for ratiometric reading of the temperature changes under a 980 nm NIR laser excitation (5.5 mW). The hybrid sensor exhibited high sensitivity of 1.53% with 0.4 K LOD. In general, the ratiometric response could be measured based on the fluorescence intensity ratio of upconversion emission peaks produced by UCNPs under NIR laser excitation. In this study the intensity ratio between green emissions at 520 nm and 540 nm was used as a temperature change indicator, because these emissions originate from the thermally coupled energy levels $^2H_{11/2}$ and $^4S_{3/2}$ of Er^{3+} activator [84].

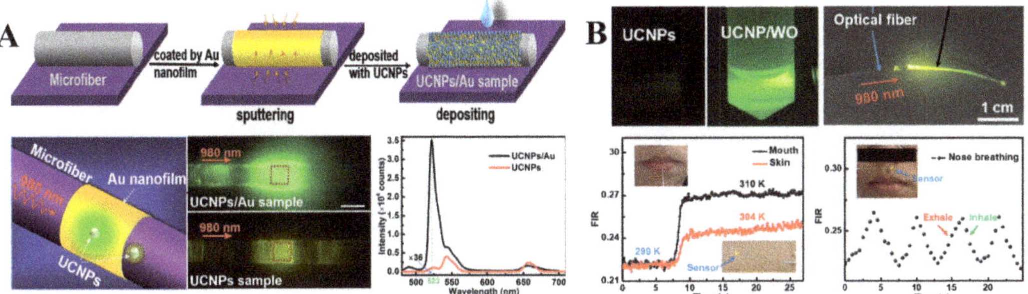

Figure 6. (**A**) Illustration of the selective plasmon-enhanced green emission upconversion nanoprobe for temperature sensing, which include: schematics of the fabrication process, optical images of the sensor upon laser excitation and comparative illustration of the upconversion photoluminescence enhancement induced by Au nanofilm. Reprinted with permission from reference [83]. (**B**) Optical images of UCNPs and UCNPs/WO solutions under a 980 nm laser illumination. Fluorescence intensity ratio changes from the air to the mouth (black line), hand skin (red line) and time-dependent breath sensor response (dotted line). Reprinted with permission from reference [84].

4. Future Directions

Continuing advances in the upconversion field opened the emerging frontiers of plasmon modulated biosensing technology, which affords background-free, rapid, well-timed and more sensitive quantitative data acquisition regarding bioanalytes compared to traditional sensing approaches. In this review, we covered the basic concepts of lanthanide upconversion and brief theoretical principles plasmon modulated upconversion, we systematically analyzed cutting edge advances in construction biosensing tools through the integration of plasmonic nanoparticles (predominantly gold nanoparticles) with upconversion nanoparticles for nanoprobing various biological analytes.

Herein, we have reviewed ongoing research progress made on the development of plasmon modulated upconversion biosensors. Thus, from the most recent experimental findings we derived the future trends and perspectives to better understand how to improve UCNPs luminescence features using plasmonic nanoparticles. In this regard, to fully realize their potential in practical plasmon modulated upconversion biosensing, the following criteria still need to be fulfilled to meet exceedingly high standards of in vivo models and their further clinical translation:

(1) The size of nanoparticles should be monodisperse and in the renal clearable range (<10 nm); however, the size should not be too small (<5 nm) to prevent too rapid excretion from the body. Another important consideration is that as the size of UCNPs decreases the brightness degrades dramatically because the quenching issue becomes more profound;

(2) the stabilization of UCNPs in biological environments with minimized toxicity, since the dissolution in aqueous media has been a general drawback of $NaYF_4$ host materials which leads to the release of cytotoxic lanthanide and fluorine ions. Moreover, plasmon modulation efficiency is sensitive to the nature and thickness of the spacer coating on UCNPs or PNPs. This aspect makes it more demanding to understand the nanoparticle functionalization process and seek alternative coatings which impart more efficient plasmon-enabled control of upconversion properties;

(3) high photoluminescence quantum yields are generally desirable for sufficient brightness under safe NIR excitation laser powers;

(4) the use of 980 nm NIR laser based photoactivation remains a subject of debate, which is unlikely to be increasingly used in the future due to tissue overheating issues. We

anticipate that the focus of the upconversion biosensing research will gradually shift toward safer and shorter NIR laser wavelengths, such as an 808 nm laser-stimulated biosensing, yet it has been restricted by a low permissible power dosage;

(5) from the perspective of temperature-sensitive luminescence of UCNPs, there is little doubt about the rationality of their complementary use, since PNPs tend to generate heat under NIR laser irradiation. It is widely acknowledged that elevated temperatures accelerate the decay rates of thermally coupled energy levels in UCNPs leading to their poor upconversion luminescence or even undesired quenching. Though from our literature survey, it is clear that plasmon enhancement does improve the upconversion efficiency and benefit its biosensing performance.

5. Conclusions

The progress in plasmon-modulated upconversion biosensing will continue to advance, which may enable to explore alternatives to gold plasmonic nanostructure that can provide better plasmonic resonances. Yet, a more profound understanding of the exact key parameters (size, geometry, thickness of the spacer, etc.) governing the efficient plasmon enhancement is therefore essential. Overall, plasmon-modulated upconversion is envisioned as exciting and very promising avenue for the future development of intelligent biosensors that could possibly benefit the healthcare system. We hope this literature survey will motivate the reader to explore solutions to tackle the remaining problems and possibly confront the conceptual challenges to overcome the current bottleneck.

Author Contributions: Conceptualization, J.-H.L. and K.S.K.; investigation, A.M., H.E.C., J.M.P. and K.S.K.; writing and original draft preparation, A.M., J.-H.L. and K.S.K.; visualization, A.M.; supervision, J.-H.L. and K.S.K.; funding acquisition, J.-H.L. and K.S.K. All authors have read and agreed to the published version of the manuscript.

Funding: This study was financially supported by the '2022 Post-Doc. Development Program' of Pusan National University. This work was implemented within the National Research Foundation of Korea (NRF) grant supported by the Ministry of Science and ICT (No. 2022R1C1C1009533, 2021M3H4A4079509, 2022R1A5A2027161), and the Korea Health Industry Development Institute (KHIDI) grant funded by the Ministry of Health & Welfare (No. HF21C0156), Republic of Korea.

Institutional Review Board Statement: Not applicable.

Informed Consent Statement: Not applicable.

Data Availability Statement: Not applicable.

Conflicts of Interest: The authors declare no conflict of interest.

Abbreviations

AChE	acetylcholinesterase
Con-A	concanavalin A
CTAB	cetyl trimethyl ammonium bromide
COVID	coronavirus disease
DNA	deoxyribonucleic acid
FRET	fluoresecence resonance energy transfer
GSH	gluthathione
HBV	Hepatitis B virus
Ln	Lanthanide
LOD	limit of detection
NAAO	nanoporous alumina
NCs	nanoclusters
NDs	nanodots
NPs	nanoparticle
NIR	near-infrared
NSs	nanosheets

PAA	polyacrylic acid
PAMAM	polyamidoamine
PEI	polyethylene imine
PLA	polyacrylic acid
PNPs	plasmonic nanoparticles
PSA	polystyrene-co-acrylic acid
RNA	ribonucleic acid
SPR	surface plasmon resonance
TEM	transmission electron microscopy
UCNPs	upconversion nanoparticles

References

1. Zhu, X.; Zhang, J.; Liu, J.; Zhang, Y. Recent Progress of Rare-Earth Doped Upconversion Nanoparticles: Synthesis, Optimization, and Applications. *Adv. Sci.* **2019**, *6*, 1901358. [CrossRef]
2. Sun, L.; Wei, R.; Feng, J.; Zhang, H. Tailored Lanthanide-Doped Upconversion Nanoparticles and Their Promising Bioapplication Prospects. *Coord. Chem. Rev.* **2018**, *364*, 10–32. [CrossRef]
3. Wang, F.; Liu, X. Recent Advances in the Chemistry of Lanthanide-Doped Upconversion Nanocrystals. *Chem. Soc. Rev.* **2009**, *38*, 976–989. [CrossRef] [PubMed]
4. Ghazy, A.; Safdar, M.; Lastusaari, M.; Savin, H.; Karppinen, M. Advances in Upconversion Enhanced Solar Cell Performance. *Sol. Energy Mater. Sol. Cells* **2021**, *230*, 111234. [CrossRef]
5. Yang, X.; Jin, X.; Zheng, A.; Duan, P. Dual Band-Edge Enhancing Overall Performance of Upconverted Near-Infrared Circularly Polarized Luminescence for Anticounterfeiting. *ACS Nano* **2023**, *17*, 2661–2668. [CrossRef] [PubMed]
6. Ansari, A.A.; Aldajani, K.M.; AlHazaa, A.N.; Albrithen, H.A. Recent Progress of Fluorescent Materials for Fingermarks Detection in Forensic Science and Anti-Counterfeiting. *Coord. Chem. Rev.* **2022**, *462*, 214523. [CrossRef]
7. Peltomaa, R.; Benito-Peña, E.; Gorris, H.H.; Moreno-Bondi, M.C. Biosensing Based on Upconversion Nanoparticles for Food Quality and Safety Applications. *Analyst* **2021**, *146*, 13–32. [CrossRef]
8. Xu, R.; Cao, H.; Lin, D.; Yu, B.; Qu, J. Lanthanide-Doped Upconversion Nanoparticles for Biological Super-Resolution Fluorescence Imaging. *Cell Rep. Phys. Sci.* **2022**, *3*, 100922. [CrossRef]
9. Xie, L.; Hong, Z.; Zan, J.; Wu, Q.; Yang, Z.; Chen, X.; Ou, X.; Song, X.; He, Y.; Li, J.; et al. Broadband Detection of X-Ray, Ultraviolet, and Near-Infrared Photons Using Solution-Processed Perovskite–Lanthanide Nanotransducers. *Adv. Mater.* **2021**, *33*, 1–7. [CrossRef]
10. Lee, G.; Park, Y. Il Lanthanide-Doped Upconversion Nanocarriers for Drug and Gene Delivery. *Nanomaterials* **2018**, *8*, 511. [CrossRef]
11. Lucky, S.S.; Soo, K.C.; Zhang, Y. Nanoparticles in Photodynamic Therapy. *Chem. Rev.* **2015**, *115*, 1990–2042. [CrossRef]
12. Han, S.; Beack, S.; Jeong, S.; Hwang, B.W.; Shin, M.H.; Kim, H.; Kim, S.; Hahn, S.K. Hyaluronate Modified Upconversion Nanoparticles for near Infrared Light-Triggered on-off Tattoo Systems. *RSC Adv.* **2017**, *7*, 14805–14808. [CrossRef]
13. Han, S.; Hwang, B.W.; Jeon, E.Y.; Jung, D.; Lee, G.H.; Keum, D.H.; Kim, K.S.; Yun, S.H.; Cha, H.J.; Hahn, S.K. Upconversion Nanoparticles/Hyaluronate-Rose Bengal Conjugate Complex for Noninvasive Photochemical Tissue Bonding. *ACS Nano* **2017**, *11*, 9979–9988. [CrossRef]
14. Sun, C.; Gradzielski, M. Advances in Fluorescence Sensing Enabled by Lanthanide-Doped Upconversion Nanophosphors. *Adv. Colloid Interface Sci.* **2022**, *300*, 102579. [CrossRef]
15. Zhao, J.; Jin, D.; Schartner, E.P.; Lu, Y.; Liu, Y.; Zvyagin, A.V.; Zhang, L.; Dawes, J.M.; Xi, P.; Piper, J.A.; et al. Single-Nanocrystal Sensitivity Achieved by Enhanced Upconversion Luminescence. *Nat. Nanotechnol.* **2013**, *8*, 729–734. [CrossRef] [PubMed]
16. Gargas, D.J.; Chan, E.M.; Ostrowski, A.D.; Aloni, S.; Altoe, M.V.P.; Barnard, E.S.; Sanii, B.; Urban, J.J.; Milliron, D.J.; Cohen, B.E.; et al. Engineering Bright Sub-10-Nm Upconverting Nanocrystals for Single-Molecule Imaging. *Nat. Nanotechnol.* **2014**, *9*, 300–305. [CrossRef] [PubMed]
17. Liang, L.; Qin, X.; Zheng, K.; Liu, X. Energy Flux Manipulation in Upconversion Nanosystems. *Acc. Chem. Res.* **2019**, *52*, 228–236. [CrossRef] [PubMed]
18. Maini, A.K. *Lasers and Optoelectronics: Fundamentals, Devices and Applications*; Maini, A.K., Ed.; John Wiley and Sons Ltd.: Chichester, UK, 2013; ISBN 9781118688977.
19. Molkenova, A.; Atabaev, T.S.; Hong, S.W.; Mao, C.; Han, D.-W.; Kim, K.S. Designing Inorganic Nanoparticles into Computed Tomography and Magnetic Resonance (CT/MR) Imaging-Guidable Photomedicines. *Mater. Today Nano* **2022**, *18*, 100187. [CrossRef]
20. Jares-Erijman, E.A.; Jovin, T.M. FRET Imaging. *Nat. Biotechnol.* **2003**, *21*, 1387–1395. [CrossRef]
21. Francés-Soriano, L.; Estebanez, N.; Pérez-Prieto, J.; Hildebrandt, N. DNA-Coated Upconversion Nanoparticles for Sensitive Nucleic Acid FRET Biosensing. *Adv. Funct. Mater.* **2022**, *32*, 2201541. [CrossRef]

22. Kotulska, A.M.; Pilch-Wróbel, A.; Lahtinen, S.; Soukka, T.; Bednarkiewicz, A. Upconversion FRET Quantitation: The Role of Donor Photoexcitation Mode and Compositional Architecture on the Decay and Intensity Based Responses. *Light Sci. Appl.* **2022**, *11*, 256. [CrossRef] [PubMed]

23. Das, A.; Corbella Bagot, C.; Rappeport, E.; Ba Tis, T.; Park, W. Quantitative Modeling and Experimental Verification of Förster Resonant Energy Transfer in Upconversion Nanoparticle Biosensors. *J. Appl. Phys.* **2021**, *130*, 023102. [CrossRef]

24. Wu, D.M.; García-Etxarri, A.; Salleo, A.; Dionne, J.A. Plasmon-Enhanced Upconversion. *J. Phys. Chem. Lett.* **2014**, *5*, 4020–4031. [CrossRef]

25. Jiang, W.; Yi, J.; Li, X.; He, F.; Niu, N.; Chen, L. A Comprehensive Review on Upconversion Nanomaterials-Based Fluorescent Sensor for Environment, Biology, Food and Medicine Applications. *Biosensors* **2022**, *12*, 1036. [CrossRef]

26. Jin, H.; Yang, M.; Gui, R. Ratiometric Upconversion Luminescence Nanoprobes from Construction to Sensing, Imaging, and Phototherapeutics. *Nanoscale* **2022**, *15*, 859–906. [CrossRef]

27. Auzel, F. History of Upconversion Discovery and Its Evolution. *J. Lumin.* **2020**, *223*, 116900. [CrossRef]

28. Li, Z.; Zhang, Y. An Efficient and User-Friendly Method for the Synthesis of Hexagonal-Phase NaYF4:Yb, Er/Tm Nanocrystals with Controllable Shape and Upconversion Fluorescence. *Nanotechnology* **2008**, *19*, 345606. [CrossRef]

29. Liu, G. Advances in the Theoretical Understanding of Photon Upconversion in Rare-Earth Activated Nanophosphors. *Chem. Soc. Rev.* **2015**, *44*, 1635–1652. [CrossRef]

30. Ding, Y.; Li, Z. Tuning the Photoluminescence Properties of β-NaYF4:Yb,Er by Bi^{3+} Doping Strategy. *Cryst. Res. Technol.* **2022**, *2100162*, 3–7. [CrossRef]

31. Wen, S.; Zhou, J.; Zheng, K.; Bednarkiewicz, A.; Liu, X.; Jin, D. Advances in Highly Doped Upconversion Nanoparticles. *Nat. Commun.* **2018**, *9*, 2415. [CrossRef]

32. Du, S.; Wu, J.; Wang, Y. Application of Upconversion-Luminescent Materials in Temperature Sensors. In *Upconversion Nanophosphors*; Elsevier: Amsterdam, The Netherlands, 2022; pp. 291–310.

33. Wang, L.; Li, P.; Li, Y. Down- and up-Conversion Luminescent Nanorods. *Adv. Mater.* **2007**, *19*, 3304–3307. [CrossRef]

34. Naccache, R.; Yu, Q.; Capobianco, J.A. The Fluoride Host: Nucleation, Growth, and Upconversion of Lanthanide-Doped Nanoparticles. *Adv. Opt. Mater.* **2015**, *3*, 482–509. [CrossRef]

35. Ansari, A.A.; Muthumareeswaran, M.R.; Lv, R. Coordination Chemistry of the Host Matrices with Dopant Luminescent Ln^{3+} Ion and Their Impact on Luminescent Properties. *Coord. Chem. Rev.* **2022**, *466*, 214584. [CrossRef]

36. Holmberg, R.J.; Aharen, T.; Murugesu, M. Paramagnetic Nanocrystals: Remarkable Lanthanide-Doped Nanoparticles with Varied Shape, Size, and Composition. *J. Phys. Chem. Lett.* **2012**, *3*, 3721–3733. [CrossRef] [PubMed]

37. Ayadi, H.; Fang, W.; Mishra, S.; Jeanneau, E.; Ledoux, G.; Zhang, J.; Daniele, S. Influence of Na^{+} Ion Doping on the Phase Change and Upconversion Emissions of the GdF3: Yb^{3+}, Tm^{3+} Nanocrystals Obtained from the Designed Molecular Precursors. *RSC Adv.* **2015**, *5*, 100535–100545. [CrossRef]

38. Wang, Y.; Deng, R.; Xie, X.; Huang, L.; Liu, X. Nonlinear Spectral and Lifetime Management in Upconversion Nanoparticles by Controlling Energy Distribution. *Nanoscale* **2016**, *8*, 6666–6673. [CrossRef]

39. del Rosal, B.; Rocha, U.; Ximendes, E.C.; Martín Rodríguez, E.; Jaque, D.; Solé, J.G. Nd^{3+} Ions in Nanomedicine: Perspectives and Applications. *Opt. Mater. Amst.* **2017**, *63*, 185–196. [CrossRef]

40. Chen, G.; Damasco, J.; Qiu, H.; Shao, W.; Ohulchanskyy, T.Y.; Valiev, R.R.; Wu, X.; Han, G.; Wang, Y.; Yang, C.; et al. Energy-Cascaded Upconversion in an Organic Dye-Sensitized Core/Shell Fluoride Nanocrystal. *Nano Lett.* **2015**, *15*, 7400–7407. [CrossRef]

41. Slooff, L.H.; Polman, A.; Oude Wolbers, M.P.; Van Veggel, F.C.J.M.; Reinhoudt, D.N.; Hofstraat, J.W. Optical Properties of Erbium-Doped Organic Polydentate Cage Complexes. *J. Appl. Phys.* **1998**, *83*, 497–503. [CrossRef]

42. Wiesholler, L.M.; Frenzel, F.; Grauel, B.; Würth, C.; Resch-Genger, U.; Hirsch, T. Yb,Nd,Er-Doped Upconversion Nanoparticles: 980 Nm: Versus 808 Nm Excitation. *Nanoscale* **2019**, *11*, 13440–13449. [CrossRef]

43. Boyer, J.C.; Vetrone, F.; Cuccia, L.A.; Capobianco, J.A. Synthesis of Colloidal Upconverting NaYF4 Nanocrystals Doped with Er^{3+}, Yb^{3+} and Tm^{3+}, Yb^{3+} via Thermal Decomposition of Lanthanide Trifluoroacetate Precursors. *J. Am. Chem. Soc.* **2006**, *128*, 7444–7445. [CrossRef]

44. Yang, Y.; Sun, Y.; Cao, T.; Peng, J.; Liu, Y.; Wu, Y.; Feng, W.; Zhang, Y.; Li, F. Hydrothermal Synthesis of NaLuF 4: 153Sm,Yb,Tm Nanoparticles and Their Application in Dual-Modality Upconversion Luminescence and SPECT Bioimaging. *Biomaterials* **2013**, *34*, 774–783. [CrossRef]

45. Marin, R.; Halimi, I.; Errulat, D.; Mazouzi, Y.; Lucchini, G.; Speghini, A.; Murugesu, M.; Hemmer, E. Harnessing the Synergy between Upconverting Nanoparticles and Lanthanide Complexes in a Multiwavelength-Responsive Hybrid System. *ACS Photonics* **2019**, *6*, 436–445. [CrossRef]

46. Huang, Y.; You, H.; Song, Y.; Jia, G.; Yang, M.; Zheng, Y.; Zhang, L.; Liu, K. Half Opened Microtubes of NaYF4:Yb,Er Synthesized in Reverse Microemulsion under Solvothermal Condition. *J. Cryst. Growth* **2010**, *312*, 3214–3218. [CrossRef]

47. Wang, X.; Zhuang, J.; Peng, Q.; Li, Y. A General Strategy for Nanocrystal Synthesis. *Nature* **2005**, *437*, 121–124. [CrossRef] [PubMed]

48. Huang, B.; Bergstrand, J.; Duan, S.; Zhan, Q.; Widengren, J.; Ågren, H.; Liu, H. Overtone Vibrational Transition-Induced Lanthanide Excited-State Quenching in Yb^{3+}/Er^{3+}-Doped Upconversion Nanocrystals. *ACS Nano* **2018**, *12*, 10572–10575. [CrossRef]

49. Rabouw, F.T.; Prins, P.T.; Villanueva-Delgado, P.; Castelijns, M.; Geitenbeek, R.G.; Meijerink, A. Quenching Pathways in NaYF4:Er3+,Yb3+ Upconversion Nanocrystals. *ACS Nano* **2018**, *12*, 4812–4823. [CrossRef]

50. Li, X.; Shen, D.; Yang, J.; Yao, C.; Che, R.; Zhang, F.; Zhao, D. Successive Layer-by-Layer Strategy for Multi-Shell Epitaxial Growth: Shell Thickness and Doping Position Dependence in Upconverting Optical Properties. *Chem. Mater.* **2013**, *25*, 1, 106–112. [CrossRef]
51. Zhang, D.; Dong, Y.; Li, D.; Jia, H.; Qin, W. Growth Regularity and Phase Diagrams of NaLu0.795−xYxF4 Upconversion Nanocrystals Synthesized by Automatic Nanomaterial Synthesizer. *Nano Res.* **2021**, *14*, 4760–4767. [CrossRef]
52. Willets, K.A.; Van Duyne, R.P. Localized Surface Plasmon Resonance Spectroscopy and Sensing. *Annu. Rev. Phys. Chem.* **2007**, *58*, 267–297. [CrossRef]
53. Qin, X.; Carneiro Neto, A.N.; Longo, R.L.; Wu, Y.; Malta, O.L.; Liu, X. Surface Plasmon-Photon Coupling in Lanthanide-Doped Nanoparticles. *J. Phys. Chem. Lett.* **2021**, *12*, 1520–1541. [CrossRef] [PubMed]
54. Yin, Z.; Zhou, D.; Xu, W.; Cui, S.; Chen, X.; Wang, H.; Xu, S.; Song, H. Plasmon-Enhanced Upconversion Luminescence on Vertically Aligned Gold Nanorod Monolayer Supercrystals. *ACS Appl. Mater. Interfaces* **2016**, *8*, 11667–11674. [CrossRef] [PubMed]
55. Saboktakin, M.; Ye, X.; Oh, S.J.; Hong, S.H.; Fafarman, A.T.; Chettiar, U.K.; Engheta, N.; Murray, C.B.; Kagan, C.R. Metal-Enhanced Upconversion Luminescence Tunable through Metal Nanoparticle-Nanophosphor Separation. *ACS Nano* **2012**, *6*, 8758–8766. [CrossRef] [PubMed]
56. Wu, M.; Wang, X.; Wang, K.; Guo, Z. Sequence-Specific Detection of Cytosine Methylation in DNA: Via the FRET Mechanism between Upconversion Nanoparticles and Gold Nanorods. *Chem. Commun.* **2016**, *52*, 8377–8380. [CrossRef]
57. Dong, J.; Gao, W.; Han, Q.; Wang, Y.; Qi, J.; Yan, X.; Sun, M. Plasmon-Enhanced Upconversion Photoluminescence: Mechanism and Application. *Rev. Phys.* **2019**, *4*, 100026. [CrossRef]
58. Wang, L.; Guo, S.; Liu, D.; He, J.; Zhou, J.; Zhang, K.; Wei, Y.; Pan, Y.; Gao, C.; Yuan, Z.; et al. Plasmon-Enhanced Blue Upconversion Luminescence by Indium Nanocrystals. *Adv. Funct. Mater.* **2019**, *29*, 1901242. [CrossRef]
59. Zong, H.; Mu, X.; Sun, M. Physical Principle and Advances in Plasmon-Enhanced Upconversion Luminescence. *Appl. Mater. Today* **2019**, *15*, 43–57. [CrossRef]
60. Mendez-Gonzalez, D.; Melle, S.; Calderón, O.G.; Laurenti, M.; Cabrera-Granado, E.; Egatz-Gómez, A.; López-Cabarcos, E.; Rubio-Retama, J.; Díaz, E. Control of Upconversion Luminescence by Gold Nanoparticle Size: From Quenching to Enhancement. *Nanoscale* **2019**, *11*, 13832–13844. [CrossRef]
61. Liu, B.; Tan, H.; Chen, Y. Upconversion Nanoparticle-Based Fluorescence Resonance Energy Transfer Assay for Cr(III) Ions in Urine. *Anal. Chim. Acta* **2013**, *761*, 178–185. [CrossRef]
62. Fang, A.; Chen, H.; Li, H.; Liu, M.; Zhang, Y.; Yao, S. Glutathione Regulation-Based Dual-Functional Upconversion Sensing-Platform for Acetylcholinesterase Activity and Cadmium Ions. *Biosens. Bioelectron.* **2017**, *87*, 545–551. [CrossRef]
63. Sun, L.; Wang, T.; Sun, Y.; Li, Z.; Song, H.; Zhang, B.; Zhou, G.; Zhou, H.; Hu, J. Fluorescence Resonance Energy Transfer between NH2−NaYF4:Yb,Er/NaYF4@SiO2 Upconversion Nanoparticles and Gold Nanoparticles for the Detection of Glutathione and Cadmium Ions. *Talanta* **2020**, *207*, 120294. [CrossRef]
64. Wu, S.; Duan, N.; Shi, Z.; Fang, C.; Wang, Z. Dual Fluorescence Resonance Energy Transfer Assay between Tunable Upconversion Nanoparticles and Controlled Gold Nanoparticles for the Simultaneous Detection of Pb^{2+} and Hg^{2+}. *Talanta* **2014**, *128*, 327–336. [CrossRef] [PubMed]
65. Liu, Z.; Yang, L.; Chen, M.; Chen, Q. Amine Functionalized NaY/GdF4:Yb,Er Upconversion-Silver Nanoparticles System as Fluorescent Turn-off Probe for Sensitive Detection of Cr(III). *J. Photochem. Photobiol. A Chem.* **2020**, *388*, 112203. [CrossRef]
66. Sun, C.; Gradzielski, M. Fluorescence Sensing of Cyanide Anions Based on Au-Modified Upconversion Nanoassemblies. *Analyst* **2021**, *146*, 2152–2159. [CrossRef] [PubMed]
67. Zhang, K.; Lu, F.; Cai, Z.; Song, S.; Jiang, L.; Min, Q.; Wu, X.; Zhu, J.J. Plasmonic Modulation of the Upconversion Luminescence Based on Gold Nanorods for Designing a New Strategy of Sensing MicroRNAs. *Anal. Chem.* **2020**, *92*, 11795–11801. [CrossRef]
68. Kannan, P.; Abdul Rahim, F.; Chen, R.; Teng, X.; Huang, L.; Sun, H.; Kim, D.H. Au Nanorod Decoration on NaYF4:Yb/Tm Nanoparticles for Enhanced Emission and Wavelength-Dependent Biomolecular Sensing. *ACS Appl. Mater. Interfaces* **2013**, *5*, 3508–3513. [CrossRef]
69. Xiao, Y.; Zeng, L.; Xia, T.; Wu, Z.; Liu, Z. Construction of an Upconversion Nanoprobe with Few-Atom Silver Nanoclusters as the Energy Acceptor. *Angew. Chem. Int. Ed.* **2015**, *54*, 5323–5327. [CrossRef]
70. Wiesholler, L.M.; Genslein, C.; Schroter, A.; Hirsch, T. Plasmonic Enhancement of NIR to UV Upconversion by a Nanoengineered Interface Consisting of NaYF4:Yb,Tm Nanoparticles and a Gold Nanotriangle Array for Optical Detection of Vitamin B12 in Serum. *Anal. Chem.* **2018**, *90*, 14247–14254. [CrossRef] [PubMed]
71. Hu, W.; Chen, Q.; Li, H.; Ouyang, Q.; Zhao, J. Fabricating a Novel Label-Free Aptasensor for Acetamiprid by Fluorescence Resonance Energy Transfer between NH2-NaYF4: Yb, Ho@SiO2 and Au Nanoparticles. *Biosens. Bioelectron.* **2016**, *80*, 398–404. [CrossRef] [PubMed]
72. Wang, F.; Han, Y.; Wang, S.; Ye, Z.; Wei, L.; Xiao, L. Single-Particle LRET Aptasensor for the Sensitive Detection of Aflatoxin B1 with Upconversion Nanoparticles. *Anal. Chem.* **2019**, *91*, 11856–11863. [CrossRef]
73. Jesu Raj, J.G.; Quintanilla, M.; Mahmoud, K.A.; Ng, A.; Vetrone, F.; Zourob, M. Sensitive Detection of SsDNA Using an LRET-Based Upconverting Nanohybrid Material. *ACS Appl. Mater. Interfaces* **2015**, *7*, 18257–18265. [CrossRef]
74. Chen, X.; Lan, J.; Liu, Y.; Li, L.; Yan, L.; Xia, Y.; Wu, F.; Li, C.; Li, S.; Chen, J. A Paper-Supported Aptasensor Based on Upconversion Luminescence Resonance Energy Transfer for the Accessible Determination of Exosomes. *Biosens. Bioelectron.* **2018**, *102*, 582–588. [CrossRef] [PubMed]

75. Zhang, K.; Yang, L.; Lu, F.; Wu, X.; Zhu, J.J. A Universal Upconversion Sensing Platform for the Sensitive Detection of Tumour-Related NcRNA through an Exo III-Assisted Cycling Amplification Strategy. *Small* **2018**, *14*, 1703858. [CrossRef]
76. Li, X.; Wei, L.; Pan, L.; Yi, Z.; Wang, X.; Ye, Z.; Xiao, L.; Li, H.W.; Wang, J. Homogeneous Immunosorbent Assay Based on Single-Particle Enumeration Using Upconversion Nanoparticles for the Sensitive Detection of Cancer Biomarkers. *Anal. Chem.* **2018**, *90*, 4807–4814. [CrossRef]
77. Chen, X.; Wang, J.; Yang, C.; Ge, Z.; Yang, H. Fluorescence Resonance Energy Transfer from NaYF4:Yb,Er to Nano Gold and Its Application for Glucose Determination. *Sens. Actuators B Chem.* **2018**, *255*, 1316–1324. [CrossRef]
78. Wu, S.; Kong, X.J.; Cen, Y.; Yuan, J.; Yu, R.Q.; Chu, X. Fabrication of a LRET-Based Upconverting Hybrid Nanocomposite for Turn-on Sensing of H2O2 and Glucose. *Nanoscale* **2016**, *8*, 8939–8946. [CrossRef] [PubMed]
79. Zhu, H.; Lu, F.; Wu, X.C.; Zhu, J.J. An Upconversion Fluorescent Resonant Energy Transfer Biosensor for Hepatitis B Virus (HBV) DNA Hybridization Detection. *Analyst* **2015**, *140*, 7622–7628. [CrossRef]
80. Tsang, M.K.; Ye, W.; Wang, G.; Li, J.; Yang, M.; Hao, J. Ultrasensitive Detection of Ebola Virus Oligonucleotide Based on Upconversion Nanoprobe/Nanoporous Membrane System. *ACS Nano* **2016**, *10*, 598–605. [CrossRef]
81. Li, L.; Song, M.; Lao, X.; Pang, S.Y.; Liu, Y.; Wong, M.C.; Ma, Y.; Yang, M.; Hao, J. Rapid and Ultrasensitive Detection of SARS-CoV-2 Spike Protein Based on Upconversion Luminescence Biosensor for COVID-19 Point-of-Care Diagnostics. *Mater. Des.* **2022**, *223*, 111263. [CrossRef] [PubMed]
82. Chen, M.; Yan, Z.; Han, L.; Zhou, D.; Wang, Y.; Pan, L.; Tu, K. Upconversion Fluorescence Nanoprobe-Based FRET for the Sensitive Determination of Shigella. *Biosensors* **2022**, *12*, 795. [CrossRef]
83. Zhang, W.; Li, J.; Lei, H.; Li, B. Plasmon-Induced Selective Enhancement of Green Emission in Lanthanide-Doped Nanoparticles. *ACS Appl. Mater. Interfaces* **2017**, *9*, 42935–42942. [CrossRef] [PubMed]
84. Zhang, W.; Huang, X.; Liu, W.; Gao, Z.; Zhong, L.; Qin, Y.; Li, B.; Li, J. Semiconductor Plasmon Enhanced Upconversion toward a Flexible Temperature Sensor. *ACS Appl. Mater. Interfaces* **2023**, *15*, 4469–4476. [CrossRef] [PubMed]
85. Kravets, V.; Almemar, Z.; Jiang, K.; Culhane, K.; Machado, R.; Hagen, G.; Kotko, A.; Dmytruk, I.; Spendier, K.; Pinchuk, A. Imaging of Biological Cells Using Luminescent Silver Nanoparticles. *Nanoscale Res. Lett.* **2016**, *11*, 1–9. [CrossRef] [PubMed]
86. Zhao, Q.; Du, P.; Wang, X.; Huang, M.; Sun, L.D.; Wang, T.; Wang, Z. Upconversion Fluorescence Resonance Energy Transfer Aptasensors for H5N1 Influenza Virus Detection. *ACS Omega* **2021**, *6*, 15236–15245. [CrossRef]
87. Alexaki, K.; Kyriazi, M.E.; Greening, J.; Taemaitree, L.; El-Sagheer, A.H.; Brown, T.; Zhang, X.; Muskens, O.L.; Kanaras, A.G. A SARS-Cov-2 Sensor Based on Upconversion Nanoparticles and Graphene Oxide. *RSC Adv.* **2022**, *12*, 18445–18449. [CrossRef]
88. Guo, J.; Chen, S.; Tian, S.; Liu, K.; Ni, J.; Zhao, M.; Kang, Y.; Ma, X.; Guo, J. 5G-Enabled Ultra-Sensitive Fluorescence Sensor for Proactive Prognosis of COVID-19. *Biosens. Bioelectron.* **2021**, *181*, 113160. [CrossRef]
89. Martiskainen, I.; Talha, S.M.; Vuorenpää, K.; Salminen, T.; Juntunen, E.; Chattopadhyay, S.; Kumar, D.; Vuorinen, T.; Pettersson, K.; Khanna, N.; et al. Upconverting Nanoparticle Reporter–Based Highly Sensitive Rapid Lateral Flow Immunoassay for Hepatitis B Virus Surface Antigen. *Anal. Bioanal. Chem.* **2021**, *413*, 967–978. [CrossRef]
90. Martiskainen, I.; Juntunen, E.; Salminen, T.; Vuorenpää, K.; Bayoumy, S.; Vuorinen, T.; Khanna, N.; Pettersson, K.; Batra, G.; Talha, S.M. Double-Antigen Lateral Flow Immunoassay for the Detection of Anti-HIV-1 and -2 Antibodies Using Upconverting Nanoparticle Reporters. *Sensors* **2021**, *21*, 330. [CrossRef]
91. Huang, C.; Wei, Q.; Hu, Q.; Wen, T.; Xue, L.; Li, S.; Zeng, X.; Shi, F.; Jiao, Y.; Zhou, L. Rapid Detection of Severe Fever with Thrombocytopenia Syndrome Virus (SFTSV) Total Antibodies by up-Converting Phosphor Technology-Based Lateral-Flow Assay. *Luminescence* **2019**, *34*, 162–167. [CrossRef]
92. Sheena, B.S.; Hiebert, L.; Han, H.; Ippolito, H.; Abbasi-Kangevari, M.; Abbasi-Kangevari, Z.; Abbastabar, H.; Abdoli, A.; Abubaker Ali, H.; Adane, M.M.; et al. Global, Regional, and National Burden of Hepatitis B, 1990–2019: A Systematic Analysis for the Global Burden of Disease Study 2019. *Lancet Gastroenterol. Hepatol.* **2022**, *7*, 796–829. [CrossRef]

biosensors

MDPI

Review

Improving Biosensors by the Use of Different Nanomaterials: Case Study with Microcystins as Target Analytes

Hanbin Park [1], Gahyeon Kim [1], Yoseph Seo [1], Yejin Yoon [1], Junhong Min [2,*], Chulhwan Park [1,*] and Taek Lee [1,*]

[1] Department of Chemical Engineering, Kwangwoon University, Seoul 01897, Korea; binwlal@naver.com (H.P.); 1497rg@hanmail.net (G.K.); akdldytpq12@gmail.com (Y.S.); wjwj0131@naver.com (Y.Y.)
[2] School of Integrative Engineering, Chung-Ang University, Seoul 06974, Korea
* Correspondence: junmin@cau.ac.kr (J.M.); chpark@kw.ac.kr (C.P.); tlee@kw.ac.kr (T.L.); Tel.: +82-2-820-8320 (J.M.); +82-2-940-5173 (C.P.); +82-2-940-5771 (T.L.)

Abstract: The eutrophication of lakes and rivers without adequate rainfall leads to excessive growth of cyanobacterial harmful algal blooms (CyanoHABs) that produce toxicants, green tides, and unpleasant odors. The rapid growth of CyanoHABs owing to global warming, climate change, and the development of rainforests and dams without considering the environmental concern towards lakes and rivers is a serious issue. Humans and livestock consuming the toxicant-contaminated water that originated from CyanoHABs suffer severe health problems. Among the various toxicants produced by CyanoHABs, microcystins (MCs) are the most harmful. Excess accumulation of MC within living organisms can result in liver failure and hepatocirrhosis, eventually leading to death. Therefore, it is essential to precisely detect MCs in water samples. To date, the liquid chromatography–mass spectrometry (LC–MS) and enzyme-linked immunosorbent assay (ELISA) have been the standard methods for the detection of MC and provide precise results with high reliability. However, these methods require heavy instruments and complicated operation steps that could hamper the portability and field-readiness of the detection system. Therefore, in order for this goal to be achieved, the biosensor has been attracted to a powerful alternative for MC detection. Thus far, several types of MC biosensor have been proposed to detect MC in freshwater sample. The introduction of material is a useful option in order to improve the biosensor performance and construct new types of biosensors. Introducing nanomaterials to the biosensor interface provides new phenomena or enhances the sensitivity. In recent times, different types of nanomaterials, such as metallic, carbon-based, and transition metal dichalcogenide-based nanomaterials, have been developed and used to fabricate biosensors for MC detection. This study reviews the recent advancements in different nanomaterial-based MC biosensors.

Keywords: microcystin; nanoparticle; biosensor; cyanobacterial harmful algal bloom

Citation: Park, H.; Kim, G.; Seo, Y.; Yoon, Y.; Min, J.; Park, C.; Lee, T. Improving Biosensors by the Use of Different Nanomaterials: Case Study with Microcystins as Target Analytes. *Biosensors* **2021**, *11*, 525. https://doi.org/10.3390/bios11120525

Received: 10 September 2021
Accepted: 15 December 2021
Published: 20 December 2021

Publisher's Note: MDPI stays neutral with regard to jurisdictional claims in published maps and institutional affiliations.

1. Introduction

Cyanobacterial harmful algal blooms (CyanoHABs) are toxic algal blooms that float on living organisms, freshwater systems, and water supply sources during the summer and produce toxicants [1,2]. The eutrophication of lakes and rivers without adequate rainfall leads to excessive growth of CyanoHABs that produce toxicants, green tides, and unpleasant odors. The massive growth of CyanoHABs owing to global climate change and global warming has led to rapid eutrophication in water bodies [3–5].

The development of rainforests without considering the environment and construction of dams in rivers could result in the overgrowth of CyanoHABs [6]. Additionally, the industrialization of rural areas located near rivers and lakes is accelerating the growth of CyanoHABs, which can cause serious economic, health, and ecological problems [7]. Furthermore, humans, livestock, and aquatic animals could suffer from damage to the liver and other organs by drinking large amounts of CyanoHAB-contaminated water, leading to death [8,9]. Moreover, due to the growing amounts of CyanoHABs, the water bodies turn green and produce a stench smell, which is not esthetically pleasing. Particularly during the summer, the overgrowth of CyanoHABs produces algal blooms in waterbodies such as rivers, lakes, and ponds [10].

CyanoHABs produce harmful cyanotoxins such as microcystin, anatoxin, saxitoxin, nodularin, and cylindrospermopsin [11,12]. When CyanoHABs undergo damage or die, they release cyanotoxins that cause hepatotoxicity and neural toxicity to humans, livestock, and other wildlife [13,14]. Microcystin, nodularin, and cylindrospermopsin are classified as hepatotoxins, whereas anatoxin-a and saxitoxin are classified as neurotoxins [15,16].

Microcystins (MCs) are produced by various cyanobacteria genera, such as Microcystis, Planktothrix, Anabaena, Nostoc, Aphanizomenon, and Limnothrix, and are the most commonly found toxins from CyanoHABs [17]. MCs comprises seven amino acids having circular forms, and over 90 different types of MCs have been reported worldwide [18]. MC was classified with amino acid composition such as MC-LR, MC-RR, MC-YR, MC-LR, and MC-LF. These different MC species showed the different toxicities. It is reported that MC-LC showed the highest toxicity in comparison to the other MC types [19]. When MCs are exposed to living organisms, they cause inhibition of protein phosphatase in liver cells, protein kinase activation malfunction, and over-phosphorylation of proteins, thereby resulting in several acute diseases [20]. Furthermore, excess accumulation of MC in liver cells results in apoptosis of cells due to cytoskeletal disruption and control loss of p53 gene regulation. The structural stability of MC is determined in cells that do not decompose for 2–3 months [21]. Therefore, the precise detection of MC in freshwater is essential for humans and wildlife.

Liquid chromatography–mass spectrometry (LC–MS) and enzyme-linked immunosorbent assay (ELISA) are two of the conventional methods used for the detection of MC [5,19,22–24]. The United States Environmental Protection Agency (USEPA) recommends using these techniques to quantify the MC and nodularins in water samples as official methodologies "Method 546" and "Method 544" [23–25]. While LC–MS provides precise data and results, it requires heavy analytical apparatus and expensive equipment. Comparatively, while ELISA does not require heavy and expensive equipment, it does require a complicated detection step and trained researchers with extensive analytical time. Therefore, these techniques cannot meet the requirements of simple field-ready detection methods.

In the meantime, biosensors can detect various molecules, including toxins, and can be used as an alternative to solve the above problems. Several biosensors have been devised to detect various analytes such as viruses, pathogens, diseases, toxicants, and microorganisms [26–30]. The biosensors are designed to meet specific goals according to various detection platforms such as electrochemical [26], electrical [31], optical [32], and spectroscopic platforms [33]. The biosensor can provide the useful platform with small size, portability and easy-to-handle nature, and field-ready measurement [34]. In constructing a biosensor, the target recognition bioreceptors should be required. Generally, two types of bioreceptors are used to bind with specific targets. The antibody is a gold standard for immunosensor construction. It provides specific target binding, low Kd constant and high selectivity. However, the manufacturing cost of antibody is expensive, and production of antibody requires animal experiments. In last 20 years, the aptamer was regarded as a powerful alternative for biosensor construction. Aptamers are short RNA or DNA strands that can bind with high specificity and affinity to target materials such as proteins, lipids, ions, and whole cells by forming a unique 3D structure by folding. Aptamers are inexpensive, easy to synthesize, and small in size, exhibiting excellent chemical stability [35]. In addition, owing to their unique structural properties, aptamers can reportedly be enhanced through systematic evolution of ligands by exponential enrichment (SELEX) to significantly improve sensitivity and selectivity when binding to target substances. Compared to antibodies, it can be chemically synthesized, reducing the manufacturing cost and being free from animal experiments. However, the aptamer showed less selectivity and stability that should be solved by aptamer-based biosensor (aptasensor) commercialization.

Various types of biosensors can detect the toxicant small molecule effectively [36–38]. In particular, several reviews on biosensors for microcystin detection have been published [39–41]. Cunha et al. discussed the aptasensor for aquatic phycotoxins and cyanotoxins [39]. In this review, they mainly focused of application of aptamer for toxin detection. In addition, Bertani and Lu recently introduced the cyanobacterial toxin biosensors for environmental monitoring and protection [40]. They introduced various cyanotoxin biosensors, including saxitoxin, microcystin, and cylindrospermopsin. Moreover, this review focused on the portability of biosensor for field-ready application. Massey et al. recently summarized the microcystin detection methods [41]. They focused not on biosensors but on ELISA and HPLC-based MC detection. Thus, these reviews discussed MC detection methods using biosensor or conventional methods; however, these reviews did not explain the usefulness of introduction of nanomaterial for MC biosensor construction.

Meanwhile, the introduction of nanomaterial for the construction of biosensors provides detection sensitivity and selectivity, as well as new detection platforms [42,43].

Several nanomaterials have been synthesized for application in fields of energy, medicine, science, and engineering [44–47]. Among them, three types of nanomaterials: noble metal-based, carbon-based, and transition metal dichalcogenide (TMD)-based nanomaterial, are useful for the construction of toxicant-detecting biosensors [48–50]. A suitable platform for detecting CyanoHABs and algal toxins can be achieved by combining adequate bioreceptors (antibody, aptamer, and nucleic acid) with the above-mentioned nanomaterials. The present review discusses the recent progress in the four types of nanoparticles and bioreceptors hybrid material-based MC biosensors.

2. Metal Nanoparticle-Based MC Biosensor

Metal-based nanoparticles with various physical and chemical properties according to their composition and shape have also been developed [48–51]. Metal nanoparticles are widely used in batteries, materials, devices, and for the treatment of cancer [52,53]. Furthermore, noble metal nanoparticles such as gold, silver, and rhodium nanoparticles exhibit superior conductivity and high stability and are used in electrochemical biosensors for the detection of microcystin [48,54,55].

Owing to the electrical or electrochemical properties of conductive nanoparticles, they can be used in MC biosensors to improve the detection sensitivity of the sensor by increasing the surface area of the interface between the target and the bioreceptors electrode. Furthermore, conductive nanoparticles can be used in electrochemical and electricity-based biosensors by pairing them with an antibody or an aptamer to achieve high detection sensitivity.

Electrochemical measurement type can be applied to immunosensor for MC-LR detection with high selectivity and sensitivity. The analytical approach exhibits acceptable precision, stability, and accuracy. Zhang et al. [56] developed electrochemical immunosensors using functional *PPy* microspheres (AuNP/PpyMS) comprising gold nanoparticles for the detection of microcystin-LR. The fabricated biosensor was composed of antibody–AuNP/PPyMS complex for generating electrochemical signal enhancement through AuNP. AuNP was prepared by depositing silver on polypyrrole microspheres that were synthesized by chemical oxidation polymerization to act as electrochemical catalysts for signal amplification. As shown in Figure 1A, AuNPs were incorporated into polypyrrole microspheres as signal antibodies, and the detection signal was enhanced by increasing the surface area. Furthermore, the electrodes of the MC-LR immunosensor were modified using carbon nanotubes (CNTs), and the excellent fixation of MC-LR antigens to the modified electrodes was due to the stabilization of the antigen binding site of the polyethylene glycol (PEG) film. Moreover, the coated MC-LRs were subjected to MC-LR antibodies to form antigen–antibody complexes for competitive immunolysis, which reacted with the AuNP/PPyMS-labeled signal antibodies to generate electrochemical signals under the silver catalyst.

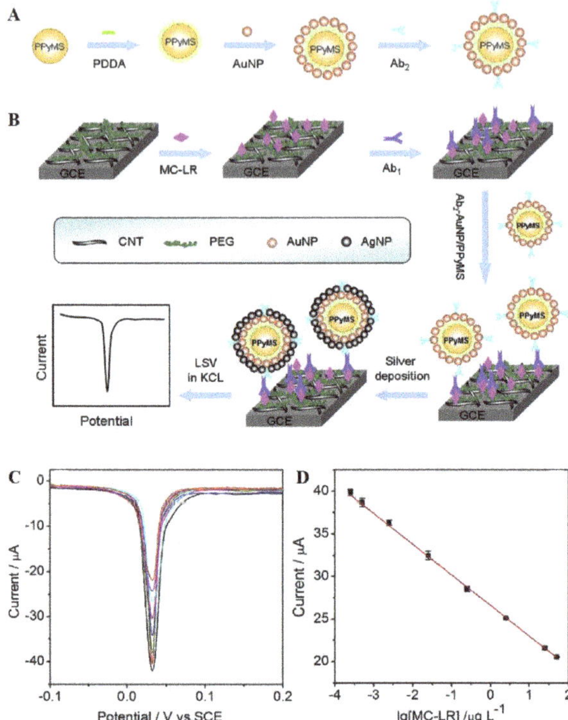

Figure 1. (**A**) Ab2-AuNP/PPyMS particle preparation method. (**B**) Schematic representation of the MC-LR immunosensor fabrication and competition immunoassay procedure. (**C**) Linear sweep stripping voltammetry curves of silver nanoparticles deposited on the immunosensors at 1.0 M KCl after incubation with the lowest peak currents in the MCLR having corresponding concentration ranges (0.00025, 0.0005, 0.0025, 0.025, 0.25, 2.5, 25, and 50 µg/L). (**D**) Calibration curve for MC-LR immunoassay. Reproduced with permission from [56], published by Elsevier, 2017.

The detection performance of MC-LR was evaluated using linear sweep voltammetry (LSV) peak currents, wherein the current of the working electrode is measured, while the potential between the working and reference electrodes changes linearly with time. Figure 1B shows the LSV peak current measured in the concentration range of 0.0005–100 µg/L, which could be used for immunoanalysis, owing to the tendency of the current to decrease as the concentration of MC-LR antigens used in peak competitive immune responses increased. The antigen–antibody immune complex reacts with Ab2-AuNP/PPyMS and induces reduction of silver ions in the dissociated Cl^- in KCl. Although the linearization curve (Figure 1C,D) of MC-LR immunolysis had a low detection limit of 0.2 ng/L in the corresponding linear range, the proposed analytical approach showed excellent stability and high precision, exhibiting potential applications for the detection of other toxins.

Secondly, noble metal nanomaterials arising from nanoscale phenomena such as localized surface plasmon resonance (LSPR) or enhancement of Raman signals are also used in the fabrication of MC biosensors. The vibrations occur at the surface of the metal nanoparticles, which have localized surfaces, in the surface plasmon resonance (SPR) sensor. This results in LSPR referred to as the resonance formed by combining the electromagnetic field (EM) with spatially limited free electrons [57]. LSPR can be excited resonantly around nanoparticles of metal surfaces or thin films of metal. Furthermore, the wavelength intensity and length of LSPR can be altered by combining the bioreceptors and target material on the surface of the nanoparticles, which was confirmed by the LSPR results [58]. Moreover, LSPR-based biosensors have a simple structure, are easy to operate, and are portable for detection in the field [59].

One of the most essential characteristics of metal nanoparticles such as Au and Ag is that LSPR is generated in metal nanoparticles. The optical plasmonic properties of metal NPs are highly dependent on the grain boundary distance between the NPs, which are small or large aggregates of NP pairs, compared to individual and well-spaced NPs. As the interparticle distance decreased, a strong overlap was observed between the plasmon fields of the surrounding particles, owing to which the intensity increased, causing a redshift in the LSPR band, making it easier to observe changes in the color solution. AuNPs and AgNPs exhibit excellent LSPR properties with strong and distinct colors and color changes between individual NPs. Compared to aggregated NPs, widely spaced NPs are easier with UV-visible spectroscopy. This can be visualized or confirmed [60].

Wang et al. fabricated an LSPR-based immune sensor using AuNP–aptamer assays for MC-LR detection [61]. As shown in Figure 2A, a target molecule-specific aptamer was used as the linker to prepare the AuNP dimer. Then, the AuNP dimer was degraded in the presence of the target molecule, and the color of the solution changed from blue to red. Furthermore, a new peak appeared at approximately 606 nm, and the absorbance increased at 606 nm as the linker concentration increased (Figure 2B). Conversely, in the absence of the target molecule, the aptamer acts as a linker that induces the formation of an asymmetrically altered AuNPs dimer, owing to which the solution appears blue, thereby indicating that the absorption peak of the AuNP monomer was at approximately 539 nm, similar to that before the linker was added. Compared to omnidirectional sensors based on the expansion of large aggregates into molecules, LSPR-based sensors exhibit significantly high sensitivity and stability and can be measured within a duration of 5 min.

Raman spectroscopy is used as a molecular identification tool, considering the fact that it enables the qualitative and quantitative analysis of molecules by measuring the vibrational spectrum of the sensor. Raman scattering mainly depends on the energy loss (Stokes) or gain (anti-Stokes) of inelastically scattered photons, owing to the molecular vibrational events, and reflects information about the molecular structure to enable in situ real-time sensing [62,63]. However, because the signal strength of the spectrum sensor was weak, surface-enhanced Raman spectroscopy (SERS) using specific metals was developed [64,65]. In SERS, the Raman signal of a chemical target material is amplified by the resonance between the wavelength of the incident light and the surface free electrons form when the chemical target material is close to a specific metal nanosurface. The SERS mechanism can be divided into electromagnetic and chemical enhancements. In general, the contribution of chemical enhancement is smaller than that of electromagnetic enhancement. SERS is advantageous, given that unlabeled non-destructive analytes can be detected from the spectral results, such as fingerprints and individual components of biochemical molecules and multi-component materials, respectively [66,67].

In Deyun's study, MC-LR was detected with high sensitivity using the SERS technology-based AuNP-aptasensor [68]. The overall schematic of the MC-LR detection method is the same as that shown in Figure 2C. The MC-LR aptamer and its corresponding complementary DNA fragment (cDNA) bind the gold (AuNPs) and magnetic (MNPs) nanoparticles, respectively. Furthermore, MC-LR aptamer–AuNPs and cDNA–MNP conjugates were used as signal reporter and bioreceptors, respectively. Figure 2D,E shows linearity ranging between 0.01 and 200 ng/mL with a proposed sensor detection limit (LOD) of 0.002 ng/mL. The reliability of the new approach was evaluated at different concentrations of spiked MC-LR in tap water samples.

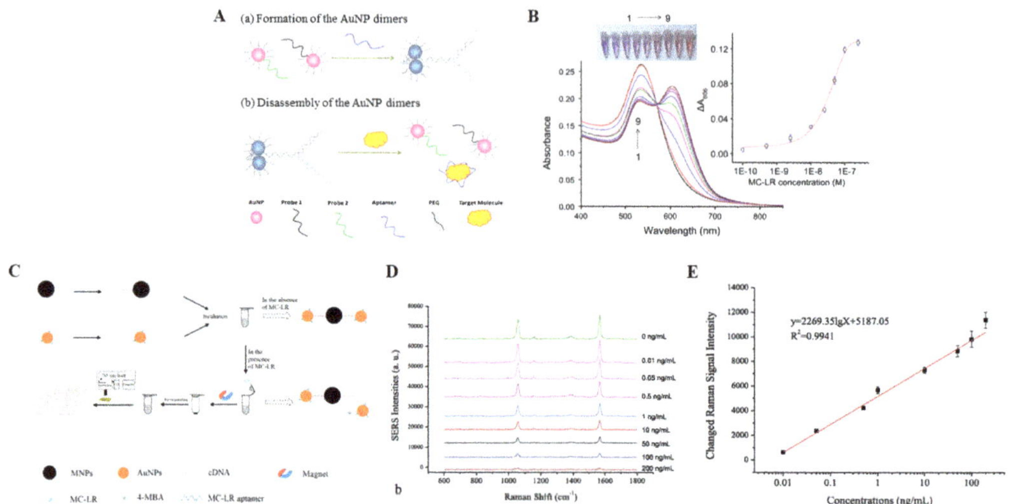

Figure 2. (**A**) Schematic representation of the formation and disassembly of AuNP dimers. (**B**) Extinction spectra of the sensors for different concentrations (0, 0.1, 0.5, 2.5, 10, 25, 50, 100, and 250 nM) of MC-LR. (**C**) Illustration of the principle behind the MC-LR-based analysis on the SERS-based aptasensor. (**D**) Typical SERS spectra with exposure to different concentrations of MC-LR (0–200 ng/mL). (**E**) Linear correlation between the changed Raman signal intensities and concentrations of MC-LR. Reproduced with permission from [61,68], published by Elsevier, 2015 and 2019, respectively.

Li et al. fabricated a sandwich SERS spectroscopic immunosensor comprising a surface-functionalized quartz substrate and a SERS tag to selectively detect MC-LR [69]. The SERS tag utilized the gold nanospike (GNS) plasmonic substrate, owing to its SERS augmentation factor, unique high-density "hotspots," and easy tuning of the LSPR band to the near-infrared (NIR) region. The GNS was identified at 750 nm using a 785 nm NIR laser excitation. NTP molecules used as Raman reporters were tightly adsorbed and immobilized on the GNS surface, which resulted in the GNS @ NTP @ SiO_2 structure. Herein, we demonstrate how the developed SERS sensor can reach a detection limit of 0.14 µg/L. A high-performing SERS immunosensor assay allows for monitoring of the dynamic generation of MC-LR.

A colorimetric sensor that uses the degree of aggregation of conductive metal nanoparticles is also an interesting approach. Metallic nanoparticles are known to have attributes such as controllable physical and chemical properties, functional flexibility, low toxicity, and excellent stability, making them ideal sensor materials [70]. Colorimetric sensors have attracted wide attention in biochemical analysis, owing to their simplicity, high sensitivity, and low cost [71]. Among the different metals, gold (AuNPs) has been used in various colorimetric sensors, owing to its wide visibility of color change [72]. However, metal nanoparticles as sensing elements can sometimes lack the ability to selectively target specific materials. Therefore, to solve this problem, recent colorimetric biosensors research has developed a nanoparticle–aptamer complex where nanoparticles are bound to the aptamer, which is an oligonucleotide with excellent selectivity for a target material used as the sensor material [73–76].

However, to date, existing analytical techniques used for the detection of ML-LR in fresh water have many limitations, such as the interference of various environmental organic pollutants, specialized technology, complex sample preparation, expensive equipment, and lengthy detection time. Therefore, to overcome these problems, recent colorimetric sensor studies have suggested using Au nanoparticle–aptamer-based colorimetric sensors, which are simple and highly sensitive, specifically for MC-LR detection [77–79].

The Au nanoparticle–aptamer-based colorimetric sensor principle used in the study by Li et al. (2016) [77] is described as follows. The aptamer and metal nanoparticles are used as recognition elements to selectively bind to MC-LR with high affinity and as sensing materials to detect the color change in the plasma resonance absorption peak when binding to a target in high-concentration sodium chloride, respectively (Figure 3A–C). When the Au nanoparticle–aptamer comes in contact with the target material (MC-LR) in the sample, the aptamer structure modifies and separates from the nanoparticles to form an MC-LR/aptamer complex. Thereafter, a concentrated salt solution was used to aggregate the released AuNPs, and the sensor was driven by a method that changes the color of the sample by interparticle plasmon coupling of metal nanoparticles. The limit of detection (LOD) of the MC-LR-specific AuNP–aptamer was estimated as 0.37 nM. In addition, it was observed for the real sample that the biosensor worked even in high-salt pond water. Therefore, given that this colorimetric sensor has low cost, excellent stability, and reproducibility and can determine the presence of MC-LR even in real freshwater, it is considered very useful for detecting MC-LR in the real environment.

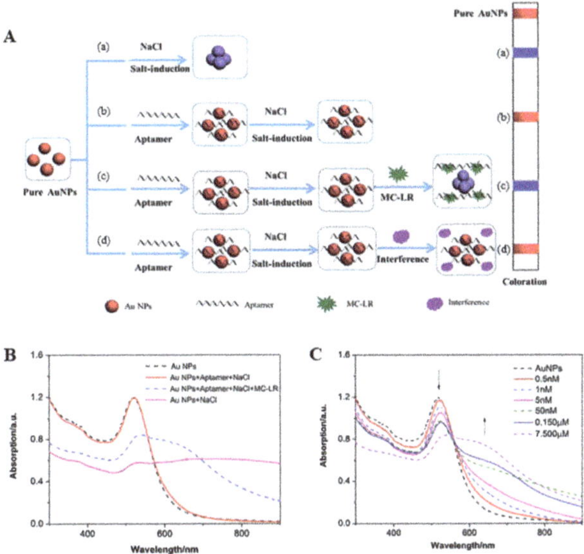

Figure 3. (**A**) Schematic illustration of the colorimetric sensor and mechanism of analytic determination of MC-LR. (**B**) UV–VIS absorption spectra of AuNPs under different experimental conditions, c(Au NPs) = 4.4 nM, c(aptamer) = 0.128 μM, c(NaCl) = 24 mM, c(MC-LR) = 750 nM, T = 298K.C. (**C**) UV–VIS absorption spectrum of Au NPs (concentration range of MC-LR: 0.5 nM–7.5 μM). Reproduced with permission from [77], published by Elsevier, 2016.

3. Carbon Nanomaterial-Based MC Biosensor

Owing to the unique structural characteristics of carbon, since 2000, carbon-based nanomaterials such as carbon nanotubes, carbon nanofibers, graphene, and graphene oxide have been used for the development of electrochemical, electrical, and spectroscopic biosensors [80–83]. Recently, a new type of structural layer called MXene was developed, which improved the scalability of the applicability of biosensors [52,84]. Recent studies have attempted using carbon-based biosensors with excellent sensor performance for MC detection.

Zhao et al. [85] developed graphene and multienzyme functions as carbon-nanosphere-based electrochemical immune sensors for MC-LR detection. The immune sensors used graphene (GSs) and chitosan (CS) to improve the electrochemical performance of the electrodes. Furthermore, a horseradish peroxidase–carbon nanosphere (CNS)–antibody system was used to amplify the electrochemical signal. Figure 4A shows a mimetic diagram of an electrochemical immune sensor comprising bio-composite nanostructures horseradish peroxidase–carbon nanosphere–antibody and microcystins–LR/graphene sheets–chitosan/glassy carbon electrode (GCE).

The above-mentioned sensor was formed by the immobilization of nanocomposite GSs-CS/CNS to the vitreous carbon electrode. The detection sensitivity was improved by using a synthesized HRP-CNS-Ab bio junction. The sensor has a three-dimensional structure that can be used as a signal reporter and interacts with MC-LR. Thereafter, hydrogen peroxide was applied to transfer electrons directly from the electrode to form an electrical signal. Owing to the electrode, the formation of free-state HRP-CNS-Ab/MC-LR/GSs-CS/GCE decreased as the concentration of the MC-LR samples increased.

Figure 4B shows the measurements of redox peaks using the cyclic voltammetry (CV) method, a method of obtaining the current potential curve from a solid electrode at each stationary stage of the sensor. The periodic changes to the electrode potential were obtained using a triangular wave $[Fe(CN)_6]^{3-/4-}$ as the redox species. The applied voltages were measured, ranging between -0.2 and 0.6 V. CV was obtained when the GCE (curve b) electrodes were modified into GS-CS, thereby resulting in a 38% increase in current of the over bare GCE (curve b) electrodes, indicating that the electrodes modified with GS-CS exhibited increased conductivity and a larger surface area. However, the MC-LR/GSs-CS/GCE (curve c) showed a decrease of 30.8% in the current value from curve b, which indicated that the MC-LR was fixed to the electrode surface. This reduction can be attributed to the inhibition of the electron transfer process by the MC-LR, which is nonconductive. The insulation properties of the electrode surface increased further after incubation with HRP-CNS-Ab (curve d), thereby reducing the peak current response.

Figure 4C,D shows the results of the evaluation of the sensor performance using the DPV measurement method. A linear range of 0.05–15 µg/L microcystin-LR with a detection limit of 0.016 µg/L was observed. Furthermore, the DPV measurements under optimized conditions (buffer: 0.2 M PBS (pH7.4), applied voltage: -0.5–0 V, amplitude: 50 mV pulse width: 0.01 s) were proportional to the concentration of MC-LR and decreased linearly as the MC-LR concentration increased in the 0.05–15 µg/L concentration range. The linear range obtained for MC-LR was much wider than the range of immune sensors (0.06–0.65 µg/L) derived from antibodies labeled directly with HRP.

In addition, various studies have attempted to detect MC-LR using carbon materials that exhibit excellent conductivity. For example, Zhang et al. (2011) [86] fabricated an immunosensor by coating AuNPs onto nitrogen-doped carbon nanotubes. Furthermore, Zhao et al. (2013) [85] fabricated a graphene-based immunosensor using a horseradish peroxidase–carbon nanosphere–antibody system for signal amplification to detect microcystine-LR.

Among the various carbon materials, graphene separated from crystalline graphite is one of the representative biosensor materials. Graphene is an allotrope of carbon comprising a single atom-thick and a flat monolayer comprising a two-dimensional sheet of honeycomb lattice [87], having unique optical, electronic, thermal, and chemical properties. Owing to this, graphene and its derivatives are emerging as the new carbon materials and are attracting wide attention in different fields such as biological detection [88], nanocomposite synthesis [89], and microelectronic device fabrication [90]. However, considering the limitations of existing physical approaches [91], the chemical modification and functionalization of graphene has attracted attention from many researchers. In particular, graphene oxide (GO) prepared by oxidizing graphite has abundant hydrophilic groups (hydroxyl, epoxide, carboxyl groups, etc.) on its surface, which indicates that it can be well dispersed in water [92,93]. Furthermore, GO maintains a delocalized π–electron system that provides a strong affinity to the carbon-based ring structure of graphene, being known to have superior value as a sensor material compared to graphene [94].

The work of Shi et al. [95] introduced GO to biosensors in their study (Figure 4E). The sensor substrate was constructed by exposing a coated glass slide surface to APTES vapor.

Graphene oxide was adsorbed through electrostatic force using a graphene oxide array. Here, crosslinking agents EDC and sulfo-NHS were added to activate the exposed carboxyl group of GO through incubation. Furthermore, a carboxyl–amine group covalent bond was formed on the graphene oxide surface through additional incubation after adding an NH2-MC antibody solution. In the detection step, the AuNP–ssDNA complex was formed with MCs in the sample by poly (A) ssDNA ($5'$-AAA AAA AAA AAA AAA-$3'$), and the residual complex, which was not bound to the MCs, was eliminated by poly (T) ssDNA ($5'$-TTT TTT TTT TTT TTT-$3'$). The bound MCs were recognized immunologically by the antibodies (NH2-MCs) adsorbed onto the GO surface. The GO and AuNPs act as–donor–acceptor pairs to induce quenching of GO fluorescence through the FRET phenomenon. Therefore, when the AuNP–ssDNA–MC complex is formed in GO, MCs are detected using a mechanism that lowers the intensity of fluorescence.

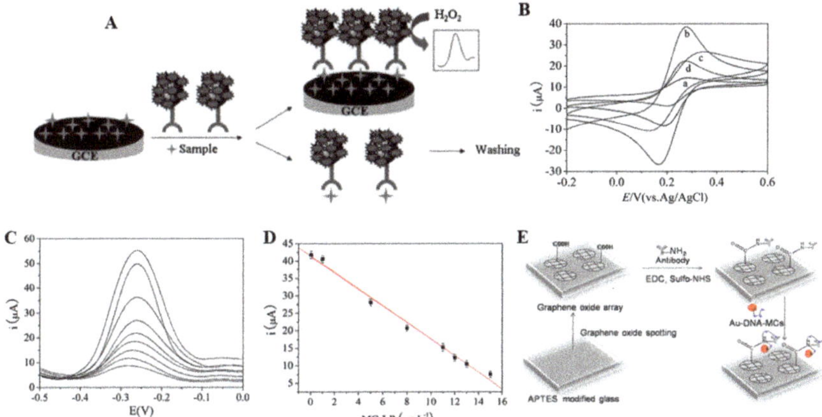

Figure 4. (**A**) Schematic illustration of the detection principles of the immunosensor using microcystins–LR/graphene sheets–chitosan/GCE and horseradish peroxidase–carbon nanosphere–antibody bioconjugates. (**B**) Cyclic voltammograms of (a) Bare GCE, (b) GSs-CS/GCE, (c) MC-LR/GSs-CS/GCE, (d) HRP-CNSs-Ab/MC-LR/GSs-CS/GCE, (e) in 0.1 M KCl containing 2.5 mM $Fe(CN)_6^{3-}$/$Fe(CN)_6^{4-}$ mixture (1:1 molar ratio). Scan rate: 60 mV s^{-1}. (**C**) In 0.2 M PBS (pH 7.4), DPV measurement results at MC-LR concentration shifts (0.05 to 15 µg/L). The current response of the immunosensor after incubation with amplitude: 50 mV, pulse width (top to bottom) containing 7.0 mM H_2O_2. The DPV measurements were performed from -0.5 V–0 V, with an amplitude of 50 mV and a pulse width of 0.01 s. (**D**) The calibration curve of the current responses vs. MC-LR concentrations and liner fit for microcystin-LR concentrations. (**E**) Illustration of GO-based fluorescence biosensor. Reproduced with permission from [85,95] published by Elsevier, 2013 and 2012.

The limit of detections of MC-LR and MC-RR in the sensor were 0.5 mg/L and 0.3 mg/L, respectively, which satisfied the strictest standards of the World Health Organization (WHO). Furthermore, we obtained significant fluorescence quenching signals from MCs in real lake water. The antibody was able to recognize the Adda ((all-S, all-E)-3-amino-9-methoxy-2,6,8-trimethyl-10-phenyldeca-4,6-dienoic acid) group in the MC structure, which is a conservative part of MCs, to sensitively and selectively detect MCs. Results showed that the GO-based sensor exhibits high sensitivity, acceptable stability, and high reproducibility, thereby indicating the possibility of using GO-based sensors for the detection of MC-LR in environmental samples.

4. Transition Metal Dichalcogenides Nanoparticle-Based MC Biosensor

Because the first paper describing the properties of transition metal dichalcogenides (TMDs) was published almost a decade ago, one-dimensional nanomaterials have become one of the most vibrant areas of study in the field of materials science [96,97]. Two-dimensional TMD nanostructures such as molybdenum disulfide (MoS_2) and bismuth selenide (Bi_2Se_3) exhibit remarkable optical, electrical, magnetic, and mechanical properties and have attracted significant attention, owing to their potential applications in silicon-based devices and various material applications [98–101]. In particular, MoS_2 has been studied extensively for the storage and conversion of electrochemical energy in the form of electrocatalyst for hydrogen evolution reactions, electrode material in lithium-ion batteries, and supercapacitors, owing to its good anti-corrosion, catalytic abilities, and biosensor [102–104]. Owing to these interesting characteristics, several groups have reported the application of MoS_2 for the development of biosensors [105–107].

Liu et al. [108] developed a sensitive aptasensor based on a dual-signal amplification system that uses both horseradish peroxidase (HRP) and a trilobe nanocomposite (AuNP@MoS_2-TiONB nanocomposite) for the measurement of MC-LR. MoS_2 nanosheets that can cover spherical TiO_2 surfaces have a large surface area, which increases the chances of binding to biomolecules. The HRP also amplifies the sensing signal of the sensor. The DPV method was used to evaluate the analytical performance of MC-LR under scanning conditions from 0.22 to 0.23 V at a pulse amplitude of 50 mV every 0.1 s. The peak current value decreased as the concentration of MC-LR increased in the concentration range of 0.005–200 nM, which can be attributed to the decrease in the binding of biotin–cDNA, which reduced the binding of avidin–HRP. Therefore, the electrocatalyst current of HRP was confirmed to be inversely proportional to MC-LR.

Zhang et al. [109] developed a sensitive electrochemical immunosensor based on molybdenum disulfide (MoS_2) and gold nanorod (AuNR) composites for the detection of MC-LR. The immunosensor was constructed by immobilizing an MC-LR antibody on a gold electrode, which was modified using a MoS_2/AuNRs nanocomposite with a large surface area and excellent biocompatibility. The detection method used had a competitive immunoassay format, wherein the coated MC-LR antibody competed with the added target MC-LR for the MC-LR antigen to form an antibody–antigen immunocomplex. Then, horseradish peroxidase-labeled anti-MC-LR antibody (HRP-Ab$_2$) was used to detect MC-LR. Figure 5A shows a 20 mM PBS (pH 7.5, 0.1 M KCl) buffer containing 5 mM $[Fe(CN)_6]^{3-/4-}$ and 0.1 M $KClO_4$ in the frequency range of 0.1 Hz to 100 kHz at a bias potential of 0.2 V. The Nyquist plots, a diagram comprising a high-frequency semicircle and low-frequency linear section corresponding to the electron transfer resistance (R_{et}), of electrochemical impedance spectroscopy (EIS) for the electrodes were plotted at each modification step. It can be seen that, unlike other electrodes, the bare electrode (curve a) has a low electron transfer resistance (R_{et}), indicated by a straight line. After modifying the AuNR in MoS_2/Au (curve c), the R_{et} value decreased, indicating that AuNR promoted $[Fe(CN)_6]^{3-/4-}$ ion transport. In the other process, it was confirmed that the R_{et} value increased, which indicated that MoS_2, AuNR, anti-MC-LR, MC-LR, and HRP–anti-MC-LR were successfully immobilized on the surface of the gold electrode. The DPV measurement method was used to evaluate the electrochemical performance of the MC-LR measurement immune sensor. Under optimal conditions, the immunosensor showed a linear response to MC-LR in the range of 0.01–20 µg/L with a detection limit of 5 ng (Figure 5B,C). This electrochemical immune sensor showed excellent potential for monitoring routine water quality for various toxins.

A MoS_2–quantum dot (MoS_2 QD) complex system was synthesized and applied to a fluorescent biosensor as a fluorophore and a quencher, respectively. MoS_2 QDs exhibit excellent stability, low toxicity, and low manufacturing costs. Furthermore, it has strong resonant light absorption, excellent photoluminescence, and excellent potential as a fluorometric sensor material [110,111]. Moreover, when N-acetyl-L-cysteine (NAC) is used as a capping agent in the synthesis of MoS_2 QDs, in near-infrared absorption caused by NAC, abnormal upconversion photoluminescence of MoS_2 QDs occurs, owing to the two successive energy transfers into the hexagonal MoS_2 QD structure, which appears as green fluorescence under UV and NIR irradiation [112]. This upconversion photoluminescence by

MoS–QDs under low-energy NIR irradiation is known to effectively eliminate background interferences [113].

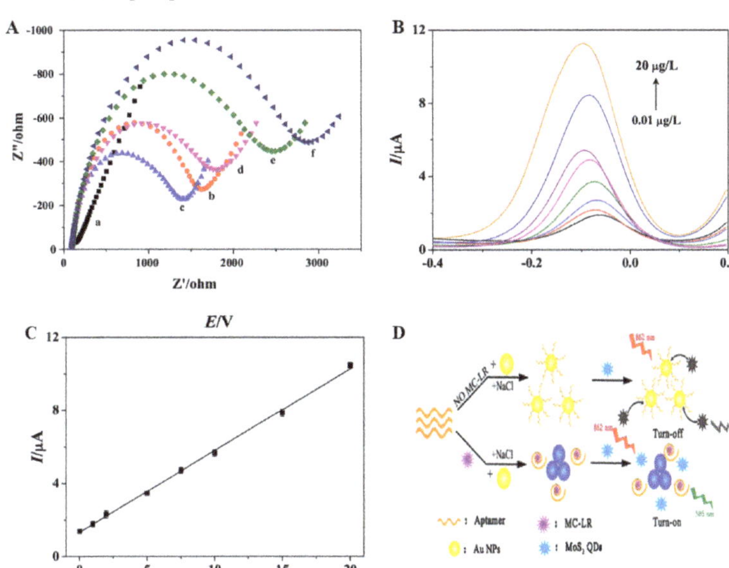

Figure 5. (**A**) Nyquist plots of (a) bare Au electrode, (b) MoS2/Au, (c) AuNRs/MoS2/Au, (d) anti-MC-LR/AuNRs/MoS2/Au, (e) MC-LR/anti-MC-LR/AuNRs/MoS2/Au, and (f) HRP–anti-MC-LR/MC-LR/anti-MC-LR/AuNRs/MoS2/Au in 20 mM PBS (pH 7.5, 0.10 M KCl) containing 5.0 mM $[Fe(CN)_6]^{3-/4-}$ and 0.10 M NaClO4 within frequency range of 0.1 Hz to 100 kHz, amplitude of 0.05 V, and applied potential of 0.20 V. (**B**) DPV responses of the immunosensor in 20 mM PBS (pH 7.5, 100 mM $NaClO_4$ and 100 mM KCl) containing 0.8 mM HQ and 2.0 mM H_2O_2 after incubation with $(0.01 \sim 20.0 \ \mu g \ L^{-1})$ MC-LR. (**C**) Calibration curve for MC-LR immunoassay. The DPV measurements were performed from -0.4 V to 0.2 V, with an amplitude of 50 mV and a pulse width of 50 ms. (**D**) Schematic representation of the detection strategy of MC-LR according to MoS_2 QDs using up conversion fluorescence and aptamer–AuNPs. Reproduced with permission from [109] published by Elsevier, 2017, and [113] published by American Chemical Society, 2020.

On the basis of the unique luminescence of MoS_2 QDs, Cao et al. (2020) [113] presented a method for detecting MC-LR according to MoS_2 QDs using upconversion fluorescence generated by NAC. Furthermore, this method used MoS_2 QDs as a signal-sensing molecule and gold nanoparticle–aptamer as a recognition factor for a target material (Figure 5D). The working principle of this study is similar to that of the colorimetric sensor proposed by Li et al. (2016) [77]. However, MoS_2 QDs cannot bind to aggregated AuNPs in the post-separation process of AuNP–aptamer in the presence of MC-LR. Therefore, the sensor measures the upconversion fluorescence from the exposed MoS_2 QDs. This method has a lower limit of detection (0.01 nM) as compared to a fluorescent sensor that uses a fluorophore and a quencher. In addition, the ability to eliminate background noise in complex environmental samples by emitting visible photoluminescence of MoS_2 QDs with upconversion fluorescence indicates that the MoS– QD-based fluorescence sensor is highly effective in accurately detecting the presence of MC-LR in the environment.

5. MC Biosensors with Other Nanomaterials

For the fabrication of field-ready biosensors, it is essential to pretreat the cyanobacterial sample and detect MC simultaneously. Dos et al. [114] developed a good example of this system by developing a portable microfluidic sensing platform for simultaneously detecting

MC-LRs in and out of Microcystis aeruginosa cells. The filter in the chip filtered the MC toxins that discriminated the sample and quantitatively detected MCs. Figure 6A shows the scheme of the manufactured sensor, wherein the sample processing module uses a micro-tank (μR), microfluidic mixer (μFM), and ultra-filter methods for pretreating the samples. The detection principle and functionalization of monetary poles follows electrochemical impedance spectroscopy (EIS) using an electrochemical cell chip (ECC). In-sample toxins competitively inhibit the binding of anti-MC-LR antibodies to the electrode surface where MC-LR is immobilized. The analytical performance was tested by characterizing the surface functionalization of ECC by increasing the concentration of MC-LR. Therefore, this process simultaneously detects the total MC-LR content (in and out of the cells) and concentration of MC-LR toxin (out of cells) in a water-free state.

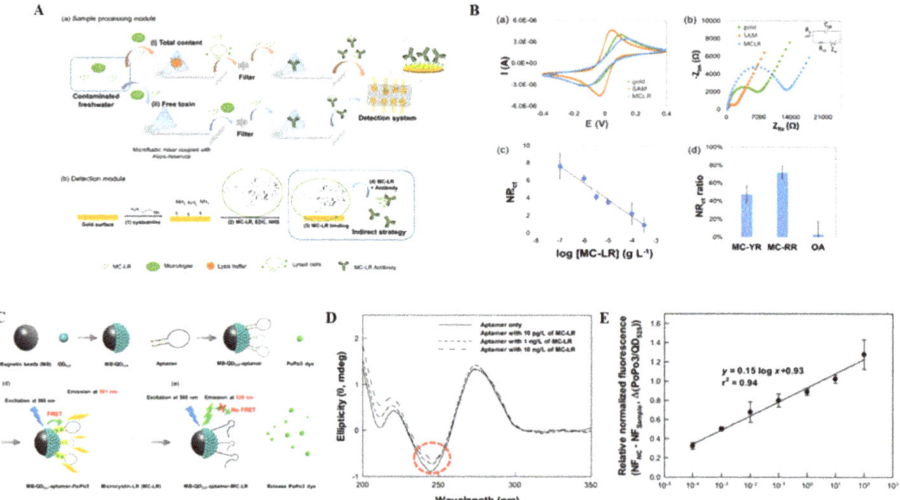

Figure 6. (**A**) Schematic representation of the overall MC-LR detection system. A sample processing system comprising (**a**) sample processing and (**b**) detection modules. Consists of (i) two microfluidic mixers coupled and (ii) reservoirs (μFM-μR) used to process freshwater cyanobacterial samples without lysis buffer. (**b**) Immunosensor fabrication; (1) diagram of SAM with cysteamine, (2) MC-LR activation, and (3) MC-LR binding to a SAM-modified surface. (**B**) Nyquist plots for (**a**) cyclic voltammetry and (**b**) surface functionalization steps; (inset) Randles equivalent circuit model used for EIS fitting. (**c**) Normalized signal of R_{ct} versus the MC-LR concentration (concentration of antibody 10^{-2} g/L). (**d**) Percentage of NRct signals obtained for the same concentration (10^{-4} g/L) of different toxins (MC-YR, MC-RR, OA). (**C**) Schematic representation of the MC-LR detection procedure of the FRET-based QD–aptasensor. (**D**) Circular dichroism (CD) spectral changes of the aptamer with MC-LR. (**E**) Sensitivity validation for MC-LR detection after 6 h of reaction. Reproduced with permission from [114,115], published by Elsevier, 2019.

Figure 6B shows the result of the performance analysis at each stage of electrode fabrication using CV and EIS methods. The surface functionalization stage CV measurements showed that the Faraday current of CV increase due to the immobilization of cysteamine in gold electrodes, which can be attributed to the electrostatic attraction between the positively charged monolayer (surface pK_a = 6.7) and negatively charged solutions, while the toxin bonded to the surface. The Nyquist plot was suitable for Randles equivalent circuit models. Furthermore, the formation of cysteamine SAM in gold electrodes reduced the R_{ct} value (1825 $\pm$ 306), whereas the combination of MC-LR on the SAM gold surface resulted in a higher R_{ct} signal (17,007 $\pm$ 4161) from MC-LR compared to SAM. The EIS measurements for MC-LR quantification were performed on custom electrochemical impedance portable platforms (EPP) (under conditions of 0.5–100,000 Hz, sinusoidal perturbation: 0.005 V). As a result, the N R_{ct} signal showed an increase in the sample concentration in the MC-LR concentration range (3.3×10^{-4}–10^{-4} g/L). In addition, the similarity (MC-YR and MC-RR) between MC-LR with a concentration of 10^{-4} g/L at (d) and the NR_{ct} signals at OA verified the detection selectivity of the sensor.

In addition, various types of nanoparticles were introduced to fabricate the MC biosensor. Among them, core-shell structured nanoparticles and upconversion nanoparticles (UCNPs) were applied to MC biosensors. Lee et al. developed a highly sensitive fluorescence resonance energy transfer (FRET)-based quantum dot (QD)–aptasensor for the detection of MC-LR during the budding phase [115] (Figure 6C). Figure 6D shows the UV–VIS measurement results. A difference was visible at 245 nm, which can be attributed to the decrease in the intensity of the negative circular dichroism (CD) band when the target molecule is inserted perpendicular to the helical DNA axis, thereby indicating that the peak at approximately 245 nm is the only negative CD band and the intensity of that peak. It is assumed that the MC-LR molecule is inserted perpendicular to the base pair and binds to the back of the aptamer. The fret-based QD-aptasensor had a measured detection limit of 10^{-4} µg/L in the range of 10^{-4} to 10^{2} µg/L (ppb or nmol/L) (Figure 6E). MC-LR is selectively detected in different homologues of MC-LR such as microcystin-YR, microcystin-LY, microcystin-LW, microcystin-RR, microcystin-LF, microcystin-LA, and nojularin. Furthermore, the laboratory culture resulted in changes in the intracellular MC-LR concentration along the bacterial growth curve. In the early stages of fixation, QD–Aputasensa detected MC-LR in bacterial cultures of 12.7–15.8 µg/L. For environmental samples, MC-LR corresponded to 1.0 and 7.2 µg of MC-LR/L-water measured at 2.7×10^{8} and 6.6×10^{10} cells/L-water, respectively, indicating that microcystin-LR can be quantified in both laboratory cultures and eutrophic conditions [115].

Wu et al. developed an MC-LR sensor that uses green and red UCNP luminescence as donors and two quenchers, black hole quencher-1 (BHQ1), and black hole quencher-3 (BHQ3-). The two donor–acceptor pairs were constructed by hybridizing the aptamer with the corresponding complementary DNA. Results showed that the overlapping spectrum of green and red UCNP emissions can be extinguished by the bioreceptor. Aptamers preferentially bind to the corresponding analyte in the presence of MC-LR and okadaic acid (OA) and dehybridize with complementary DNA. The detection limits for MC-LR and OA were found to be 0.025 and 0.05 ng/mL, respectively. Experiments showed that the relative luminescence intensity increased as the algal toxin concentration increases, which promoted the quantification of MC-LR and OA [116].

6. Future Perspectives

Although developed countries are attempting to control the growth of cyanobacteria by limiting the construction of lake areas and improving the operation of water purification systems, developing countries are unable to ensure such control, and the sparse development is further accelerated. Furthermore, the indiscriminate increase in cyanobacteria owing to global warming is expected to increase in the future. The WHO has set a concentration value of 1 µg/L or less as the standard concentration of cyanobacteria in drinking water and recommends countries to comply with this. However, because this situation is relatively well observed in developed countries, there is still a long way to go. Therefore, the development of low-cost and highly reliable biosensors for the detection of MCs is expected to increase enormously.

Electrochemical and optical devices satisfy the above requirements in terms of ease of use, portability, and detection sensitivity. In particular, nanobiotechnology has led to the development of various sensor electrodes by combining different nanomaterials and biomaterials, which can detect the threshold concentration by increasing the surface area of the electrode on the basis of the roughness of the electrode. This resulted in the development of new types of sensors such as LSPR and SERS for the amplification of nanoparticle conductivity and spectral power at the nano level. Here, the present review discussed the recent progress of MC biosensors composed of various nanoparticle and bioreceptors. Table 1 showed the recent MC biosensors in terms of nanoparticle types. Thus, the introduction of nanomaterial provides sensitivity and detection method widely. Moreover, those nanomaterials have the different characteristics for MC biosensor construction. Table 2 displays the pros and cons of each nanomaterial in terms of biosensor fabrication. Noble metal-based nanomaterials provide high conductivity, durability, and stability. These characteristics are essential to MC biosensor because MC should be detected in the river or freshwater samples. The high salt, precipitate, and other microorganisms erode the fabricated MC biosensor electrode substrate. However, it is still expensive for manufacturing. The strategy for cost down with noble nanomaterial should be solved for commercialization. The use of carbon-based nanomaterial can be reduced the fabrication cost compared to noble nanomaterial. Moreover, it showed the stable semiconducting property that can bring the chance to fabricate new type of MC biosensor. However, it can be easily oxidized, hampering the sensing performance. Finally, the TMD-based nanomaterial provides the unique conductivity as it called topological insulator; furthermore, the unique chemical property of TMD can provide an opportunity for new concept of MC detection system. However, it still requires a research test for biosensor construction because it usually forms a two-dimensional shape and is hard to particulate.

Table 1. MC biosensors composed of various nanomaterials and bioreceptors.

Materials	Bioreceptors	Nanoparticle	Detection Method	Linear Range (μg/L)	LOD (μg/L)	References
Metal-NP	Aptamer	AuNS	CV	0–99.5	9.95×10^{-4}	[55]
	Antibody	AuNP/PPyMS	LSV	0.25–50	0.1	[56]
	Aptamer	AuNP/MNP	SERS	0.01–200	0.002	[68]
	Aptamer	CuNC	Fluorescence	0.005–1200	0.003	[73]
	Aptamer	AuNP	Fluorescence	0.25–19.90	0.83	[79]
	Antibody	CNS	DPV	0.05–15	0.016	[85]
Carbon-NP	Antibody	CNx-MWNT/AuNP	DPV	0.01–2	0.004	[86]
	Antibody	GO/AuNP	FRET	$10^{-4} \times 2.5$	0.5	[95]
	Antibody	MWCNT	EIS	0.05–20	0.04	[117]
	Antibody	CNF/AuNP	DPV	0.0025–5	0.00168	[118]
TMD-NP	Aptamer	MoS_2	DPV	0–199	1.99×10^{-3}	[108]
	Antibody/antibody	MoS_2/AuNRs	DPV	0.01–20	0.005	[109]
	Aptamer	MoS_2 QD	Fluorescence	$19.90–43.8 \times 10^3$	9.95×10^{-3}	[113]
	Aptamer	MoS_2	Fluorescence	0.01–50	0.02	[119]
	Antibody/antibody	MoS_2/AuNCs	DPV	0.001–1000	3×10^{-4}	[120]
Others	Aptamer	-	Fluorescence	10–100	0.110	[76]
	Aptamer	fluorescence resonance energy transfer based QD	Fluorescence	10^{-4}–100	10.4	[114]
	Aptamer	UCNP	Fluorescence	0.1–50	25×10^{-6}	[115]
	Antibody	Ag@MSN	CA	$0.5–30 \times 10^3$	0.2	[121]
	Antibody	Cds QD	ECL	0.01–50	0.0028	[122]

Table 2. Features and disadvantages of each nanomaterial for MC biosensor construction.

Materials	Features	Disadvantages
Noble metal-based nanomaterial	High stability and durability High conductivity (Ex 1. Ag (6.30×10^7 S/m et 20 °C)) (Ex 2. Au (4.11×10^7 S/m et 20 °C)) [123]	Expensive cost (Ex. Au nanopowder (USD 446 per 1 g))
Carbon-based nanomaterial	Cheap cost Stable semiconducting property (Ex. Carbon nanopowder (USD 11 per 1 g))	Easy to oxidize Low conductivity (Ex 1. carbon-nanotube) (3×10^7 S/m) (Ex 2. graphene (1.72×10^7 S/m)) [124]
Transition metal dichalcogenide-based nanomaterial	Unique conductivity (topological insulator) Unique chemical property	Hard to particulate Low conductivity MoS_2 (10^{-1} to 10^1 S/m) [125]

Although conventional ELISA is mainly based on antibodies and is widely performed to MC detection, the development of aptamers for electrochemical and optical sensors, as well as a new type of ELISA, has become a powerful substitute for antibodies. Aptamers exhibit similar detection capabilities as that of existing antibodies, and because they can be chemically synthesized, they have a low production cost and can be produced more ethically than antibodies. This combination of bioreceptors and nanomaterials can be applied not only for the detection of MC but also to detect other toxic substances such as saxitoxin, anatoxin, and cylindrospermopsin.

However, in order for an actual MC portable sensor to be utilized as a portable biosensor, the following conditions must be met. Although most studies described above were performed on actual samples, most extracted and used only one type of sample to detect toxic substances, that is, blue-green algae. Therefore, an appropriate pretreatment system that can be applied directly in the field should be developed simultaneously. In addition, consistent MC detection ability should be obtained, even for mixed samples such as various blue-green algae or freshwater samples. If miniature equipment is used, the instrument (electrochemical or spectroscopic equipment) performance tends to decrease. Therefore, an optimal size should be considered for MC detection equipment. Finally, the reaction time between the current MC and manufactured bioreceptors/nanomaterial is naturally reversible according to the thermodynamic entropy, but it takes several hours of target binding capacity. Therefore, in order for the detection time to be reduced, an external voltage can be applied to construct an effective portable sensor. If these factors are achieved, an ideal system that can detect the toxic concentration of cyanobacteria directly at the site of occurrence can be developed, rather than conducting a precise analysis in the laboratory.

Author Contributions: T.L. and J.M. conceived and designed the manuscript; H.P. and G.K. wrote the metal nanoparticle-based biosensor section; G.K. and Y.S. wrote the carbon-based biosensor section; C.P., Y.Y. and T.L. wrote the TMD-based biosensor. All authors have read and agreed to the published version of the manuscript.

Funding: This work was supported by the National Research Foundation of Korea (NRF) grant funded by the Korean government (MSIT) (no. 2021R1C1C1005583) and by the Korea Environment Industry and Technology Institute (KEITI) through the program for the management of aquatic ecosystem health, funded by the Korea Ministry of Environment (MOE) (2020003030001) and by the Industrial Core Technology Development Program (20009121, Development of early diagnostic system of peritoneal fibrosis by multiplex detection of exosomal nucleic acids and protein markers) funded by the Ministry of Trade, Industry and Energy (MOTIE, Korea).

Institutional Review Board Statement: Not applicable.

Informed Consent Statement: Not applicable.

Data Availability Statement: Not applicable.

Conflicts of Interest: The authors declare no conflict of interest.

References

1. Heisler, J.; Glibert, P.M.; Burkholder, J.M.; Anderson, D.M.; Cochlan, W.; Dennison, W.C.; Dortch, Q.; Gobler, C.J.; Heil, C.A.; Humphries, E. Eutrophication and harmful algal blooms: A scientific consensus. *Harmful Algae* **2008**, *8*, 3–13. [CrossRef]
2. Anderson, D.M. Approaches to monitoring, control and management of harmful algal blooms (HABs). *Ocean Coast. Manag.* **2009**, *52*, 342–347. [CrossRef]
3. Abdallah, M.F.; Van Hassel, W.H.; Andjelkovic, M.; Wilmotte, A.; Rajkovic, A. Cyanotoxins and Food Contamination in Developing Countries: Review of Their Types, Toxicity, Analysis, Occurrence and Mitigation Strategies. *Toxins* **2021**, *13*, 786. [CrossRef] [PubMed]
4. Steffen, M.M.; Belisle, B.S.; Watson, S.B.; Boyer, G.L.; Wilhelm, S.W. Status, causes and controls of cyanobacterial blooms in Lake Erie. *J. Great Lakes Res.* **2014**, *40*, 215–225. [CrossRef]
5. Weller, M.G. Immunoassays and biosensors for the detection of cyanobacterial toxins in water. *Sensors* **2013**, *13*, 15085–15112. [CrossRef]
6. Harke, M.J.; Steffen, M.M.; Gobler, C.J.; Otten, T.G.; Wilhelm, S.W.; Wood, S.A.; Paerl, H.W. A review of the global ecology, genomics, and biogeography of the toxic cyanobacterium, *Microcystis* spp. *Harmful Algae* **2016**, *54*, 4–20. [CrossRef]

7.   Magalhaes, V.d.; Marinho, M.M.; Domingos, P.; Oliveira, A.C.; Costa, S.M.; Azevedo, L.d.; Azevedo, S.M. Microcystins (cyanobacteria hepatotoxins) bioaccumulation in fish and crustaceans from Sepetiba Bay (Brasil, RJ). *Toxicon* **2003**, *42*, 289–295. [CrossRef]
8.   Lee, S.; Kim, J.; Choi, B.; Kim, G.; Lee, J. Harmful algal blooms and liver diseases: Focusing on the areas near the four major rivers in South Korea. *J. Environ. Sci. Health C* **2019**, *37*, 356–370. [CrossRef]
9.   Paerl, H.W.; Huisman, J. Climate change: A catalyst for global expansion of harmful cyanobacterial blooms. *Environ. Microbiol. Rep.* **2009**, *1*, 27–37. [CrossRef]
10. Wu, D.; Li, R.; Zhang, F.; Liu, J. A review on drone-based harmful algae blooms monitoring. *Environ. Monit. Assess.* **2019**, *191*, 211. [CrossRef] [PubMed]
11. Rastogi, R.P.; Sinha, R.P.; Incharoensakdi, A. The cyanotoxin-microcystins: Current overview. *Rev. Environ. Sci. Biotechnol.* **2014**, *13*, 215–249. [CrossRef]
12. Mehinto, A.C.; Smith, J.; Wenger, E.; Stanton, B.; Linville, R.; Brooks, B.W.; Sutula, M.A.; Howard, M.D. Synthesis of ecotoxicological studies on cyanotoxins in freshwater habitats–Evaluating the basis for developing thresholds protective of aquatic life in the United States. *Sci. Total Environ.* **2021**, *795*, 148864. [CrossRef] [PubMed]
13. Sorlini, S.; Abbà, A.; Collivignarelli, M.C. CONTROL MEASURES FOR Cyanobacteria AND Cyanotoxins IN DRINKING WATER. *Environ. Eng. Manag. J.* **2018**, *17*, 2455–2463. [CrossRef]
14. Kaur, S.; Srivastava, A.; Ahluwalia, A.S.; Mishra, Y. Cyanobacterial blooms and Cyanotoxins: Occurrence and Detection. In *Algae*; Springer: Singapore, 2021; pp. 339–352.
15. Khomutovska, N.; Sandzewicz, M.; Łach, Ł.; Suska-Malawska, M.; Chmielewska, M.; Mazur-Marzec, H.; Cegłowska, M.; Niyatbekov, T.; Wood, S.A.; Puddick, J. Limited microcystin, anatoxin and cylindrospermopsin production by cyanobacteria from microbial mats in cold deserts. *Toxins* **2020**, *12*, 244. [CrossRef] [PubMed]
16. Metcalf, J.S.; Codd, G.A. Cyanobacteria, neurotoxins and water resources: Are there implications for human neurodegenerative disease? *Amyotroph. Lat. Scler.* **2009**, *10*, 74–78. [CrossRef]
17. Catherine, A.; Bernard, C.; Spoof, L.; Bruno, M. Microcystins and Nodularins. In *Handbook of Cyanobacterial Monitoring and Cyanotoxin Analysis*; Meriluoto, J., Lisa Spoof, L., Codd, G.A., Eds.; John Wiley & Sons, Ltd.: Chichester, UK, 2017; pp. 107–126.
18. McElhiney, J.; Lawton, L.A. Detection of the cyanobacterial hepatotoxins microcystins. *Toxicol. Appl. Pharmacol.* **2005**, *203*, 219–230. [CrossRef]
19. Zurawell, R.W.; Chen, H.; Burke, J.M.; Prepas, E.E. Hepatotoxic cyanobacteria: A review of the biological importance of microcystins in freshwater environments. *J. Toxicol. Environ. Health Part B* **2005**, *8*, 1–37. [CrossRef] [PubMed]
20. Singh, S.; Srivastava, A.; Oh, H.; Ahn, C.; Choi, G.; Asthana, R.K. Recent trends in development of biosensors for detection of microcystin. *Toxicon* **2012**, *60*, 878–894. [CrossRef]
21. Abe, T.; Lawson, T.; Weyers, J.D.; Codd, G.A. Microcystin-LR inhibits photosynthesis of Phaseolus vulgaris primary leaves: Implications for current spray irrigation practice. *New Phytol.* **1996**, *133*, 651–658. [CrossRef]
22. Pang, P.; Lai, Y.; Zhang, Y.; Wang, H.; Conlan, X.A.; Barrow, C.J.; Yang, W. Recent advancement of biosensor technology for the detection of microcystin-LR. *Bull. Chem. Soc. Jpn.* **2020**, *93*, 637–646. [CrossRef]
23. ISO 20179: 2005 Water Quality-Determination of Microcystins-Method Using Solid Phase extraction (SPE) and High Performance Liquid Chromatography (HPLC) with Ultraviolet (UV) Detection. 2005. Available online: https://www.iso.org/standard/34098.html (accessed on 16 December 2021).
24. He, X.; Stanford, B.D.; Adams, C.; Rosenfeldt, E.J.; Wert, E.C. Varied influence of microcystin structural difference on ELISA cross-reactivity and chlorination efficiency of congener mixtures. *Water Res.* **2017**, *126*, 515–523. [CrossRef]
25. Khreich, N.; Lamourette, P.; Renard, P.; Clavé, G.; Fenaille, F.; Créminon, C.; Volland, H. A highly sensitive competitive enzyme immunoassay of broad specificity quantifying microcystins and nodularins in water samples. *Toxicon* **2009**, *53*, 551–559. [CrossRef]
26. Cesewski, E.; Johnson, B.N. Electrochemical biosensors for pathogen detection. *Biosens. Bioelectron.* **2020**, *159*, 112214. [CrossRef]
27. Vidic, J.; Manzano, M.; Chang, C.; Jaffrezic-Renault, N. Advanced biosensors for detection of pathogens related to livestock and poultry. *Vet. Res.* **2017**, *48*, 1–22. [CrossRef] [PubMed]
28. Dhara, K.; Mahapatra, D.R. Review on electrochemical sensing strategies for C-reactive protein and cardiac troponin I detection. *Microchem. J.* **2020**, *156*, 104857. [CrossRef]
29. Lee, Y.; Choi, J.; Han, H.; Park, S.; Park, S.Y.; Park, C.; Baek, C.; Lee, T.; Min, J. Fabrication of ultrasensitive electrochemical biosensor for dengue fever viral RNA Based on CRISPR/Cpf1 reaction. *Sens. Actuators B Chem.* **2021**, *326*, 128677. [CrossRef]
30. Raja, I.S.; Vedhanayagam, M.; Preeth, D.R.; Kim, C.; Lee, J.H.; Han, D.W. Development of Two-Dimensional Nanomaterials Based Electrochemical Biosensors on Enhancing the Analysis of Food Toxicants. *Int. J. Mol. Sci.* **2021**, *22*, 3277. [CrossRef] [PubMed]
31. Lee, T.; Mohammadniaei, M.; Zhang, H.; Yoon, J.; Choi, H.K.; Guo, S.; Guo, P.; Choi, J. Single Functionalized pRNA/Gold Nanoparticle for Ultrasensitive MicroRNA Detection Using Electrochemical Surface-Enhanced Raman Spectroscopy. *Adv. Sci.* **2020**, *7*, 1902477. [CrossRef]
32. Mitchell, M.J.; Billingsley, M.M.; Haley, R.M.; Wechsler, M.E.; Peppas, N.A.; Langer, R. Engineering precision nanoparticles for drug delivery. *Nat. Rev. Drug Discov.* **2021**, *20*, 101–124. [CrossRef] [PubMed]
33. Minea, A.A. A review on electrical conductivity of nanoparticle-enhanced fluids. *Nanomaterials* **2019**, *9*, 1592. [CrossRef] [PubMed]
34. Senf, B.; Yeo, W.; Kim, J. Recent advances in portable biosensors for biomarker detection in body fluids. *Biosensors* **2020**, *10*, 127. [CrossRef]

35. Mehlhorn, A.; Rahimi, P.; Joseph, Y. Aptamer-based biosensors for antibiotic detection: A review. *Biosensors* **2018**, *8*, 54. [CrossRef] [PubMed]
36. Hara, T.O.; Singh, B. Electrochemical Biosensors for Detection of Pesticides and Heavy Metal Toxicants in Water: Recent Trends and Progress. *ACS EST Water* **2021**, *1*, 462–478. [CrossRef]
37. Wang, X.; Wang, S. Sensors and biosensors for the determination of small molecule biological toxins. *Sensors* **2008**, *8*, 6045–6054. [CrossRef]
38. Buckova, M.; Licbinsky, R.; Jandova, V.; Krejci, J.; Pospichalova, J.; Huzlik, J. Fast ecotoxicity detection using biosensors. *Water Air Soil Pollut.* **2017**, *228*, 166. [CrossRef]
39. Cunha, I.; Biltes, R.; Sales, M.G.F.; Vasconcelos, V. Aptamer-based biosensors to detect aquatic phycotoxins and cyanotoxins. *Sensors* **2018**, *18*, 2367. [CrossRef]
40. Bertani, P.; Lu, W. Cyanobacterial toxin biosensors for environmental monitoring and protection. *Med. Nov. Technol. Devices* **2021**, *10*, 100059. [CrossRef]
41. Massey, I.Y.; Wu, P.; Wei, J.; Luo, J.; Ding, P.; Wei, H.; Yang, F. A mini-review on detection methods of microcystins. *Toxins* **2020**, *12*, 641. [CrossRef] [PubMed]
42. Naresh, V.; Lee, N. A Review on Biosensors and Recent Development of Nanostructured Materials-Enabled Biosensors. *Sensors* **2021**, *21*, 1109. [CrossRef]
43. Sharma, P.; Pandey, V.; Sharma, M.M.M.; Patra, A.; Singh, B.; Mehta, S.; Husen, A. A Review on Biosensors and Nanosensors Application in Agroecosystems. *Nanoscale Res. Lett.* **2021**, *16*, 1–24. [CrossRef]
44. Ma, J.; Du, M.; Wang, C.; Xie, X.; Wang, H.; Zhang, Q. Advances in airborne microorganisms detection using biosensors: A critical review. *Front. Environ. Sci. Eng.* **2021**, *15*, 47. [CrossRef] [PubMed]
45. Yu, H.; Yu, J.; Li, L.; Zhang, Y.; Xin, S.; Ni, X.; Sun, Y.; Song, K. Recent Progress of the Practical Applications of the Platinum Nanoparticle-Based Electrochemistry Biosensors. *Front. Chem.* **2021**, *9*, 677876. [CrossRef]
46. Chen, C.; Wang, J. Optical biosensors: An exhaustive and comprehensive review. *Analyst* **2020**, *145*, 1605–1628. [CrossRef]
47. Gergeroglu, H.; Yildirim, S.; Ebeoglugil, M.F. Nano-carbons in biosensor applications: An overview of carbon nanotubes (CNTs) and fullerenes (C 60). *Appl. Sci.* **2020**, *2*, 603. [CrossRef]
48. Kim, C.; Jung, J.; Yoon, K.R.; Youn, D.; Park, S.; Kim, I. A high-capacity and long-cycle-life lithium-ion battery anode architecture: Silver nanoparticle-decorated SnO2/NiO nanotubes. *ACS Nano* **2016**, *10*, 11317–11326. [CrossRef]
49. Hu, Y.; Huang, Y.; Tan, C.; Zhang, X.; Lu, Q.; Sindoro, M.; Huang, X.; Huang, W.; Wang, L.; Zhang, H. Two-dimensional transition metal dichalcogenide nanomaterials for biosensing applications. *Mater. Chem. Front.* **2017**, *1*, 24–36. [CrossRef]
50. Khan, I.; Saeed, K.; Khan, I. Nanoparticles: Properties, applications and toxicities. *Arab. J. Chem.* **2019**, *12*, 908–931. [CrossRef]
51. Cao, X.; Ye, Y.; Liu, S. Gold nanoparticle-based signal amplification for biosensing. *Anal. Biochem.* **2011**, *417*, 1–16. [CrossRef] [PubMed]
52. Sinha, A.; Zhao, H.; Huang, Y.; Lu, X.; Chen, J.; Jain, R. MXene: An emerging material for sensing and biosensing. *Trends Anal. Chem.* **2018**, *105*, 424–435. [CrossRef]
53. Laurent, S.; Mahmoudi, M. Superparamagnetic iron oxide nanoparticles: Promises for diagnosis and treatment of cancer. *Int. J. Mol. Epidemiol. Genet.* **2011**, *2*, 367.
54. Loiseau, A.; Asila, V.; Boitel-Aullen, G.; Lam, M.; Salmain, M.; Boujday, S. Silver-based plasmonic nanoparticles for and their use in biosensing. *Biosensors* **2019**, *9*, 78. [CrossRef] [PubMed]
55. Lee, T.; Lee, Y.; Park, S.Y.; Hong, K.; Kim, Y.; Park, C.; Chung, Y.; Lee, M.; Min, J. Fabrication of electrochemical biosensor composed of multi-functional DNA structure/Au nanospike on micro-gap/PCB system for detecting troponin I in human serum. *Colloids Surf. B.* **2019**, *175*, 343–350. [CrossRef]
56. Zhang, J.; Xiong, Z.; Chen, Z. Ultrasensitive electrochemical microcystin-LR immunosensor using gold nanoparticle functional polypyrrole microsphere catalyzed silver deposition for signal amplification. *Sens. Actuators B Chem.* **2017**, *246*, 623–630. [CrossRef]
57. Hutter, E.; Fendler, J.H. Exploitation of localized surface plasmon resonance. *Adv Mater.* **2004**, *16*, 1685–1706. [CrossRef]
58. Willets, K.A.; Van Duyne, R.P. Localized surface plasmon resonance spectroscopy and sensing. *Annu. Rev. Phys. Chem.* **2007**, *58*, 267–297. [CrossRef]
59. Mayer, K.M.; Hafner, J.H. Localized surface plasmon resonance sensors. *Chem. Rev.* **2011**, *111*, 3828–3857. [CrossRef] [PubMed]
60. Chen, Y.; Ming, H. Review of surface plasmon resonance and localized surface plasmon resonance sensor. *Photonic Sens.* **2012**, *2*, 37–49. [CrossRef]
61. Wang, F.; Liu, S.; Lin, M.; Chen, X.; Lin, S.; Du, X.; Li, H.; Ye, H.; Qiu, B.; Lin, Z.; et al. Colorimetric detection of microcystin-LR based on disassembly of orient-aggregated gold nanoparticle dimers. *Biosens. Bioelectron.* **2015**, *68*, 475–480. [CrossRef]
62. Sun, X.; Wu, L.; Ji, J.; Jiang, D.; Zhang, Y.; Li, Z.; Zhang, G.; Zhang, H. Longitudinal surface plasmon resonance assay enhanced by magnetosomes for simultaneous detection of Pefloxacin and Microcystin-LR in seafoods. *Biosens. Bioelectron.* **2013**, *47*, 318–323. [CrossRef]
63. Stiles, P.L.; Dieringer, J.A.; Shah, N.C.; Van Duyne, R.P. Surface-enhanced Raman spectroscopy. *Annu. Rev. Anal. Chem.* **2008**, *1*, 601–626. [CrossRef] [PubMed]
64. Haynes, C.L.; McFarland, A.D.; Van Duyne, R.P. Surface-enhanced Raman spectroscopy. *Anal. Chem.* **2005**, *77*, 338–346. [CrossRef]
65. Moskovits, M. Surface-enhanced Raman spectroscopy: A brief retrospective. *J. Raman Spectrosc.* **2005**, *36*, 485–496. [CrossRef]

66. Le Ru, E.; Etchegoin, P. *Principles of Surface-Enhanced Raman Spectroscopy: And Related Plasmonic Effects*; Elsevier: Amsterdam, The Netherlands, 2008.
67. Tian, Z.Q. Surface-enhanced Raman spectroscopy: Advancements and applications. *J. Raman Spectrosc.* **2005**, *36*, 466–470. [CrossRef]
68. He, D.; Wu, Z.; Cui, B.; Jin, Z. A novel SERS-based aptasensor for ultrasensitive sensing of microcystin-LR. *Food Chem.* **2019**, *278*, 197–202. [CrossRef] [PubMed]
69. Li, M.; Paidi, S.K.; Sakowski, E.; Preheim, S.; Barman, I. Ultrasensitive detection of hepatotoxic microcystin production from cyanobacteria using surface-enhanced Raman scattering immunosensor. *ACS Sens.* **2019**, *4*, 1203–1210. [CrossRef] [PubMed]
70. Yu, L.; Li, N. Noble metal nanoparticles-based colorimetric biosensor for visual quantification: A mini review. *Chemosensors* **2019**, *7*, 53. [CrossRef]
71. Kanjanawarut, R.; Su, X. Colorimetric detection of DNA using unmodified metallic nanoparticles and peptide nucleic acid probes. *Anal. Chem.* **2009**, *81*, 6122–6129. [CrossRef]
72. Sun, J.; Lu, Y.; He, L.; Pang, J.; Yang, F.; Liu, Y. Colorimetric sensor array based on gold nanoparticles: Design principles and recent advances. *TrAC Trends Anal. Chem.* **2020**, *122*, 115754. [CrossRef]
73. Zhang, Y.; Zhu, Z.; Teng, X.; Lai, Y.; Pu, S.; Pang, P.; Wang, H.; Yang, C.; Barrow, C.J.; Yang, W. Enzyme-free fluorescent detection of microcystin-LR using hairpin DNA-templated copper nanoclusters as signal indicator. *Talanta* **2019**, *202*, 279–284. [CrossRef]
74. Ruscito, A.; DeRosa, M.C. Small-molecule binding aptamers: Selection strategies, characterization, and applications. *Front. Chem.* **2016**, *4*, 14. [CrossRef] [PubMed]
75. Zhou, W.; Ding, J.; Liu, J. A highly specific sodium aptamer probed by 2-aminopurine for robust Na sensing. *Nucleic Acids Res.* **2016**, *44*, 10377–10385. [CrossRef]
76. Alhadrami, H.A.; Chinnappan, R.; Eissa, S.; Rahamn, A.A.; Zourob, M. High affinity truncated DNA aptamers for the development of fluorescence based progesterone biosensors. *Anal. Biochem.* **2017**, *525*, 78–84. [CrossRef]
77. Li, X.; Cheng, R.; Shi, H.; Tang, B.; Xiao, H.; Zhao, G. A simple highly sensitive and selective aptamer-based colorimetric sensor for environmental toxins microcystin-LR in water samples. *J. Hazard. Mater.* **2016**, *304*, 474–480. [CrossRef]
78. He, F.; Liang, L.; Zhou, S.; Xie, W.; He, S.; Wang, Y.; Tlili, C.; Tong, S.; Wang, D. Label-free sensitive detection of microcystin-LR via aptamer-conjugated gold nanoparticles based on solid-state nanopores. *Langmuir* **2018**, *34*, 14825–14833. [CrossRef]
79. Xie, W.; He, S.; Fang, S.; Liang, L.; Shi, B.; Wang, D. Visualizing of AuNPs Protection Aptamer from DNase I enzyme digestion Based on Nanopipette and Its Use for Microcystin-LR Detection. *Anal. Chim. Acta* **2021**, *1173*, 338698. [CrossRef]
80. Liang, X.; Li, N.; Zhang, R.; Yin, P.; Zhang, C.; Yang, N.; Liang, K.; Kong, B. Carbon-based SERS biosensor: From substrate design to sensing and bioapplication. *NPG Asia Mater.* **2021**, *13*, 1–36. [CrossRef]
81. Hwang, H.S.; Jeong, J.W.; Kim, Y.A.; Chang, M. Carbon nanomaterials as versatile platforms for biosensing applications. *Micromachines* **2020**, *11*, 814. [CrossRef]
82. Krishnan, S.K.; Singh, E.; Singh, P.; Meyyappan, M.; Nalwa, H.S. A review on graphene-based nanocomposites for electrochemical and fluorescent biosensors. *RSC Adv.* **2019**, *9*, 8778–8881. [CrossRef]
83. Bai, Y.; Xu, T.; Zhang, X. Graphene-based biosensors for detection of biomarkers. *Micromachines* **2020**, *11*, 60. [CrossRef] [PubMed]
84. Yoon, J.; Shin, M.; Lim, J.; Lee, J.; Choi, J. Recent Advances in MXene Nanocomposite-Based Biosensors. *Biosensors* **2020**, *10*, 185. [CrossRef] [PubMed]
85. Zhao, H.; Tian, J.; Quan, X. A graphene and multienzyme functionalized carbon nanosphere-based electrochemical immunosensor for microcystin-LR detection. *Colloids Surf. B.* **2013**, *103*, 38–44. [CrossRef] [PubMed]
86. Zhang, J.; Lei, J.; Pan, R.; Leng, C.; Hu, Z.; Ju, H. In situ assembly of gold nanoparticles on nitrogen-doped carbon nanotubes for sensitive immunosensing of microcystin-LR. *Chem. Commun.* **2011**, *47*, 668–670. [CrossRef]
87. Novoselov, K.S.; Geim, A.K.; Morozov, S.V.; Jiang, D.; Zhang, Y.; Dubonos, S.V.; Grigorieva, I.V.; Firsov, A.A. Electric field effect in atomically thin carbon films. *Science* **2004**, *306*, 666–669. [CrossRef]
88. Zhao, X.; Kong, R.; Zhang, X.; Meng, H.; Liu, W.; Tan, W.; Shen, G.; Yu, R. Graphene–DNAzyme based biosensor for amplified fluorescence "turn-on" detection of Pb2 with a high selectivity. *Anal. Chem.* **2011**, *83*, 5062–5066. [CrossRef] [PubMed]
89. Li, Y.; Zhou, W.; Wang, H.; Xie, L.; Liang, Y.; Wei, F.; Idrobo, J.; Pennycook, S.J.; Dai, H. An oxygen reduction electrocatalyst based on carbon nanotube–graphene complexes. *Nat. Nanotechnol.* **2012**, *7*, 394–400. [CrossRef] [PubMed]
90. Gilje, S.; Han, S.; Wang, M.; Wang, K.L.; Kaner, R.B. A chemical route to graphene for device applications. *Nano Lett.* **2007**, *7*, 3394–3398. [CrossRef]
91. Stankovich, S.; Dikin, D.A.; Dommett, G.H.; Kohlhaas, K.M.; Zimney, E.J.; Stach, E.A.; Piner, R.D.; Nguyen, S.T.; Ruoff, R.S. Graphene-based composite materials. *Nature* **2006**, *442*, 282–286. [CrossRef] [PubMed]
92. Hirata, M.; Gotou, T.; Horiuchi, S.; Fujiwara, M.; Ohba, M. Thin-film particles of graphite oxide 1:: High-yield synthesis and flexibility of the particles. *Carbon* **2004**, *42*, 2929–2937. [CrossRef]
93. Szabó, T.; Szeri, A.; Dékány, I. Composite graphitic nanolayers prepared by self-assembly between finely dispersed graphite oxide and a cationic polymer. *Carbon* **2005**, *43*, 87–94. [CrossRef]
94. Pan, S.; Chen, X.; Li, X.; Cai, M.; Shen, H.; Zhao, Y.; Jin, M. In situ controllable synthesis of graphene oxide-based ternary magnetic molecularly imprinted polymer hybrid for efficient enrichment and detection of eight microcystins. *J. Mater. Chem. A.* **2015**, *3*, 23042–23052. [CrossRef]

95. Shi, Y.; Wu, J.; Sun, Y.; Zhang, Y.; Wen, Z.; Dai, H.; Wang, H.; Li, Z. A graphene oxide based biosensor for microcystins detection by fluorescence resonance energy transfer. *Biosens. Bioelectron.* **2012**, *38*, 31–36. [CrossRef] [PubMed]
96. Manzeli, S.; Ovchinnikov, D.; Pasquier, D.; Yazyev, O.V.; Kis, A. 2D transition metal dichalcogenides. *Nat. Rev. Mater.* **2017**, *2*, 1–15. [CrossRef]
97. Chowdhury, T.; Sadler, E.C.; Kempa, T.J. Progress and Prospects in Transition-Metal Dichalcogenide Research Beyond 2D. *Chem. Rev.* **2020**, *120*, 12563–12591. [CrossRef]
98. Zhang, X.; Teng, S.Y.; Loy, A.C.M.; How, B.S.; Leong, W.D.; Tao, X. Transition metal dichalcogenides for the application of pollution reduction: A review. *Nanomaterials* **2020**, *10*, 1012. [CrossRef] [PubMed]
99. Kaushik, S.; Tiwari, U.K.; Deep, A.; Sinha, R.K. Two-dimensional transition metal dichalcogenides assisted biofunctionalized optical fiber SPR biosensor for efficient and rapid detection of bovine serum albumin. *Sci. Rep.* **2019**, *9*, 6987. [CrossRef] [PubMed]
100. Wang, Y.; Huang, K.; Wu, X. Recent advances in transition-metal dichalcogenides based electrochemical biosensors: A review. *Biosens. Bioelectron.* **2017**, *97*, 305–316. [CrossRef]
101. Gracia-Abad, R.; Sangiao, S.; Bigi, C.; Kumar Chaluvadi, S.; Orgiani, P.; De Teresa, J.M. Omnipresence of Weak Antilocalization (WAL) in Bi2Se3 Thin Films: A Review on Its Origin. *Nanomaterials* **2021**, *11*, 1077. [CrossRef]
102. Yang, M.; Hwang, S.; Jeong, J.; Huh, Y.S.; Choi, B.G. Nitrogen-doped carbon-coated molybdenum disulfide nanosheets for high-performance supercapacitor. *Synth. Met.* **2015**, *209*, 528–533. [CrossRef]
103. Song, I.; Park, C.; Choi, H.C. Synthesis and properties of molybdenum disulphide: From bulk to atomic layers. *RSC Adv.* **2015**, *5*, 7495–7514. [CrossRef]
104. Shin, M.; Yoon, J.; Yi, C.; Lee, T.; Choi, J. Flexible HIV-1 biosensor based on the Au/MoS2 nanoparticles/Au nanolayer on the PET substrate. *Nanomaterials* **2019**, *9*, 1076. [CrossRef] [PubMed]
105. Yoon, J.; Lee, S.N.; Shin, M.K.; Kim, H.; Choi, H.K.; Lee, T.; Choi, J. Flexible electrochemical glucose biosensor based on GOx/gold/MoS2/gold nanofilm on the polymer electrode. *Biosens. Bioelectron.* **2019**, *140*, 111343. [CrossRef]
106. Barua, S.; Dutta, H.S.; Gogoi, S.; Devi, R.; Khan, R. Nanostructured MoS2-based advanced biosensors: A review. *ACS Appl. Nano Mater.* **2017**, *1*, 2–25. [CrossRef]
107. Lee, J.; Dak, P.; Lee, Y.; Park, H.; Choi, W.; Alam, M.A.; Kim, S. Two-dimensional layered MoS 2 biosensors enable highly sensitive detection of biomolecules. *Sci. Rep.* **2014**, *4*, 7352. [CrossRef]
108. Liu, X.; Tang, Y.; Liu, P.; Yang, L.; Li, L.; Zhang, Q.; Zhou, Y.; Khan, M.Z.H. A highly sensitive electrochemical aptasensor for detection of microcystin-LR based on a dual signal amplification strategy. *Analyst* **2019**, *144*, 1671–1678. [CrossRef]
109. Zhang, Y.; Chen, M.; Li, H.; Yan, F.; Pang, P.; Wang, H.; Wu, Z.; Yang, W. A molybdenum disulfide/gold nanorod composite-based electrochemical immunosensor for sensitive and quantitative detection of microcystin-LR in environmental samples. *Sens. Actuators B Chem.* **2017**, *244*, 606–615. [CrossRef]
110. Ji, Q.; Zhang, Y.; Gao, T.; Zhang, Y.; Ma, D.; Liu, M.; Chen, Y.; Qiao, X.; Tan, P.; Kan, M. Epitaxial monolayer MoS$_2$ on mica with novel photoluminescence. *Nano Lett.* **2013**, *13*, 3870–3877. [CrossRef]
111. Guo, Y.; Li, J. MoS2 quantum dots: Synthesis, properties and biological applications. *Mater. Sci. Eng. C* **2020**, *109*, 110511. [CrossRef]
112. Huang, H.; Du, C.; Shi, H.; Feng, X.; Li, J.; Tan, Y.; Song, W. Water-soluble monolayer molybdenum disulfide quantum dots with upconversion fluorescence. *Part. Part. Syst. Charact.* **2015**, *32*, 72–79. [CrossRef]
113. Cao, H.; Dong, W.; Wang, T.; Shi, W.; Fu, C.; Wu, Y. Aptasensor Based on MoS2 Quantum Dots with Upconversion Fluorescence for Microcystin-LR Detection via the Inner Filter Effect. *ACS Sustain. Chem. Eng.* **2020**, *8*, 10939–10946. [CrossRef]
114. Dos Santos, M.B.; Queirós, R.B.; Geraldes, Á.; Marques, C.; Vilas-Boas, V.; Dieguez, L.; Paz, E.; Ferreira, R.; Morais, J.; Vasconcelos, V. Portable sensing system based on electrochemical impedance spectroscopy for the simultaneous quantification of free and total microcystin-LR in freshwaters. *Biosens. Bioelectron.* **2019**, *142*, 111550. [CrossRef] [PubMed]
115. Lee, E.; Son, A. Fluorescence resonance energy transfer based quantum dot-Aptasensor for the selective detection of microcystin-LR in eutrophic water. *Chem. Eng. J.* **2019**, *359*, 1493–1501. [CrossRef]
116. Wu, S.; Duan, N.; Zhang, H.; Wang, Z. Simultaneous detection of microcysin-LR and okadaic acid using a dual fluorescence resonance energy transfer aptasensor. *Anal. Bioanal. Chem.* **2015**, *407*, 1303–1312. [CrossRef]
117. Han, C.; Doepke, A.; Cho, W.; Likodimos, V.; de la Cruz, A.A.; Back, T.; Heineman, W.R.; Halsall, H.B.; Shanov, V.N.; Schulz, M.J.; et al. A multiwalled-carbon-nanotube-based biosensor for monitoring microcystin-LR in sources of drinking water supplies. *Adv. Funct. Mater.* **2013**, *23*, 1807–1816. [CrossRef]
118. Zhang, J.; Sun, Y.; Dong, H.; Zhang, X.; Wang, W.; Chen, Z. An electrochemical non-enzymatic immunosensor for ultrasensitive detection of microcystin-LR using carbon nanofibers as the matrix. *Sens. Actuators B Chem.* **2016**, *233*, 624–632. [CrossRef]
119. Lv, J.; Zhao, S.; Wu, S.; Wang, Z. Upconversion nanoparticles grafted molybdenum disulfide nanosheets platform for microcystin-LR sensing. *Biosens. Bioelectron.* **2017**, *90*, 203–209. [CrossRef]
120. Pang, P.; Teng, X.; Chen, M.; Zhang, Y.; Wang, H.; Yang, C.; Yang, W.; Barrow, C.J. Ultrasensitive enzyme-free electrochemical immunosensor for microcystin-LR using molybdenum disulfide/gold nanoclusters nanocomposites as platform and Au@ Pt core-shell nanoparticles as signal enhancer. *Sens. Actuators B Chem.* **2018**, *266*, 400–407. [CrossRef]
121. Zhao, C.; Hu, R.; Liu, T.; Liu, Y.; Bai, R.; Zhang, K.; Yang, Y. A non-enzymatic electrochemical immunosensor for microcystin-LR rapid detection based on Ag@ MSN nanoparticles. Colloids Surf. Physicochem. *Eng. Asp.* **2016**, *490*, 336–342. [CrossRef]

122. Zhang, J.; Kang, T.; Hao, Y.; Lu, L.; Cheng, S. Electrochemiluminescent immunosensor based on CdS quantum dots for ultrasensitive detection of microcystin-LR. *Sens. Actuators B Chem.* **2015**, *214*, 117–123. [CrossRef]
123. Raymond, A. *Serway. Principles of Physics*, 2nd ed.; Saunders College Pub.: Fort Worth, TX, USA; London, UK, 1998; p. 602.
124. Graphene Fiber, Method for Manufacturing Same and Use There of_(WO2011099761A2). Available online: https://patents.google.com/patent/WO2011099761A2/en (accessed on 16 December 2021).
125. Yakovleva, G.E.; Berdinsky, A.S.; Romanenko, A.I.; Khabarov, S.P.; Fedorov, V.E. The conductivity and TEMF of MoS_2 with Mo_2S_3 additive. In Proceedings of the 2015 38th International Convention on Information and Communication Technology, Electronics and Microelectronics (MIPRO), Opatija, Croatia, 25–29 May 2015; IEEE: Manhattan, NY, USA, 2015; pp. 12–14.

Review

Enzymatic Electrochemical/Fluorescent Nanobiosensor for Detection of Small Chemicals

Hye Kyu Choi [1] and Jinho Yoon [2,*]

[1] Department of Chemistry and Chemical Biology, Rutgers, The State University of New Jersey, Piscataway, NJ 08854, USA; hye.choi@rutgers.edu
[2] Department of Biomedical-Chemical Engineering, The Catholic University of Korea, 43 Jibong-ro, Bucheon-si 14662, Gyeonggi-do, Republic of Korea
* Correspondence: jyoon@catholic.ac.kr

Abstract: The detection of small molecules has attracted enormous interest in various fields, including the chemical, biological, and healthcare fields. In order to achieve such detection with high accuracy, up to now, various types of biosensors have been developed. Among those biosensors, enzymatic biosensors have shown excellent sensing performances via their highly specific enzymatic reactions with small chemical molecules. As techniques used to implement the sensing function of such enzymatic biosensors, electrochemical and fluorescence techniques have been mostly used for the detection of small molecules because of their advantages. In addition, through the incorporation of nanotechnologies, the detection property of each technique-based enzymatic nanobiosensors can be improved to measure harmful or important small molecules accurately. This review provides interdisciplinary information related to developing enzymatic nanobiosensors for small molecule detection, such as widely used enzymes, target small molecules, and electrochemical/fluorescence techniques. We expect that this review will provide a broad perspective and well-organized roadmap to develop novel electrochemical and fluorescent enzymatic nanobiosensors.

Keywords: nanobiosensor; enzymatic biosensor; electrochemistry; fluorescence; small molecules

Citation: Choi, H.K.; Yoon, J. Enzymatic Electrochemical/ Fluorescent Nanobiosensor for Detection of Small Chemicals. *Biosensors* **2023**, *13*, 492. https:// doi.org/10.3390/bios13040492

Received: 8 March 2023
Revised: 13 April 2023
Accepted: 18 April 2023
Published: 19 April 2023

1. Introduction

The accurate detection and quantification of small molecules have become increasingly important in various fields, such as chemistry, biology, medicine, and environmental science. Small molecules are ubiquitous in nature and play important roles in various biological processes, including gene expression, metabolism, and cell signaling [1,2]. For instance, neurotransmitters such as dopamine (DA), serotonin, and norepinephrine, are representative small molecules that regulate the mood, behavior, and cognitive function of organisms, and their accurate detection and quantification are essential for understanding neurological disorders such as depression [3,4] and anxiety. In addition, small molecules can be used as biomarkers for various diseases [5]. Therefore, the detection of small molecules can be critical for early disease diagnosis and treatment. For example, glucose is a small molecule that is used as a biomarker for diabetes, and its accurate quantification is necessary for monitoring blood glucose levels and adjusting insulin therapy [6]. Other important molecules, such as prostate-specific antigen (PSA) and carcinoembryonic antigen (CEA), are also used as significant biomarkers for prostate and colon cancer, respectively, in the biomedical field [7,8]. The accurate detection of PSA and CEA can be critical for early cancer diagnosis and treatment. In addition to their biomedical contribution, small molecules can also be used as indicators of environmental pollution and food contamination, which help to ensure public health and safety. Heavy metal ions, such as lead, cadmium, and mercury, are small molecules that can be present in industrial waste, and monitoring the level of these small molecules is necessary to confirm that these contaminants do not affect the environment and harm human health [9]. As described above, small molecules play critical roles in various fields, and their accurate detection and quantification are essential for disease diagnosis, drug discovery, environmental monitoring, and food safety. Therefore, the development of sensitive and selective biosensors for small molecules has the potential to advance these fields and improve human health and safety.

Among the different types of biosensors, enzymatic biosensors have shown improved sensing due to their highly specific enzymatic reactions with small chemical molecules [10]. Enzymatic biosensors typically consist of an enzyme that catalyzes a reaction with the target molecule and a transducer that converts the reaction into a measurable signal [11]. This transducer is developed based on various detection principles, such as electrochemical, optical, or mass-based techniques [12]. Enzymatic biosensors can provide reliable and accurate measurements of various analytes under different conditions. One of the advantages of enzymatic biosensors is their stability under different temperatures and pH levels, making them suitable for use in a variety of settings. Enzymatic biosensors can typically operate at a wide range of temperatures, which allows the biosensors to be used in various applications where the temperature may fluctuate, such as in industrial processes or the outdoors. Additionally, many enzymatic biosensors are stable at room temperature, making them convenient to use and store without the need for special refrigeration or temperature-controlled storage [13–15]. Their ability to function reliably under different conditions ensures that they can provide accurate measurements of analytes in various settings, from the laboratory to the field. Compared to non-enzymatic or enzyme-free biosensors, enzymatic biosensors have several advantages, including high specificity and sensitivity, low detection limits, and a wide detection range [16,17]. In addition, enzymatic biosensors can also be tailored to detect specific molecules, making them suitable for a wide range of applications, including clinical diagnosis, environmental monitoring, and food safety. Enzymatic biosensors offer several advantages over antibody- and aptamer-based biosensors. Firstly, enzymes are more cost-effective than antibodies, which can be expensive to purchase [18]. Additionally, in electrochemical biosensors, enzymes capable of undergoing redox reactions can be directly utilized as redox signal probes, whereas antibody-based biosensors typically require additional redox-active molecules such as methylene blue or ferritin [19]. Further, there are limitations to developing biosensors for detecting various small chemicals using aptamers unless a specific aptamer sequence is designed via complex techniques, such as the systematic evolution of ligands by exponential enrichment (SELEX) [20].

To achieve accurate detection of small molecules, electrochemical and fluorescence techniques have been mainly used. Electrochemical techniques are based on the measurement of current or potential changes resulting from the enzymatic reaction, while fluorescence techniques involve the detection of changes in the fluorescence intensity of the reaction product. Both techniques offer high sensitivity, selectivity, and specificity, making them suitable for the detection of a wide range of small molecules [21,22].

The incorporation of nanotechnologies into enzymatic biosensors has further enhanced the performance of enzymatic biosensors, improving their ability to detect harmful or important small molecules with high accuracy. Nanomaterials with unique properties, such as high surface area, conductivity, and catalytic properties, have been utilized to improve the sensitivity, selectivity, and stability of enzymatic biosensors [23]. For example, metal nanoparticles (NPs) and carbon-based nanomaterials (e.g., carbon nanotubes, graphene) have been used to improve the performance of enzymatic biosensors [24,25]. Using these nanomaterials, researchers have increased the surface area available for immobilizing enzymes, created a more favorable microenvironment for enzymatic reactions, and improved signal amplification [26]. The resulting improvements have led to enzymatic biosensors with enhanced sensitivity and selectivity for detecting small molecules. Additionally, the incorporation of nanomaterials has also facilitated the development of hybrid biosensors that combine the advantages of both electrochemical and optical detection techniques, leading to even better detection performance [27,28]. The incorporation of nanotechnologies has opened up new avenues for the development of enzymatic nanobiosensors, allowing researchers to fabricate biosensors that are highly selective and accurate in their detection of small molecules. With ongoing research and development, enzymatic nanobiosensors hold great promise for a wide range of applications in fields such as healthcare, environmental monitoring, and food safety.

Therefore, this review aims to provide interdisciplinary information on the development of enzymatic nanobiosensors for small molecule detection, including the widely used enzymes, electrochemical/fluorescence techniques, and electrochemical/fluorescent biosensors (Figure 1). Through a well-organized roadmap, this review will offer a broad perspective on the development of novel electrochemical and fluorescent enzymatic nanobiosensors for small molecule detection.

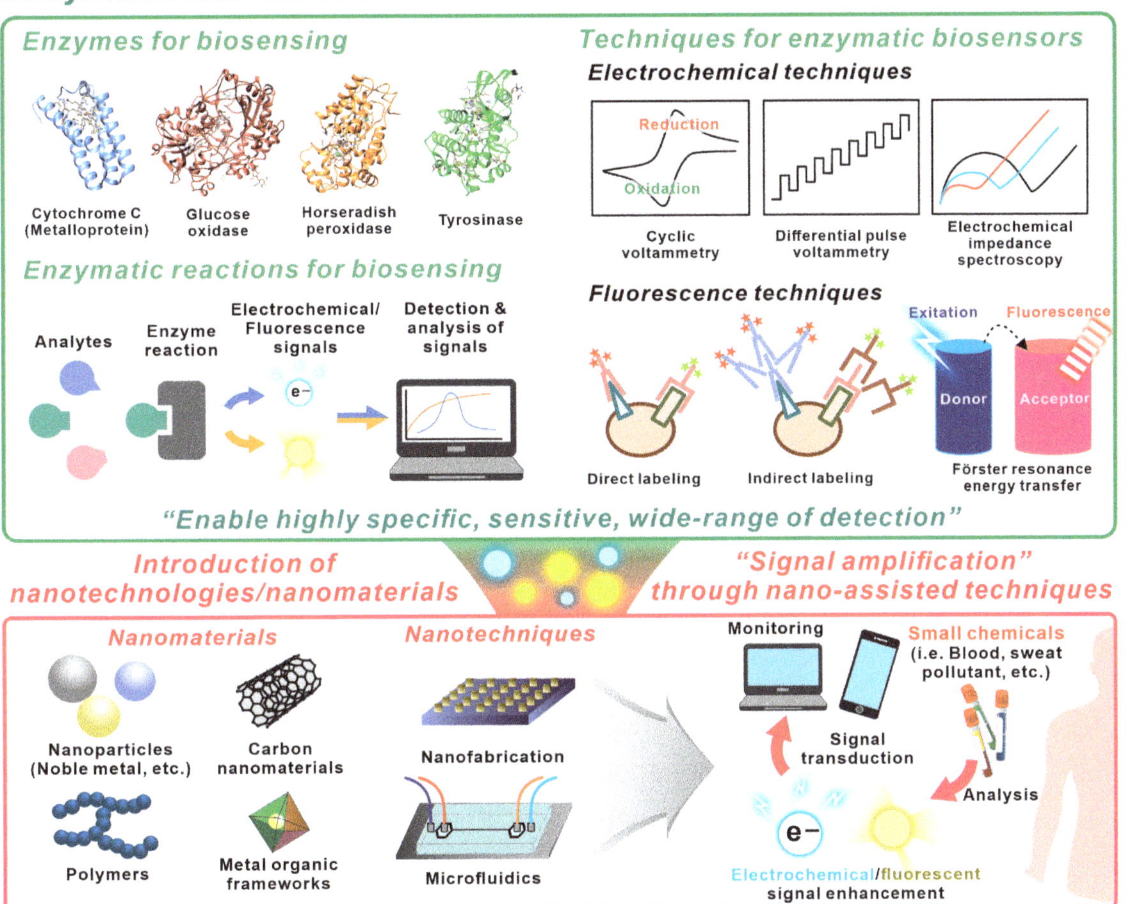

Figure 1. Schematic diagram of enzymatic electrochemical/fluorescent nanobiosensor for detection of small chemicals.

2. Components of Enzymatic Biosensors

2.1. Enzymes

Among various enzymes, metalloenzymes are mainly used in the development of enzymatic biosensors. Since metalloenzymes possess metal ions in their core structure, they can form enzymatic reactions with certain small molecules using these metal ions and have the potential to develop biosensors [29,30]. For example, metalloenzymes have been used to develop biosensors to detect small chemicals that are harmful or require sophisticated measurements in the environment. Representative examples of such enzymes include cytochrome c (Cyt C), myoglobin (Mb), hemoglobin (Hb), and nitrogenase (Nase). Using iron ions in their core, both Mb and Hb act as oxygen carriers. Cyt c possesses an iron ion that can react with hydrogen peroxide (H_2O_2) in an enzymatic manner and has been hugely applied to develop H_2O_2 biosensors [31,32]. Moreover, as shown in the reaction diagram of Mb in Figure 2A, Mb can also enzymatically react with nitric oxide (NO) [33]. Some examples of enzymatic reactions are discussed in more detail in the next subsection. In addition, Nase reacts with nitrogen and can thus be used to measure nitrogen, which is important in chemistry [34].

In addition, metalloenzymes and some enzymes capable of performing redox reactions are being used to diagnose chemical substances that play important roles in living organisms. Glucose oxidase (GOx) is one of the most widely used enzymes to develop the glucose biosensor. Since glucose plays a large role in diabetes, a metabolic disorder related to the regulation of blood glucose levels, the accurate monitoring of glucose is important in the healthcare field [35]. To achieve this goal, directly monitoring glucose in the blood using an implantable biosensor is more accurate than a traditional blood test. The implantable glucose biosensor uses poly (3,4-ethylenedioxythiophene) (PEDOT) modified carbon fiber (CF) as an electrode [36]. Here, GOx was introduced as an enzymatic probe to detect glucose based on the enzymatic reaction between flavin adenine dinucleotide (FAD) in the active site of GOx and glucose. During this enzymatic reaction, FAD in GOx converts glucose to gluconolactone, and the generated electrons are measured and quantified using the biosensor. GOx is used as a key component not only in the development of the glucose biosensor at the laboratory level but also in the development of commercially available glucose biosensors. Similarly to this, urease, which contains nickel ions, is used to detect urea in healthcare monitoring applications [37].

Recently, in line with stem cell research, some metalloenzymes are also being used for cell differentiation monitoring in the biomedical field. DA is an important chemical neurotransmitter that plays an important role in signal transmission in the nervous system. Horseradish peroxidase (HRP) is an enzyme that is widely used to detect DA [38]. In addition, tyrosinase (Tyr), containing copper ions, can oxidase catechol, a toxic small chemical compound [39]. Using the copper ions in its core, Tyr can react with catechol and convert it to *o*-quinone via oxidation and can also convert phenol to *o*-quinone via hydroxylation and oxidation enzymatic reactions, as shown in Figure 2B. In addition to the enzymes discussed here, there are numerous enzymes that can be used to detect small chemicals (e.g., laccase). Biosensors using such enzymes have high selectivity and specificity, but highly sensitive detection is difficult due to the intrinsic characteristics of biomolecules. In order to address this issue, various nanomaterials and nanotechnologies are being grafted to enzymatic biosensors. For instance, metal nanoparticles have been utilized in enzymatic biosensors for significant enhancement of signal amplification and sensitivity due to their large surface area for higher biorecognition sites and immobilization of biomolecules [40]. In addition, various nanomaterials and nanotechnologies, including 2D nanomaterials (e.g., graphene and carbon dots) and nanofabrication techniques (e.g., electron beam lithography), have been incorporated in enzymatic biosensors to improve sensitivity and stability [41–43].

2.2. Enzymatic Reactions for Biosensing Applications

In biosensing applications, enzymes are typically immobilized on the surface (electrode or substrate) of the sensor and react with a specific target molecule to produce a detectable signal. The processes of enzyme reactions in biosensing applications typically involve the following steps.

At first, the enzymes recognize and bind to the target molecule, typically through specific binding sites or active sites on the enzyme [44]. Next, the enzymes catalyze chemical reactions with the target molecules that lead to the production of detectable electrochemical/fluorescence signals. Finally, the signal produced by the enzyme reaction is detected and quantified using sensing techniques, such as electrochemical and fluorescence techniques.

To explain in detail, electrochemical biosensing of small molecules can be achieved by immobilizing the enzyme on an electrode surface, where it catalyzes a redox reaction with the target molecule, producing an electrical current that can be measured. For example, the detection of H_2O_2 can be achieved through an enzymatic reaction between metalloenzymes possessing iron ions (such as Mb) and the H_2O_2 that results in the iron ion of the metalloenzyme being oxidized from a Fe^{2+} state to a Fe^{3+} state accompanying the production of H_2O via the reduction of H_2O_2. During these redox enzymatic reactions, produced or transferred electrons can be measured easily using an electrochemical method. To mimic the efficient enzymatic reaction of metalloenzymes, several artificial metalloenzymes have been developed [45,46]. Likewise, for the detection of glucose, GOx can be immobilized on an electrode surface to catalyze the oxidation of glucose to gluconic acid with H_2O_2, producing an electrical current from a redox reaction that is proportional to the glucose concentration [47]. Similar to GOx, several oxidase enzymes (e.g., glutamate oxidase, lactate oxidase, and ascorbate oxidase) have been used as components for electrochemical enzymatic biosensors [48] (Figure 2C). Utilizing an enzyme that produces or modifies a fluorescent molecule upon reaction with the target molecule, fluorescent biosensing can also be conducted. For example, acetylcholinesterase (Ache) can be used to catalyze the hydrolysis of acetylcholine (Ach), producing choline and acetate. This reaction can be coupled to a fluorescent signal using a fluorescent probe that binds to the choline produced in the reaction, producing a measurable fluorescent signal [49]. In addition to the examples mentioned earlier, numerous other enzyme reactions can be used in biosensing applications. For instance, enzymes, such as peroxidase and catalase, can be used to catalyze the oxidation of various substrates using H_2O_2 as a co-substrate, producing detectable signals, such as color changes or oxygen gas bubbles [50]. Similarly, enzymes, such as alcohol dehydrogenase and lactate dehydrogenase, can be used to catalyze redox reactions with coenzymes, such as NAD+ and NADH, which can be measured electrochemically or optically [51]. In such reactions, the enzyme facilitates the transfer of electrons between the substrate and the coenzyme, resulting in a change in the redox state of the coenzyme that can be detected. Other enzymes, such as proteases and nucleases, can be used to cleave specific peptide or nucleic acid sequences, generating fragments that can be detected using various techniques, such as mass spectrometry or fluorescence. The enzymes play a crucial role in catalyzing the reactions and facilitating the formation of detectable signals. Using appropriate co-substrates, coenzymes, and reaction conditions can ensure that the redox reactions occur catalytically and produce robust signals. As seen in this section, enzymes have the potential to develop electrochemical or fluorescent biosensors for the detection of small molecules.

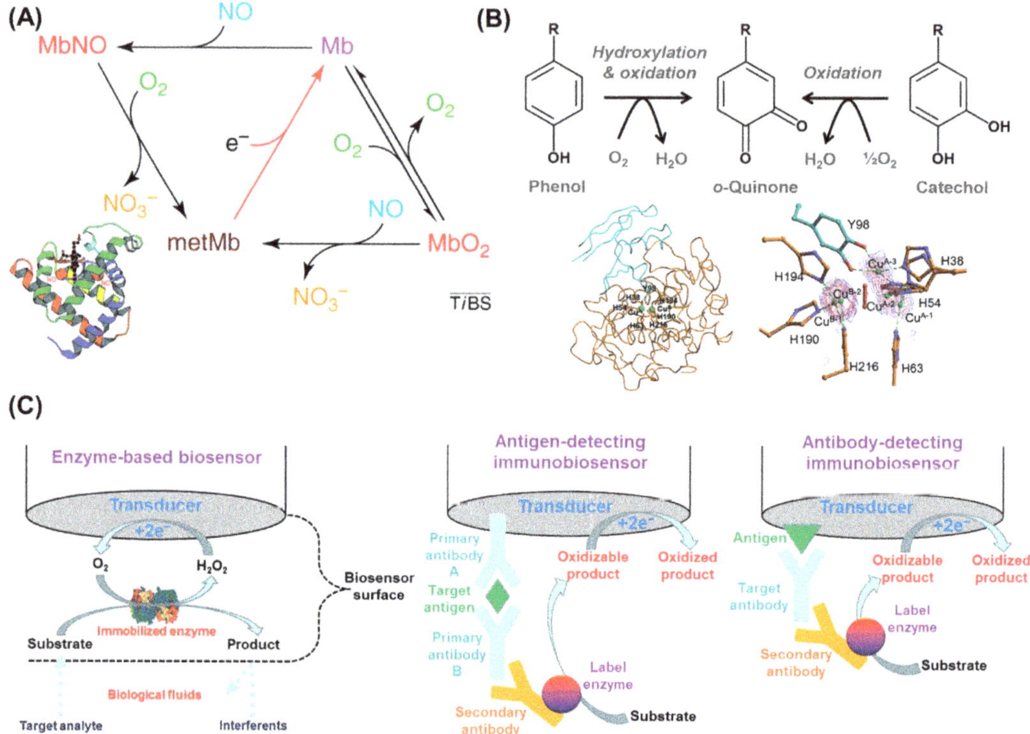

Figure 2. (**A**) Structure of Mb for reaction with NO, and its enzymatic reaction process for detection of NO by regulation of its states, Reprinted with permission from ref. [33]. Copyright © 2001, Elsevier Science Ltd., (**B**) Schematic diagram of enzymatic reactions of the Tyr using its copper ions for conversion of catechol and phenol to *o*-quinone, and its structure containing copper ions, Reprinted with permission from ref. [39]. Copyright © 2018 by the authors, (**C**) Schematic representation of enzyme-based biosensors. Examples of conventional enzymatic biosensors (left) and immune biosensors that employ an enzyme as a labeling component for the indirect detection of a target antigen (middle) or a target antibody (right), Reprinted with permission from ref. [48]. Copyright © 2016 by the authors, Licensee MDPI.

2.3. Techniques Used in Enzymatic Biosensing

2.3.1. Electrochemical Technique

Among various techniques applicable to developing biosensors, an electrochemical technique is the most actively used technique. The electrochemical method has characteristics required for biosensor development, such as rapid response, inherent miniaturization, high sensitivity, feasibility in practical use, and portability [52,53]. Particularly, the electrochemical technique is advantageous in that many different electrochemical techniques can be applied to develop biosensors that can be used in various ways according to targets and measurement environments [54]. Generally, electrochemical biosensors use electrochemical signal changes from the detection of target molecules using probe molecules, such as the production or removal of electrons or change of impedance. Therefore, to generate electrochemical signal changes, electroactive molecules, such as ferrocene, methylene blue, or the other redox molecules dissolved in the electrolyte, are introduced normally for electrochemical biosensing. Here, metalloenzymes can exclude the additional tagging process of the electroactive molecules. By using its inherent redox properties from metal ions in its structure directly, a metalloenzyme can be considered the best candidate as an electrochemical probe for developing electrochemical biosensors. The amperometry technique, such as amperometric I-T measurement and chronoamperometry, is one of the most broadly used electrochemical techniques to develop metalloenzyme-based electrochemical biosensors using the direct electron transfer reactions between the metalloenzyme and target small molecule [55]. During these electron transfer reactions, the amperometry technique measures the produced or removed electrons directly, and it is normally plotted as stair-like stepwise graph shapes. For instance, the electrochemical Tyr-based biosensor is prepared on a chitosan NPs (ChitNPs)-modified carbon electrode to detect catecholamines composed of catechol and amine side chains [56]. Here, in the presence of oxygen, Tyr catalyzes the oxidization of catecholamine, and the oxidized catecholamine is recovered to its original state by reduction. At this moment, the electrochemical reduction of the oxidized catecholamine can be measured directly by amperometry through the increase in the cathodic signals, as shown in Figure 3A.

In addition, cyclic voltammetry (CV) is also widely utilized to develop electrochemical enzymatic biosensors. Particularly, this technique can investigate the redox activity of metalloenzymes, even in the absence of an enzymatic reaction between metalloenzyme and target small molecules [57]. For example, the redox property of metal ions intercalated in the mismatched nucleic acids was investigated by CV to develop an electrochemical SARS-CoV-2 RNA biosensor for the determination of single-point mutation occurrence [58]. In this study, since certain metal ions can be intercalated in the mismatched nucleic acid sequence to stabilize the overall mismatched nucleic acid structure, the authors designed the probe DNA sequence capable of hybridizing with target SARS-CoV-2 RNA and forming intended mismatched sequences. Using this metal ion–nucleic acid biocomplex, the authors successfully determined the single-point mutation occurrence (Figure 3B). In the presence of pure SARS-CoV-2 RNA, only one redox peak pair was measured, but two different redox peak pairs were measured when the mutation occurred due to the formation of additional metal ion–nucleic acid biocomplexes. In addition, differential pulse voltammetry (DPV) is a broadly used electrochemical technique. In this technique, the measured current difference at two points immediately before and after the potential applied to the biosensor rises is expressed as a graph of voltage [59]. Since this technique measures the current difference obtained before and after the potential rises, it is less disturbed by current measurement disturbances such as electrical double layers. Using DPV, the enzymatic glutathione biosensor was developed based on glutathione peroxidase and graphene oxide (GO) [60]. The electrochemical technique can also be used for the development of electrochemical biosensors using redox-free proteins, enzymes, and antibodies. Electrochemical impedance spectroscopy (EIS) measures the electrical resistance or capacitance of the layers formed through the capture of targets such as antigens by probes like an antibody, so it does not require the inherent redox properties from biomolecules or target molecules [61,62].

However, despite the many advantages of electrochemical techniques, it may sometimes be difficult to sensitively and reliably measure the redox signals derived only from biomolecular reactions because of the inherent limitations of biomolecules. Further, to achieve high sensitivity and selectivity, electrochemical biosensors inevitably require high conductivity to measure the redox signals produced by target detection. To grant sufficient conductivity to biosensors and to solve the intrinsic limitations of biomolecules (e.g., low signal production and low stability), with the introduction of various nanomaterials and nanotechnologies to enzymatic electrochemical biosensors, the studies for developing functional electrodes with high conductivity or the increase in electron transfer of the enzymatic reactions are becoming parallel.

2.3.2. Fluorescence Technique

The fluorescence technique is a widely used technique for developing enzymatic biosensors because it can provide sensitive and specific detection of biomolecules. The basic principle of fluorescence biosensing is that when a molecule is excited by light of a certain wavelength, it absorbs the energy and emits light at a longer wavelength [63].

In a fluorescent biosensor, the target molecule is labeled with a fluorescent molecule or a fluorescent molecule is labeled with the sensing probe conversely, that can be excited by a specific wavelength of light. When the labeled molecule interacts with the biosensor, it changes the local environment of the fluorescent molecule, which alters its emission properties. By measuring the change in fluorescence, the presence and concentration of the target molecule can be detected. To design and develop the fluorescent biosensor, several approaches can be applied. First, a direct labeling method can be utilized to detect the target molecule using the fluorescence technique. To detect ions or small molecules, single directly labeled proteins or peptide-based biosensors have been developed [64]. For example, D-myo-inositol-1,4,5-trisphosphate (IP$_3$), an important signaling molecule that modulates cellular Ca^{2+} concentrations, can be detected by signal transduction from a cysteine mutation and alkylation of the active site (pleckstrin homology domain) of phospholipase C, such as 6F106 and DAN106 domains [65] (Figure 3C). The direct labeling method allows simple and straightforward sensing, but it can be challenging to label the target molecule without any damage or effect on the function and structure of the target molecules [66,67].

The indirect labeling method has been used to address the limitations of direct labeling fluorescence methods. In indirect labeling, a probe molecule that is labeled with a fluorescent molecule can specifically bind to the target molecule, and the fluorescence signal is altered from the conjugation of the probe and target molecules. For instance, amantadine can be quantitatively detected using an indirect competitive enzyme-linked immunosorbent assay [68]. Here, amantadine and ovalbumin are conjugated as a coating antigen and coated on a 96-well plate. GOx can also be employed as an enzyme label for immunoassays for GOx/glucose-mediated H$_2$O$_2$ production, resulting in a fluorescence response at 540 nm (Figure 3D). The Förster resonance energy transfer (FRET) is a phenomenon in which energy is transferred between two fluorescent molecules when they are in close distance proximity [69]. In a FRET-based biosensor, the biosensor contains two fluorescent molecules, a donor and an acceptor. The donor is excited by light and transfers its energy to the acceptor if it is in close proximity. The distance between the two fluorescent molecules changes when the target molecule binds to the biosensor, which alters the FRET signal. Several fluorescence techniques have also been utilized in biosensors, such as luminescence resonance energy transfer, photoluminescence, and electrochemiluminescence [70–72]. Overall, fluorescent biosensors have been used in many applications, including the detection of biomarkers for diseases, the monitoring of enzyme activity, and the study of protein–protein interactions. Fluorescent biosensors are versatile and sensitive tools that can provide real-time, non-invasive detection of biomolecules in complex biological environments.

While fluorescent biosensors offer many advantages over traditional enzymatic biosensors, such as high sensitivity, selectivity, and the ability to detect multiple analytes simultaneously, there are also some limitations to their use. One major limitation of fluorescent enzymatic biosensors is the requirement for an external excitation source, such as a laser or UV light, to produce a measurable signal. The necessity of an external source can limit the portability and ease of use of the biosensor in further applications [73]. In addition, the use of external excitation sources can lead to the production of background fluorescence from biological samples, which can interfere with the accuracy of the measurements. Another limitation of fluorescent enzymatic biosensors is the photobleaching of fluorescent dyes over time. The unexpected photobleaching can reduce the sensitivity and stability of the biosensor, making it less reliable over extended periods of time [74]. The bleaching process can be accelerated by exposure to light, temperature changes, and other environmental factors. In addition, the design and optimization of fluorescent enzymatic biosensors can be a complex and time-consuming process [75]. The choice of the fluorescent molecule, the method of attachment to the enzyme or substrate, and the selection of appropriate detection parameters all play important roles in determining the sensitivity and specificity of the biosensor. The optimization of the parameters requires extensive experimentation and validation, adding to the cost and time required to develop and manufacture the biosensor. Despite these limitations, fluorescent enzymatic biosensors remain a valuable tool for a wide range of applications, including medical diagnostics, environmental monitoring, and food safety testing. Current research and development in the field are focused on improving the sensitivity, stability, and ease of use of these biosensors to make them more practical and accessible for a variety of applications with the incorporation of nanomaterials and nanotechnologies.

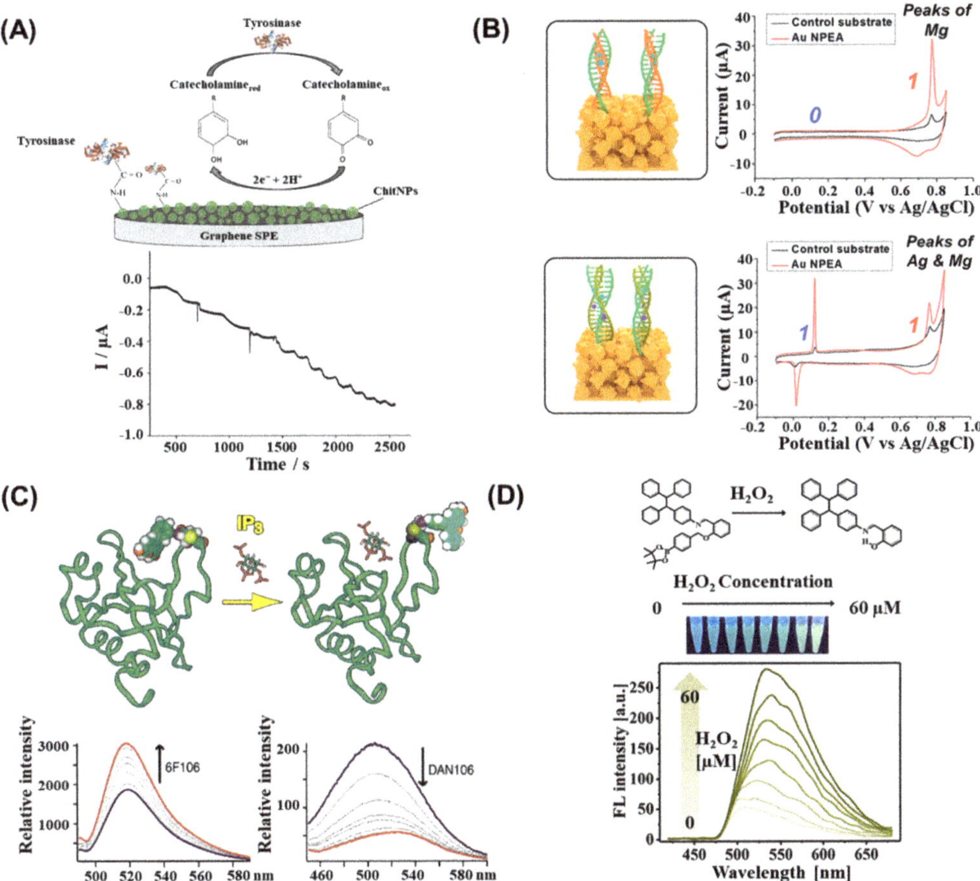

Figure 3. (**A**) Schematic image of amperometric Tyr-based biosensor on ChitNPs for catecholamine detection and its amperometric detection of catecholamine, Reprinted with permission from ref. [56]. Copyright © 2022 by the authors. Licensee MDPI, (**B**) Redox characteristics investigation of metal ions (Mg and Ag) of metal ion–nucleic acid biocomplex using CV, Reprinted with permission from ref. [58]. Copyright © 2022, American Chemical Society, (**C**) A schematic illustration of the structure of IP$_3$ complex (top) and Emission spectra of the fluorophore-labeled pleckstrin homology domains in phospholipase C; 6F106 domain (bottom left) and DAN106 domain (bottom right). With an increasing amount of IP$_3$, the fluorophore labeled-6F106 domain showed an increase in fluorescence intensity and DAN106 domain showed a decrease in fluorescence intensity, Reprinted with permission from ref. [65]. Copyright © 2002, American Chemical Society, (**D**) Fluorescence spectra with H$_2$O$_2$ concentrations ranging from 0 to 60 M using H$_2$O$_2$-triggered "turn-on" fluorescence biosensor, Reprinted with permission from ref. [68]. Copyright © 2019 by the authors.

3. Enzymatic Electrochemical Nanobiosensors

As discussed above, numerous biosensors have been developed so far using various enzymes, particularly metalloenzymes, and electrochemical techniques. These enzymatic biosensors have difficulty in highly sensitive diagnostics due to the intrinsic characteristics of biomolecules, such as low activity and stability, and therefore have limitations in diagnostics in real samples or harsh conditions. To overcome this limitation, in recent years, with the introduction of nanomaterials or nanotechnologies, nanobiosensors that measure targets in real samples or real environments with high sensitivity are being researched [76,77]. In terms of electrochemical biosensors, nanomaterials and nanotechnologies are being studied to accelerate enzymatic reactions and promote electron transfer or fabricate functional conductive electrodes [25,78]. In one study, carbon nanomaterials and nanodiamonds (ND) were employed to develop the Tyr-based amperometric nanobiosensor for detecting small pollutant chemicals (Figure 4A) [79]. Here, as the electrode materials of an enzymatic electrochemical nanobiosensor, carbon nanotubes (CNTs), NDs, and starch were used to develop a nanobiocomposite on a glassy carbon electrode (GCE). To fabricate this, NDs combined with soluble starch (ND-SS) were prepared on the GCE first, and then, the Tyr-trapped CNTs were anchored on the ND-SS using glutaraldehyde (Glu). The developed nanobiocomposite provided enhanced load capacity for Tyr, improved electron transfer efficacy from CNTs, and provided the biocompatibility for Tyr via the introduction of ND-SS. Consequently, the redox signals from this nanobiocomposite-based nanobiosensor during the detection of small pollutant chemicals (phenolic compounds) exhibited sufficiently enhanced signals in various real water samples with a 2.9 nM limit of detection (LoD) and long-term activity in an amperometric manner. In addition to this study, there are some recent studies developing enzymatic nanobiosensors by introducing nanomaterials to enzymes. For instance, a nanocomposite composed of an AuNP and titanium disulfide (TiS$_2$) nanosheet (NS) was developed and combined with uricase to develop an amperometric uric acid nanobiosensor. Due to the synergistic effects of the AuNP and TiS$_2$ NS for increasing the redox signals produced from the enzymatic reaction between uricase and uric acid, the developed nanobiosensor successfully detected uric acid prepared with interfering molecules, such as ascorbic acid, urea, and glucose, and detected uric acid dissolved in commercial human serum [80]. In another study, gold (Au) and platinum (Pt)-decorated CNTs were proposed as the nanocomposite for the immobilization of cholesterol oxidase (COx) to develop an enzymatic electrochemical cholesterol nanobiosensor [81]. In addition to the metallic nanocomposite, other materials, such as organic materials, are also being incorporated with metallic nanomaterials to develop enzymatic nanobiosensors. In one study, polyaniline (PANI), a conducting polymer, was combined with titanium dioxide (PANI@TiO$_2$) and prepared on an indium tin oxide (ITO) electrode using electrophoretic deposition to provide enhanced electron transfer kinetics for xanthine oxidase (XOs) immobilized on the PANI@TiO$_2$ electrode [82]. Using the enzymatic activity of XOs for the detection of xanthine (Xn), Xn was detected sensitively with a 100 nM LoD and rapid response (within 10 s) using the DPV technique. Additionally, various functional nanocomposite-based enzymatic nanobiosensors have been reported, such as the use of magnetic NPs (MNPs) [83,84]. For instance, the MNP, Iron (III) oxide (Fe$_3$O$_4$), encapsulated by polynorepinephrine (PNE), was developed and combined with GOx to develop an enzymatic glucose nanobiosensor operable with a smartphone (Figure 4B) [84]. Due to the encapsulation of the MNPs by PNE, PNE increased the enzyme immobilization efficiency doubly compared to bare MNPs for achieving improved enzymatic activity of GOx for glucose detection. The developed nanobiosensor successfully and rapidly detected glucose in human serum, blood, and commercially available glucose solutions using a smartphone that supports the applicability of this nanobiosensor for point-of-care testing (POCT) in broad industries. Moreover, the ability to recollect MNPs by a magnet can be further utilized in biomedical applications.

In addition to nanomaterial-assisted enzymatic electrochemical nanobiosensors, through the development of electrodes with improved conductivity via nanotechnologies, various novel enzymatic electrochemical nanobiosensors are also being studied. Nanotechnologies, such as lithography, microelectromechanical systems (MEMSs), and nanoelectromechanical system (NEMS), are being used to develop conductive functional electrodes [85,86]. For instance, the enzymatic electrochemical glucose nanobiosensor was developed by combining reduced GO (rGO), GOx, a microfluidics chip, and a 3D-printed paper electrode [86]. This nanobiosensor measured the glucose in human sweat and blood via the enzymatic catalytic reaction of GOx with glucose, and the utilized nanotechnologies offered high conductivity for this enzymatic reaction and large surface areas for GOx immobilization. This study suggested one approach for fabricating disposable nanobiosensors. In addition to this, there was a study that used micro/nanoarrays composed of Au and zinc oxide (Au-ZnO) nanocrystals developed as an electrode for the enzymatic detection of catechol (Figure 4C) [87]. In this study, to fabricate the Au-ZnO nanocrystals, the zinc nitrate $(Zn(NO_3)_2)$ precursor was prepared on the ITO electrode, and then, the ZnO nanocrystals were developed via a hydrothermal reaction. Next, through the electrodeposition of Au on the ZnO nanocrystals, the Au-ZnO nanocrystals were finally developed on the ITO electrode (Figure 4C), and laccase was immobilized on the Au-ZnO nanocrystals as a sensing probe enzyme. The developed Au-ZnO nanocrystal-assisted nanobiosensor detected catechol selectively and sensitively using the amperometry technique and also accurately measured catechol in tap water and lake water. Additionally, along with the development of smart devices, such nanotechnologies have the tremendous potential to develop flexible or wearable electrochemical nanobiosensors, which have received attention recently in the field of biosensors, particularly for the development of flexible or implementable conductive electrodes [88]. The development of wearable, flexible, or implementable nanobiosensors can enable the measurement of small chemicals directly in the body instead of using an in vitro solution. For the easy and simple fabrication of flexible electrodes as enzymatic nanobiosensors, one research group developed a flexible enzymatic electrochemical nanobiosensor composed of molybdenum disulfide (MoS_2) NPs, an Au nanolayer, and GOx to detect glucose (Figure 4D) [89]. Here, instead of relatively complex techniques, such as lithography, electrochemical deposition, and the sputtering technique were employed to fabricate the sandwiched structural nanofilm ($Au/MoS_2/Au$) on the flexible polymer electrode capable of providing a large surface area and facilitating electron transfer with sufficient bendability with GOx immobilized on the $Au/MoS_2/Au$ nanofilm via a chemical linker. The developed flexible nanobiosensor detected glucose with nanomolar sensitivity and high selectivity while retaining its flexibility. In another example, a bendable electrode composed of the silver nanowire (AgNW) and electropolymerized polymers was developed to be used for the noninvasive detection of lactate in human sweat [90]. In addition to this, the 3D printing technique, which has recently attracted attention in various scientific fields, is also being applied to develop novel enzymatic electrochemical nanobiosensors [91]. As discussed in this section, numerous enzymatic electrochemical nanobiosensors have been studied through the introduction of nanomaterials or nanotechnologies with functional enzymes to detect small chemicals. Such studies are expected to contribute to the development of enzymatic electrochemical nanobiosensors for POCT application in a simple and rapid detection manner in the near future.

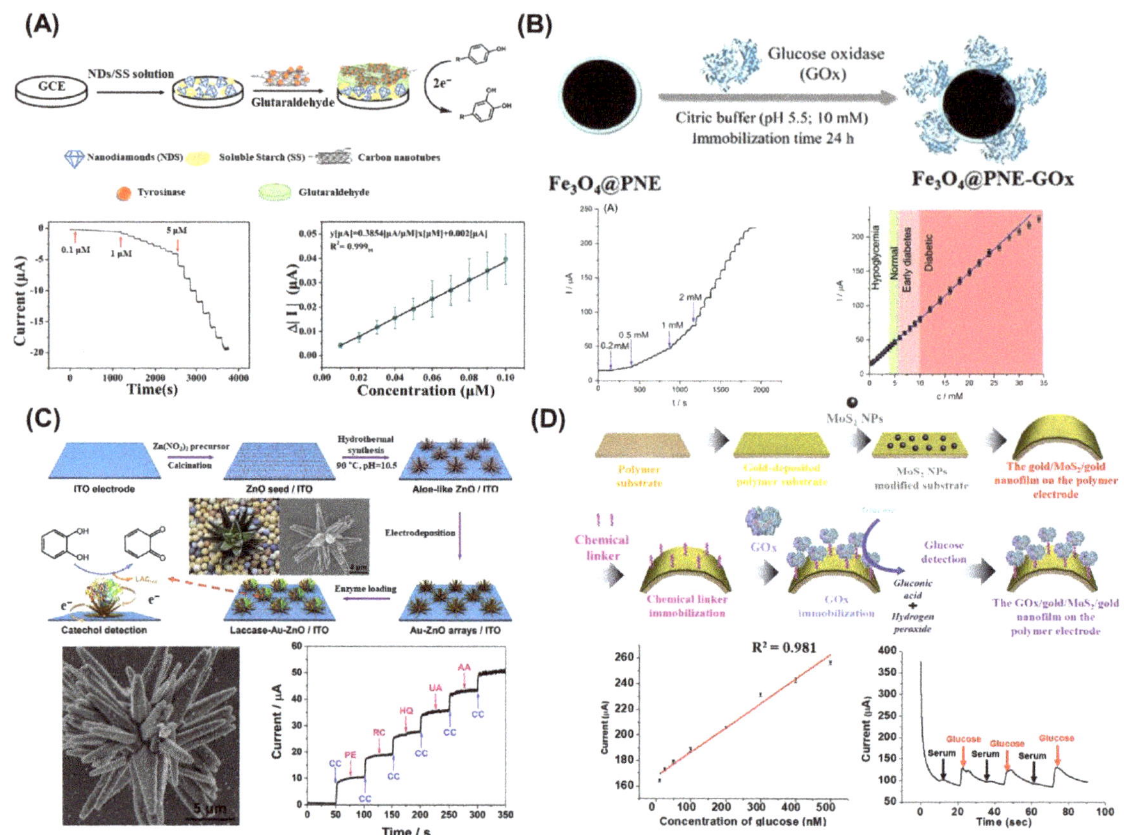

Figure 4. (**A**) Schematic illustration of the fabrication of nanobiosensor composed of Tyr, ND, and CNT, and electrochemical phenol detection results, Reprinted with permission from ref. [79]. Copyright © 2022 by the authors, (**B**) Scheme for encapsulation of the MNP by PNE and immobilization of GOx, and amperometric response and linear response plot for glucose detection in the presence of a different concentration of the glucose, Reprinted with permission from ref. [84]. Copyright © 2022 Elsevier B.V., (**C**) Schematic diagram of catechol nanobiosensor fabrication by using Au-ZnO nanocrystal arrays and Laccase, scanning electron microscopy (SEM) image of Au-ZnO nanocrystal, and its selective electrochemical catechol (CC) detection property, Reprinted with permission from ref. [87]. Copyright © 2020 Elsevier B.V., (**D**) Schematic image of flexible GOx/Au/MoS$_2$/Au nanofilm-based nanobiosensor and electrochemical results for glucose detection with high linearity prepared in the serum sample, Reprinted with permission from ref. [89]. Copyright © 2019 Elsevier B.V.

4. Enzymatic Fluorescent Nanobiosensors

Similar to enzymatic electrochemical nanobiosensors, enzymatic fluorescent nanobiosensors have gained significant attention in recent decades due to their high sensitivity and selectivity, as well as their capability to monitor analytes or targets in real time. With the integration of nanomaterials or nanotechnologies in fluorescent biosensors, the sensitive and selective detection of analytes or targets is allowed, while enzymatic reactions enhance the specificity and accuracy of the detection. One of the general approaches to developing enzymatic fluorescent nanobiosensors is the immobilization of fluorescent NPs on specific enzymes, which can result in fluorescent excitation through an enzymatic reaction between targets and enzymes. One study utilized silica-functionalized carbon dots to achieve multi-color imaging detection of DA [92]. In this study, a bioprobe was fabricated using silica-functionalized carbon dot-immobilized laccase enzyme, and the bioprobe was coated on optical fibers via the dip-coating method (Figure 5A). In the presence of DA, since the laccase enzyme contained in the bioprobe facilitates an oxidation reaction in which the DA is oxidized, DA is transformed into dopaquinone. Then, photoinduced electron transfer 12 occurs between the bioprobe and dopaquinone, with the carbon dots acting as electron donors and dopaquinone as acceptors. As a result, photoluminescence quenching of the bioprobe occurs, and DA can be quantitatively detected with a linear range of 0 to 30 μM. In addition, the developed bioprobe showed selectivity to DA only, as shown by the highest quenching efficiency compared with other interferants, such as metal ions, amino acids, macromolecules, and neurotransmitters. Among the several nano-materials, the metal–organic framework (MOF) has attracted considerable attention from researchers in the biosensor field [93,94]. The MOF has unique properties and advantages in biosensing compared with other nanomaterials, such as (i) enabling target binding with probe molecules via π-π stacking [95], (ii) high surface area with porosity [96], (iii) easy functionalization and tailoring by controlling the ratio of metal ions in the MOF [97], and (iv) stability [98]. For example, organophosphorus pesticides (OPs), which can bind with AChe resulting in the breakdown of neurotransmitters, were detected using a MOF-applied nanobiosensor [99] (Figure 5B). Herein, the Au nanocluster (AuNC)-modified zeolite-like imidazole framework structure (ZIF) composite (AuNCs@ZIF) was developed as fluores-cence material. The AuNCs@ZIF emitted intense fluorescence because ZIF-8 restricted the movement of the AuNCs, inhibited the nonradiative decay, and activated the radiation channel. The combination of AChE and choline oxidase (CHO) can hydrolyze ACh to generate H_2O_2, which can degrade the structure of ZIF and diminish the fluorescence of AuNCs@ZIF. Furthermore, the OPs can limit the activity of AChE, resulting in a decrease in the generation of H_2O_2, a weakening of ZIF degradation, and a slow fluorescence recovery. As a result, a "turn-on" fluorescence mode was developed, and the detection of OPs was achieved with a 0.4 μg/L LoD. As briefly mentioned in Section 2.2, artificial enzymes composed of nanomaterials that can act as enzymes in the presence of an analyte have been developed. These artificial enzymes are called nanozymes, and numerous studies reported several types of nanozymes. For example, the MOF, which is fabricated by modu-lating synthetic methods to combine the advantages of natural enzymes and nanomaterials, has a similar catalytic function to natural enzymes [100]. The MOFs are very flexible to biomimetic design and can accommodate a variety of enzymes. For instance, the "armor-like" exoskeleton of MOFs around enzymes can transport small molecules selectively while guaranteeing the stability of enzymes [101]. Since the nanozymes composed of MOFs can be easily tuned and can detect the target small molecules, several nanozymes that mimic ox-idase, peroxidase, catalase, and hydrolase have been developed [102–104]. As an example, the flower cluster morphology of dicopper (II) complexes (Cu-TPP MOF) was synthesized to detect DA (Figure 5C) [105]. Significantly, by imitating the active sites of a binuclear Cu (II) metal center coordinated by six nitrogen-containing coordination units in natural catechol oxidase, the Cu-TPP MOF demonstrated a high potential for simulating natural biological enzymes. Additionally, applying the flower cluster morphology, the surface area of Cu-TPP increased, which resulted in a change of contact between the DA and the substrate that was significantly enhanced. Cu-TPP MOF was utilized as a mimic of catechol oxidase in the presence of H_2O_2 to catalyze the oxidation of DA, producing the equivalent catechol derivative, o-quinone intermediate. Dihydroxynaphthalene was then used as an indicator to react with o-quinone, and a fluorescent signal was promptly generated. Even-tually, the DA detection using Cu-TPP MOF with a 2.5 nM LoD was achieved. Similarly, metal oxide- or carbon-based NPs have been utilized as nanozymes [106]. For instance, mercury metal was detected using the peroxidase-like property of polyvinylpyrrolidone Ag NPs [107]. The catalytic activity of synthesized Ag NPs oxidized the o-phenylenediamine (non-fluorescence reagent) to 2,3-diaminophenazine, a high-fluorescence reaction product. In addition, since only mercury (II) ions among the heavy metals inhibited the catalytic reaction, the fluorescence intensity was quenched in the presence of mercury (II) ions. Accordingly, mercury (II) ions were detected with a linear range of 20–2000 nM and an 8.9 nM LoD.

In addition, through the integration of microfluidic technology and enzymatic reactions, microfluidic-assisted fluorescent enzymatic biosensors have been reported to detect specific molecules or analytes in a sample [108]. The enzymatic reactions in microfluidic systems can be facilitated by immobilizing the enzyme onto a surface, such as a microfluidic channel wall or a microbead, or by incorporating the enzyme into the fluid stream. Once the enzyme is immobilized or introduced into the fluid stream, the sample is introduced into the microfluidic system, and the enzyme catalyzes a reaction with the target analyte [109]. With the incorporation of nanomaterials in microfluidic devices, various enzymatic fluorescent biosensors have been developed and offer faster response times, higher sensitivity, and the ability to analyze small sample volumes. For instance, a 3D paper-based analytical device composed of carbon dots was developed to analyze saliva samples (Figure 5D) [110]. To detect glucose and lactate, fluorescence from the carbon dots was used to quantify the H_2O_2 generated during the enzymatic oxidation of the analyte. The carbon dot dispersion had a blue emission under UV light in the absence of H_2O_2, but the intensity was reduced in the presence of H_2O_2 and HRP, allowing the measurement of glucose and lactate via the quenching of fluorescence.

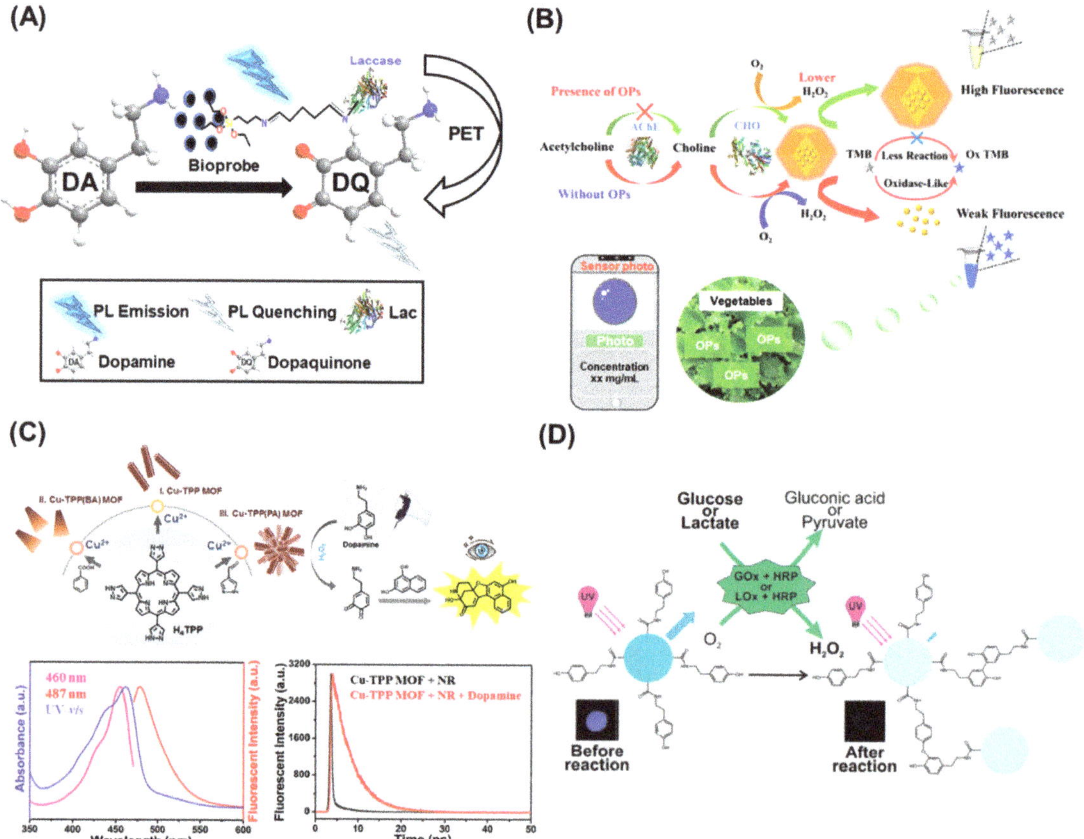

Figure 5. (**A**) Schematic representation for the plausible mechanism of photoluminescence quenching

of bioprobe by DA, Reprinted with permission from ref. [92]. Copyright © 2021, Elsevier B.V., (**B**) Schematic Diagram of the Mechanism for the Detection of OPs, Reprinted with permission from ref. [99]. Copyright © 2021, American Chemical Society, (**C**) Schematic illustration of the sensing process of Cu-TPP MOFs toward DA (top), UV vis absorption spectra at 460 nm, fluorescence excitation spectra at 460 nm, and emission spectra at 487 nm for the final product aqueous solution (bottom left), and time-resolved photoluminescence emission decay spectra of Cu-TPP MOF with and without DA, Reprinted with permission from ref. [105]. Copyright © 2021 Elsevier B.V., and (**D**) Illustration of the proposed mechanism of carbon dots fluorescence quenching when reacting with glucose in the presence of GOx and HRP or lactate in the presence of lactate oxidase and HRP., Reprinted with permission from ref. [110]. Copyright © 2020 Elsevier B.V.

In summary, enzymatic fluorescent nanobiosensors have emerged as a promising tool for the sensitive and selective detection of analytes or targets in real time. The integration of NPs or nanotechnologies into fluorescent biosensors has allowed for sensitive and selective detection, while enzymatic reactions have enhanced the specificity and accuracy of detection. The immobilization of fluorescent NPs to specific enzymes is a general approach used to develop enzymatic fluorescent nanobiosensors. Nanomaterials, such as the MOF, have unique properties and advantages in biosensing. In addition, artificial enzymes composed of nanomaterials, nanozymes, have been developed, which can mimic natural enzymes and possess a high potential for simulating natural biological enzymes. Nanotechnologies, such as microfluidics with NPs, enable the fabrication of novel fluorescent enzymatic biosensors. These advances in the development of enzymatic fluorescent nanobiosensors will provide a promising platform for the sensitive and selective detection of various analytes or targets for various applications in areas such as medical diagnostics, environmental monitoring, and food safety.

5. Conclusions and Future Perspectives

The precise detection and quantification of small molecules serve a vital role in various disciplines, including healthcare, environmental monitoring, and food safety. Enzymatic biosensors have shown exceptionally high sensing capabilities, which can be attributed to the specificity with which enzymes react with relatively small chemical molecules. In addition, enzymatic biosensors have various advantages over enzyme-free or non-enzymatic biosensors, including high specificity and sensitivity, low detection limits, and a broad detection range. To develop enzymatic biosensors, electrochemical and fluorescence techniques have been used to detect small molecules, offering high sensitivity, selectivity, and specificity. Moreover, through the incorporation of nanotechnologies, enzymatic biosensors have seen considerable performance improvements, which have led to the development of enzymatic nanobiosensors that have enhanced sensitivity and selectivity for the detection of small molecules [111–116] (Table 1). The development of enzymatic nanobiosensors has a wide range of potential applications, including but not limited to medicine, the monitoring of the environment, and food safety. In this review, interdisciplinary information on the development of enzymatic nanobiosensors for small molecule detection, including commonly used enzymes, electrochemical/fluorescence approaches, and electrochemical/fluorescent biosensors, was discussed. In addition, this review provided a comprehensive overview of the development of innovative electrochemical and fluorescent enzymatic nanobiosensors for the detection of small molecules.

Table 1. The representative electrochemical/fluorescent enzymatic nanobiosensors illustrated in this review and the other enyzamtic nanobiosensors using Surface-enhanced Raman spectroscopy (SERS), Surface plasmon resonance (SPR), and colorimetric techniques.

Sensing Technique	Enzymatic Reaction	Nano-Assistance	Target	LoD	Ref.
Electrochemical	Tyrosinase reaction	Carbon nanotubes Nanodiamonds	Phenolic compounds	2.9 nM	[79]
	Cholesterol oxidase reaction	Carbon nanotubes	Cholesterol	0.5 µM	[81]
	Glucose oxidase reaction	Polynorepinephrine grafted on magnetite nanoparticles	Glucose	6.1 µM	[84]
	Laccase reaction	Gold–Zinc oxide micro/nanoarrays	Catechol	25 nM	[87]
	Glucose oxidase reaction	Glucose oxidase/Gold/ Molybdenum disulfide/ Gold nanofilm	Glucose	10 nM	[89]
Fluorescent	Laccase reaction	Silica-functionalized carbon dots	Dopamine	41.2 nM	[92]
	Acetylcholinesterase and choline oxidase reaction	Au nanoclusters modified zeolite-like imidazole framework	Organophosphorus pesticides	1.79 nM	[99]
	Catechol oxidase reaction	Pyrazolate-based porphyrinic metal–organic framework	Dopamine	2.5 nM	[105]
	Peroxidase reaction	Polyvinylpyrrolidone stabilized silver nanoparticles	Mercury (II) ion	8.9 nM	[107]
	Glucose oxidase and lactate oxidase reaction	Carbon dots	Glucose Lactate	2.6 µM 0.8 µM	[110]
SERS	Peroxidase-mimicking reaction	Silver nanoparticles/metal organic framework	Cholesterol	0.36 µM	[111]
	Glucose oxidase-like reaction	Silver/gold nanoparticles	Glucose	50 nM	[112]
SPR	Glucose oxidase oxidation	Polystyrene nanoparticle with Manganese dioxide	Glucose	3.1 pM	[113]
	Acetylcholinesterase reaction	Molybdenum disulfide/gold nanoparticle multicore fiber	Acetylcholine	14.28 µM	[114]
Colorimetric	Uricase, glucose oxidase, choline oxidase reaction	Magnetic nanoparticles	Uric acid Glucose Choline	0.34 µM 0.59 µM 0.20 µM	[115]
	Glucose oxidase reaction	Acrylamide based-copolymer hydrogel	H_2O_2 Glucose	8.9 µM 1.6 mM	[116]

While enzymatic nanobiosensors have shown great promise for the detection of small molecules, there are still some limitations that need to be addressed. One major challenge is the stability and reproducibility of these nanobiosensors, particularly when they are exposed to complex biological or environmental samples [117]. Another limitation is the potential interference from other molecules in the sample, which can affect the accuracy and specificity of the nanobiosensors [118]. Additionally, the integration of nanomaterials into enzymatic biosensors normally requires a complex and expensive process, requiring specialized equipment and expertise. Molecularly imprinted polymers (MIPs) have the potential to overcome the limitations of enzymatic biosensors and provide an alternative means of analysis. Compared to enzyme-based devices, MIPs offer superior chemical stability, cost-effectiveness, and ease of fabrication, making them an attractive option for various analytical applications [119,120]. Despite these challenges, recent studies and development in the field of enzymatic nanobiosensors promise to overcome these limitations and improve their performance. Advancements in nanomaterial synthesis and fabrication techniques, as well as the development of new enzyme technologies, are likely to lead to the development of more stable, selective, and sensitive nanobiosensors in the future. Additionally, the integration of machine learning and artificial intelligence may help to overcome some of the challenges associated with interference from other molecules and improve the accuracy of small molecule detection [121]. Moreover, the integration of different nanotechnologies, such as nanofabrication and microfluidics, can facilitate the fabrication of advanced enzymatic nanobiosensors. This integration can enhance the performance and enable multiplex detection of small molecules [58,122]. The progress in the development of enzymatic nanobiosensors will have a significant impact on various fields and lead to the improvement of human health and safety.

Author Contributions: Conceptualization, H.K.C. and J.Y.; writing—original draft preparation, H.K.C. and J.Y.; supervision, H.K.C. and J.Y.; project administration, H.K.C. and J.Y.; funding acquisition, H.K.C. and J.Y. All authors have read and agreed to the published version of the manuscript.

Funding: This work was supported by the National Research Foundation of Korea (NRF) grant funded by the Korea government (MSIT) (2022R1C1C2005003), and by the NIH T32 Postdoctoral Training Program in Translational Research in Regenerative Medicine, under Award Number T32EB005583.

Institutional Review Board Statement: Not applicable.

Informed Consent Statement: Not applicable.

Data Availability Statement: Not applicable.

Conflicts of Interest: The authors declare no conflict of interest.

References

1. Wu, L.; Liu, J.; Li, P.; Tang, B.; James, T.D. Two-photon small-molecule fluorescence-based agents for sensing, imaging, and therapy within biological systems. *Chem. Soc. Rev.* **2021**, *50*, 702–734. [CrossRef] [PubMed]
2. Han, H.-H.; Tian, H.; Zang, Y.; Sedgwick, A.C.; Li, J.; Sessler, J.L.; He, X.-P.; James, T.D. Small-molecule fluorescence-based probes for interrogating major organ diseases. *Chem. Soc. Rev.* **2021**, *50*, 9391–9429. [CrossRef] [PubMed]
3. Liu, R.; Feng, Z.-Y.; Li, D.; Jin, B.; Lan, Y.; Meng, L.-Y. Recent trends in carbon-based microelectrodes as electrochemical sensors for neurotransmitter detection: A review. *TrAC Trends Anal. Chem.* **2022**, *148*, 116541. [CrossRef]
4. Stolz, R.M.; Kolln, A.F.; Rocha, B.C.; Brinks, A.; Eagleton, A.M.; Mendecki, L.; Vashisth, H.; Mirica, K.A. Epitaxial Self-Assembly of Interfaces of 2D Metal–Organic Frameworks for Electroanalytical Detection of Neurotransmitters. *ACS Nano* **2022**, *16*, 13869–13883. [CrossRef]
5. Ryu, J.; Lee, E.; Kang, C.; Lee, M.; Kim, S.; Park, S.; Lee, D.; Kwon, Y. Rapid Screening of Glucocorticoid Receptor (GR) Effectors Using Cortisol-Detecting Sensor Cells. *Int. J. Mol. Sci.* **2021**, *22*, 4747. [CrossRef]
6. Dong, Q.; Ryu, H.; Lei, Y. Metal oxide based non-enzymatic electrochemical sensors for glucose detection. *Electrochim. Acta* **2021**, *370*, 137744. [CrossRef]
7. Shamsazar, A.; Asadi, A.; Seifzadeh, D.; Mahdavi, M. A novel and highly sensitive sandwich-type immunosensor for prostate-specific antigen detection based on MWCNTs-Fe3O4 nanocomposite. *Sens. Actuators B Chem.* **2021**, *346*, 130459. [CrossRef]
8. Huang, L.; Zeng, Y.; Liu, X.; Tang, D. Pressure-Based Immunoassays with Versatile Electronic Sensors for Carcinoembryonic Antigen Detection. *ACS Appl. Mater. Interfaces* **2021**, *13*, 46440–46450. [CrossRef]
9. Ding, Q.; Li, C.; Wang, H.; Xu, C.; Kuang, H. Electrochemical detection of heavy metal ions in water. *Chem. Commun.* **2021**, *57*, 7215–7231. [CrossRef]
10. Xianyu, Y.; Lin, Y.; Chen, Q.; Belessiotis-Richards, A.; Stevens, M.M.; Thomas, M.R. Iodide-Mediated Rapid and Sensitive Surface Etching of Gold Nanostars for Biosensing. *Angew. Chem. Int. Ed.* **2021**, *60*, 9891–9896. [CrossRef]
11. Cavalcante, F.T.T.; de A. Falcão, I.R.; da S. Souza, J.E.; Rocha, T.G.; de Sousa, I.G.; Cavalcante, A.L.G.; de Oliveira, A.L.B.; de Sousa, M.C.M.; dos Santos, J.C.S. Designing of Nanomaterials-Based Enzymatic Biosensors: Synthesis, Properties, and Applications. *Electrochem* **2021**, *2*, 149–184. [CrossRef]
12. Sohal, N.; Maity, B.; Shetti, N.P.; Basu, S. Biosensors Based on MnO$_2$ Nanostructures: A Review. *ACS Appl. Nano Mater.* **2021**, *4*, 2285–2302. [CrossRef]
13. Yoon, J.; Chung, Y.-H.; Lee, T.; Kim, J.H.; Kim, J.; Choi, J.-W. A biomemory chip composed of a myoglobin/CNT heterolayer fabricated by the protein-adsorption-precipitation-crosslinking (PAPC) technique. *Colloids Surf. B Biointerfaces* **2015**, *136*, 853–858. [CrossRef] [PubMed]

14. Yoon, J.; Lee, T.; Bapurao, G.B.; Jo, J.; Oh, B.-K.; Choi, J.-W. Electrochemical H_2O_2 biosensor composed of myoglobin on MoS_2 nanoparticle-graphene oxide hybrid structure. *Biosens. Bioelectron.* **2017**, *93*, 14–20. [CrossRef] [PubMed]
15. Lee, T.; Chung, Y.-H.; Chen, Q.; Min, J.; Choi, J.-W. Fatigue Test of Cytochrome C Self-Assembled on a 11-MUA Layer Based on Electrochemical Analysis for Bioelectronic Device. *J. Nanosci. Nanotechnol.* **2015**, *15*, 5537–5542. [CrossRef] [PubMed]
16. Raymundo-Pereira, P.A.; Silva, T.A.; Caetano, F.R.; Ribovski, L.; Zapp, E.; Brondani, D.; Bergamini, M.F.; Marcolino, L.H.; Banks, C.E.; Oliveira, O.N.; et al. Polyphenol oxidase-based electrochemical biosensors: A review. *Anal. Chim. Acta* **2020**, *1139*, 198–221. [CrossRef] [PubMed]
17. Kucherenko, I.S.; Soldatkin, O.O.; Kucherenko, D.Y.; Soldatkina, O.V.; Dzyadevych, S.V. Advances in nanomaterial application in enzyme-based electrochemical biosensors: A review. *Nanoscale Adv.* **2019**, *1*, 4560–4577. [CrossRef] [PubMed]
18. Irfan Azizan, M.A.; Taufik, S.; Norizan, M.N.; Abdul Rashid, J.I. A review on surface modification in the development of electrochemical biosensor for malathion. *Biosens. Bioelectron. X* **2023**, *13*, 100291. [CrossRef]
19. Wei, X.; Wang, S.; Zhan, Y.; Kai, T.; Ding, P. Sensitive Identification of Microcystin-LR via a Reagent-Free and Reusable Electrochemical Biosensor Using a Methylene Blue-Labeled Aptamer. *Biosensors* **2022**, *12*, 556. [CrossRef] [PubMed]
20. Gan, Z.; Roslan, M.A.M.; Abd Shukor, M.Y.; Halim, M.; Yasid, N.A.; Abdullah, J.; Md Yasin, I.S.; Wasoh, H. Advances in Aptamer-Based Biosensors and Cell-Internalizing SELEX Technology for Diagnostic and Therapeutic Application. *Biosensors* **2022**, *12*, 922. [CrossRef] [PubMed]
21. Pan, C.; Wei, H.; Han, Z.; Wu, F.; Mao, L. Enzymatic electrochemical biosensors for in situ neurochemical measurement. *Curr. Opin. Electrochem.* **2020**, *19*, 162–167. [CrossRef]
22. Choi, J.-H.; Choi, J.-W. Metal-Enhanced Fluorescence by Bifunctional Au Nanoparticles for Highly Sensitive and Simple Detection of Proteolytic Enzyme. *Nano Lett.* **2020**, *20*, 7100–7107. [CrossRef] [PubMed]
23. Lee, M.; Kim, D. Exotic carbon nanotube based field effect transistor for the selective detection of sucrose. *Mater. Lett.* **2020**, *268*, 127571. [CrossRef]
24. Si, Y.; Lee, H.J. Carbon nanomaterials and metallic nanoparticles-incorporated electrochemical sensors for small metabolites: Detection methodologies and applications. *Curr. Opin. Electrochem.* **2020**, *22*, 234–243. [CrossRef]
25. Eivazzadeh-Keihan, R.; Bahojb Noruzi, E.; Chidar, E.; Jafari, M.; Davoodi, F.; Kashtiaray, A.; Ghafori Gorab, M.; Masoud Hashemi, S.; Javanshir, S.; Ahangari Cohan, R.; et al. Applications of carbon-based conductive nanomaterials in biosensors. *Chem. Eng. J.* **2022**, *442*, 136183. [CrossRef]
26. Zhang, Y.; Li, X.; Li, D.; Wei, Q. A laccase based biosensor on $AuNPs-MoS_2$ modified glassy carbon electrode for catechol detection. *Colloids Surf. B Biointerfaces* **2020**, *186*, 110683. [CrossRef]
27. He, Y.; Hu, F.; Zhao, J.; Yang, G.; Zhang, Y.; Chen, S.; Yuan, R. Bifunctional Moderator-Powered Ratiometric Electrochemiluminescence Enzymatic Biosensors for Detecting Organophosphorus Pesticides Based on Dual-Signal Combined Nanoprobes. *Anal. Chem.* **2021**, *93*, 8783–8790. [CrossRef]
28. Alizadeh, N.; Ghasemi, S.; Salimi, A.; Sham, T.-K.; Hallaj, R. CuO nanorods as a laccase mimicking enzyme for highly sensitive colorimetric and electrochemical dual biosensor: Application in living cell epinephrine analysis. *Colloids Surf. B Biointerfaces* **2020**, *195*, 111228. [CrossRef]
29. Yang, Z.; Liang, G.; Xu, B. Enzymatic Hydrogelation of Small Molecules. *Acc. Chem. Res.* **2008**, *41*, 315–326. [CrossRef]
30. Singh, H.; Tiwari, K.; Tiwari, R.; Pramanik, S.K.; Das, A. Small Molecule as Fluorescent Probes for Monitoring Intracellular Enzymatic Transformations. *Chem. Rev.* **2019**, *119*, 11718–11760. [CrossRef]
31. Zhang, K.; Cai, R.; Chen, D.; Mao, L. Determination of hemoglobin based on its enzymatic activity for the oxidation of o-phenylenediamine with hydrogen peroxide. *Anal. Chim. Acta* **2000**, *413*, 109–113. [CrossRef]
32. Bickar, D.; Bonaventura, J.; Bonaventura, C. Cytochrome c oxidase binding of hydrogen peroxide. *Biochemistry* **1982**, *21*, 2661–2666. [CrossRef] [PubMed]
33. Brunori, M. Nitric oxide moves myoglobin centre stage. *Trends Biochem. Sci.* **2001**, *26*, 209–210. [CrossRef] [PubMed]
34. Hoffman, B.M.; Dean, D.R.; Seefeldt, L.C. Climbing Nitrogenase: Toward a Mechanism of Enzymatic Nitrogen Fixation. *Acc. Chem. Res.* **2009**, *42*, 609–619. [CrossRef] [PubMed]
35. Bankar, S.B.; Bule, M.V.; Singhal, R.S.; Ananthanarayan, L. Glucose oxidase—An overview. *Biotechnol. Adv.* **2009**, *27*, 489–501. [CrossRef]
36. Chen, J.; Zheng, X.; Li, Y.; Zheng, H.; Liu, Y.; Suye, S.-i. A Glucose Biosensor Based on Direct Electron Transfer of Glucose Oxidase on PEDOT Modified Microelectrode. *J. Electrochem. Soc.* **2020**, *167*, 067502. [CrossRef]
37. Özbek, O.; Berkel, C.; Isildak, Ö.; Isildak, I. Potentiometric urea biosensors. *Clin. Chim. Acta* **2022**, *524*, 154–163. [CrossRef]
38. Baluta, S.; Zając, D.; Szyszka, A.; Malecha, K.; Cabaj, J. Enzymatic Platforms for Sensitive Neurotransmitter Detection. *Sensors* **2020**, *20*, 423. [CrossRef]
39. Matoba, Y.; Kihara, S.; Bando, N.; Yoshitsu, H.; Sakaguchi, M.; Kayama, K.e.; Yanagisawa, S.; Ogura, T.; Sugiyama, M. Catalytic mechanism of the tyrosinase reaction toward the Tyr98 residue in the caddie protein. *PLoS Biol.* **2019**, *16*, e3000077. [CrossRef]
40. Malathi, S.; Pakrudheen, I.; Narayana Kalkura, S.; Webster, T.J.; Balasubramanian, S. Disposable biosensors based on metal nanoparticles. *Sens. Int.* **2022**, *3*, 100169. [CrossRef]
41. Demuru, S.; Huang, C.-H.; Parvez, K.; Worsley, R.; Mattana, G.; Piro, B.; Noël, V.; Casiraghi, C.; Briand, D. All-Inkjet-Printed Graphene-Gated Organic Electrochemical Transistors on Polymeric Foil as Highly Sensitive Enzymatic Biosensors. *ACS Appl. Nano Mater.* **2022**, *5*, 1664–1673. [CrossRef]

42. Martínez-Periñán, E.; Domínguez-Saldaña, A.; Villa-Manso, A.M.; Gutiérrez-Sánchez, C.; Revenga-Parra, M.; Mateo-Martí, E.; Pariente, F.; Lorenzo, E. Azure A embedded in carbon dots as NADH electrocatalyst: Development of a glutamate electrochemical biosensor. *Sens. Actuators B Chem.* **2023**, *374*, 132761. [CrossRef]
43. Karimian, N.; Campagnol, D.; Tormen, M.; Stortini, A.M.; Canton, P.; Ugo, P. Nanoimprinted arrays of glassy carbon nanoelectrodes for improved electrochemistry of enzymatic redox-mediators. *J. Electroanal. Chem.* **2023**, *932*, 117240. [CrossRef]
44. Rahmawati, I.; Einaga, Y.; Ivandini, T.A.; Fiorani, A. Enzymatic Biosensors with Electrochemiluminescence Transduction. *ChemElectroChem* **2022**, *9*, e202200175. [CrossRef]
45. Chen, J.; Gao, H.; Li, Z.; Li, Y.; Yuan, Q. Ferriporphyrin-inspired MOFs as an artificial metalloenzyme for highly sensitive detection of H_2O_2 and glucose. *Chin. Chem. Lett.* **2020**, *31*, 1398–1401. [CrossRef]
46. Wang, X.; Wang, C.; Pan, M.; Wei, J.; Jiang, F.; Lu, R.; Liu, X.; Huang, Y.; Huang, F. Chaperonin-Nanocaged Hemin as an Artificial Metalloenzyme for Oxidation Catalysis. *ACS Appl. Mater. Interfaces* **2017**, *9*, 25387–25396. [CrossRef] [PubMed]
47. Wong, C.M.; Wong, K.H.; Chen, X.D. Glucose oxidase: Natural occurrence, function, properties and industrial applications. *Appl. Microbiol. Biotechnol.* **2008**, *78*, 927–938. [CrossRef]
48. Rocchitta, G.; Spanu, A.; Babudieri, S.; Latte, G.; Madeddu, G.; Galleri, G.; Nuvoli, S.; Bagella, P.; Demartis, M.I.; Fiore, V.; et al. Enzyme Biosensors for Biomedical Applications: Strategies for Safeguarding Analytical Performances in Biological Fluids. *Sensors* **2016**, *16*, 780. [CrossRef]
49. Cheon, H.J.; Nguyen, Q.H.; Kim, M.I. Highly Sensitive Fluorescent Detection of Acetylcholine Based on the Enhanced Peroxidase-Like Activity of Histidine Coated Magnetic Nanoparticles. *Nanomaterials* **2021**, *11*, 1207. [CrossRef]
50. Wang, Q.; Wang, B.; Shi, D.; Li, F.; Ling, D. Cerium Oxide Nanoparticles-Based Optical Biosensors for Bio-medical Applications. *Adv. Sens. Res.* **2023**, *2*, 2200065. [CrossRef]
51. Wang, L.; Sun, P.; Yang, Y.; Qiao, H.; Tian, H.; Wu, D.; Yang, S.; Yuan, Q.; Wang, J. Preparation of ZIF@ADH/NAD-MSN/LDH Core Shell Nanocomposites for the Enhancement of Coenzyme Catalyzed Double Enzyme Cascade. *Nanomaterials* **2021**, *11*, 2171. [CrossRef]
52. Dai, Y.; Liu, C.C. Recent Advances on Electrochemical Biosensing Strategies toward Universal Point-of-Care Systems. *Angew. Chem. Int. Ed.* **2019**, *58*, 12355–12368. [CrossRef] [PubMed]
53. Huang, X.; Zhu, Y.; Kianfar, E. Nano Biosensors: Properties, applications and electrochemical techniques. *J. Mater. Res. Technol.* **2021**, *12*, 1649–1672. [CrossRef]
54. Chupradit, S.; Km Nasution, M.; Rahman, H.S.; Suksatan, W.; Turki Jalil, A.; Abdelbasset, W.K.; Bokov, D.; Markov, A.; Fardeeva, I.N.; Widjaja, G.; et al. Various types of electrochemical biosensors for leukemia detection and therapeutic approaches. *Anal. Biochem.* **2022**, *654*, 114736. [CrossRef] [PubMed]
55. Schachinger, F.; Chang, H.; Scheiblbrandner, S.; Ludwig, R. Amperometric Biosensors Based on Direct Electron Transfer Enzymes. *Molecules* **2021**, *26*, 4525. [CrossRef] [PubMed]
56. Gigli, V.; Tortolini, C.; Capecchi, E.; Angeloni, A.; Lenzi, A.; Antiochia, R. Novel Amperometric Biosensor Based on Tyrosinase/Chitosan Nanoparticles for Sensitive and Interference-Free Detection of Total Catecholamine. *Biosensors* **2022**, *12*, 519. [CrossRef]
57. Harnisch, F.; Freguia, S. A Basic Tutorial on Cyclic Voltammetry for the Investigation of Electroactive Microbial Biofilms. *Chem. Asian J.* **2012**, *7*, 466–475. [CrossRef]
58. Yoon, J.; Conley, B.M.; Shin, M.; Choi, J.-H.; Bektas, C.K.; Choi, J.-W.; Lee, K.-B. Ultrasensitive Electrochemical Detection of Mutated Viral RNAs with Single-Nucleotide Resolution Using a Nanoporous Electrode Array (NPEA). *ACS Nano* **2022**, *16*, 5764–5777. [CrossRef]
59. Baluta, S.; Meloni, F.; Halicka, K.; Szyszka, A.; Zucca, A.; Pilo, M.I.; Cabaj, J. Differential pulse voltammetry and chronoamperometry as analytical tools for epinephrine detection using a tyrosinase-based electrochemical biosensor. *RSC Adv.* **2022**, *12*, 25342–25353. [CrossRef]
60. Cheraghi, S.; Taher, M.A.; Karimi-Maleh, H.; Karimi, F.; Shabani-Nooshabadi, M.; Alizadeh, M.; Al-Othman, A.; Erk, N.; Yegya Raman, P.K.; Karaman, C. Novel enzymatic graphene oxide based biosensor for the detection of glutathione in biological body fluids. *Chemosphere* **2022**, *287*, 132187. [CrossRef]
61. Lisdat, F.; Schäfer, D. The use of electrochemical impedance spectroscopy for biosensing. *Anal. Bioanal. Chem.* **2008**, *391*, 1555–1567. [CrossRef] [PubMed]
62. Rashed, M.Z.; Kopechek, J.A.; Priddy, M.C.; Hamorsky, K.T.; Palmer, K.E.; Mittal, N.; Valdez, J.; Flynn, J.; Williams, S.J. Rapid detection of SARS-CoV-2 antibodies using electrochemical impedance-based detector. *Biosens. Bioelectron.* **2021**, *171*, 112709. [CrossRef] [PubMed]
63. Gaviria-Arroyave, M.I.; Cano, J.B.; Peñuela, G.A. Nanomaterial-based fluorescent biosensors for monitoring environmental pollutants: A critical review. *Talanta Open* **2020**, *2*, 100006. [CrossRef]
64. Enander, K.; Choulier, L.; Olsson, A.L.; Yushchenko, D.A.; Kanmert, D.; Klymchenko, A.S.; Demchenko, A.P.; Mély, Y.; Altschuh, D. A Peptide-Based, Ratiometric Biosensor Construct for Direct Fluorescence Detection of a Protein Analyte. *Bioconjug. Chem.* **2008**, *19*, 1864–1870. [CrossRef]
65. Morii, T.; Sugimoto, K.; Makino, K.; Otsuka, M.; Imoto, K.; Mori, Y. A New Fluorescent Biosensor for Inositol Trisphosphate. *J. Am. Chem. Soc.* **2002**, *124*, 1138–1139. [CrossRef] [PubMed]

66. Scheller, F.W.; Wollenberger, U.; Warsinke, A.; Lisdat, F. Research and development in biosensors. *Curr. Opin. Biotechnol.* **2001**, *12*, 35–40. [CrossRef]

67. Bhirde, A.; Xie, J.; Swierczewska, M.; Chen, X. Nanoparticles for cell labeling. *Nanoscale* **2011**, *3*, 142–153. [CrossRef]

68. Yu, W.; Li, Y.; Xie, B.; Ma, M.; Chen, C.; Li, C.; Yu, X.; Wang, Z.; Wen, K.; Tang, B.Z.; et al. An Aggregation-Induced Emission-Based Indirect Competitive Immunoassay for Fluorescence "Turn-On" Detection of Drug Residues in Foodstuffs. *Front. Chem.* **2019**, *7*, 228. [CrossRef]

69. Zhang, X.; Hu, Y.; Yang, X.; Tang, Y.; Han, S.; Kang, A.; Deng, H.; Chi, Y.; Zhu, D.; Lu, Y. FÖrster resonance energy transfer (FRET)-based biosensors for biological applications. *Biosens. Bioelectron.* **2019**, *138*, 111314. [CrossRef]

70. Zhang, Y.; Xu, S.; Li, X.; Zhang, J.; Sun, J.; Tong, L.; Zhong, H.; Xia, H.; Hua, R.; Chen, B. Improved LRET-based detection characters of Cu^{2+} using sandwich structured $NaYF_4@NaYF_4:Er_{3+}/Yb_{3+}@NaYF_4$ nanoparticles as energy donor. *Sens. Actuators B Chem.* **2018**, *257*, 829–838. [CrossRef]

71. Syshchyk, O.; Skryshevsky, V.A.; Soldatkin, O.O.; Soldatkin, A.P. Enzyme biosensor systems based on porous silicon photoluminescence for detection of glucose, urea and heavy metals. *Biosens. Bioelectron.* **2015**, *66*, 89–94. [CrossRef] [PubMed]

72. Ballesta-Claver, J.; Ametis-Cabello, J.; Morales-Sanfrutos, J.; Megía-Fernández, A.; Valencia-Mirón, M.C.; Santoyo-González, F.; Capitán-Vallvey, L.F. Electrochemiluminescent disposable cholesterol biosensor based on avidin–biotin assembling with the electroformed luminescent conducting polymer poly(luminol-biotinylated pyrrole). *Anal. Chim. Acta* **2012**, *754*, 91–98. [CrossRef] [PubMed]

73. Cheng, Y.; Wang, H.; Zhuo, Y.; Song, D.; Li, C.; Zhu, A.; Long, F. Reusable smartphone-facilitated mobile fluorescence biosensor for rapid and sensitive on-site quantitative detection of trace pollutants. *Biosens. Bioelectron.* **2022**, *199*, 113863. [CrossRef] [PubMed]

74. Koveal, D.; Rosen, P.C.; Meyer, D.J.; Díaz-García, C.M.; Wang, Y.; Cai, L.-H.; Chou, P.J.; Weitz, D.A.; Yellen, G. A high-throughput multiparameter screen for accelerated development and optimization of soluble ge-netically encoded fluorescent biosensors. *Nat. Commun.* **2022**, *13*, 2919. [CrossRef]

75. Grazon, C.; Chern, M.; Lally, P.; Baer, R.C.; Fan, A.; Lecommandoux, S.; Klapperich, C.; Dennis, A.M.; Galagan, J.E.; Grinstaff, M.W. The quantum dot vs. organic dye conundrum for ratiometric FRET-based biosensors: Which one would you chose? *Chem. Sci.* **2022**, *13*, 6715–6731. [CrossRef]

76. Yoon, J.; Shin, M.; Lee, T.; Choi, J.-W. Highly Sensitive Biosensors Based on Biomolecules and Functional Nanomaterials Depending on the Types of Nanomaterials: A Perspective Review. *Materials* **2020**, *13*, 299. [CrossRef]

77. Hashem, A.; Hossain, M.A.M.; Marlinda, A.R.; Mamun, M.A.; Simarani, K.; Johan, M.R. Nanomaterials based electrochemical nucleic acid biosensors for environmental monitoring: A review. *Appl. Surf. Sci.* **2021**, *4*, 100064. [CrossRef]

78. Kosri, E.; Ibrahim, F.; Thiha, A.; Madou, M. Micro and Nano Interdigitated Electrode Array (IDEA)-Based MEMS/NEMS as Electrochemical Transducers: A Review. *Nanomaterials* **2022**, *12*, 4171. [CrossRef]

79. Liu, Y.; Chen, Y.; Fan, Y.; Gao, G.; Zhi, J. Development of a Tyrosinase Amperometric Biosensor Based on Carbon Nanomaterials for the Detection of Phenolic Pollutants in Diverse Environments. *ChemElectroChem* **2022**, *9*, e202200861. [CrossRef]

80. Öndeş, B.; Evli, S.; Şahin, Y.; Uygun, M.; Uygun, D.A. Uricase based amperometric biosensor improved by $AuNPs-TiS_2$ nanocomposites for uric acid determination. *Microchem. J.* **2022**, *181*, 107725. [CrossRef]

81. Haritha, V.S.; Kumar, S.R.S.; Rakhi, R.B. Amperometric cholesterol biosensor based on cholesterol oxidase and Pt-Au/ MWNTs modified glassy carbon electrode. *Mater. Today Proc.* **2022**, *50*, 34–39. [CrossRef]

82. Thakur, D.; Pandey, C.M.; Kumar, D. Highly Sensitive Enzymatic Biosensor Based on Polyaniline-Wrapped Titanium Dioxide Nanohybrid for Fish Freshness Detection. *Appl. Biochem. Biotechnol.* **2022**, *194*, 3765–3778. [CrossRef] [PubMed]

83. Carinelli, S.; Fernández, I.; Luis González-Mora, J.; Salazar-Carballo, P.A. Hemoglobin-modified nanoparticles for electrochemical determination of haptoglobin: Application in bovine mastitis diagnosis. *Microchem. J.* **2022**, *179*, 107528. [CrossRef]

84. Jędrzak, A.; Kuznowicz, M.; Rębiś, T.; Jesionowski, T. Portable glucose biosensor based on polynorepinephrine@magnetite nanomaterial integrated with a smartphone analyzer for point-of-care application. *Bioelectrochemistry* **2022**, *145*, 108071. [CrossRef]

85. Li, X.; Xu, M.; Wu, Q.; Wei, W.; Liu, X. Photolithographic 3D microarray electrode-based high-performance non-enzymatic H_2O_2 sensor. *Colloids Surf. Physicochem. Eng. Asp.* **2021**, *628*, 127249. [CrossRef]

86. Cao, L.; Han, G.-C.; Xiao, H.; Chen, Z.; Fang, C. A novel 3D paper-based microfluidic electrochemical glucose biosensor based on rGO-TEPA/PB sensitive film. *Anal. Chim. Acta* **2020**, *1096*, 34–43. [CrossRef]

87. Liu, T.; Zhao, Q.; Xie, Y.; Jiang, D.; Chu, Z.; Jin, W. In situ fabrication of aloe-like Au–ZnO micro/nanoarrays for ultrasensitive biosensing of catechol. *Biosens. Bioelectron.* **2020**, *156*, 112145. [CrossRef]

88. Yoon, J.; Cho, H.-Y.; Shin, M.; Choi, H.K.; Lee, T.; Choi, J.-W. Flexible electrochemical biosensors for healthcare monitoring. *J. Mater. Chem. B* **2020**, *8*, 7303–7318. [CrossRef]

89. Yoon, J.; Lee, S.N.; Shin, M.K.; Kim, H.-W.; Choi, H.K.; Lee, T.; Choi, J.-W. Flexible electrochemical glucose biosensor based on $GOx/gold/MoS_2/gold$ nanofilm on the polymer electrode. *Biosens. Bioelectron.* **2019**, *140*, 111343. [CrossRef]

90. Zhang, Q.; Jiang, D.; Xu, C.; Ge, Y.; Liu, X.; Wei, Q.; Huang, L.; Ren, X.; Wang, C.; Wang, Y. Wearable electrochemical biosensor based on molecularly imprinted Ag nanowires for noninvasive monitoring lactate in human sweat. *Sens. Actuators B Chem.* **2020**, *320*, 128325. [CrossRef]

91. Muñoz, J.; Pumera, M. 3D-printed biosensors for electrochemical and optical applications. *TrAC Trends Anal. Chem.* **2020**, *128*, 115933. [CrossRef]

92. Sangubotla, R.; Kim, J. Fiber-optic biosensor based on the laccase immobilization on silica-functionalized fluorescent carbon dots for the detection of dopamine and multi-color imaging applications in neuroblastoma cells. *Mater. Sci. Eng. C* **2021**, *122*, 111916. [CrossRef] [PubMed]

93. Osman, D.I.; El-Sheikh, S.M.; Sheta, S.M.; Ali, O.I.; Salem, A.M.; Shousha, W.G.; El-Khamisy, S.F.; Shawky, S.M. Nucleic acids biosensors based on metal-organic framework (MOF): Paving the way to clinical laboratory diagnosis. *Biosens. Bioelectron.* **2019**, *141*, 111451. [CrossRef] [PubMed]

94. Hu, S.; Liu, J.; Wang, Y.; Liang, Z.; Hu, B.; Xie, J.; Wong, W.-L.; Wong, K.-Y.; Qiu, B.; Peng, W. A new fluorescent biosensor based on inner filter effect and competitive coordination with the europium ion of non-luminescent Eu-MOF nanosheets for the determination of alkaline phosphatase activity in human serum. *Sens. Actuators B Chem.* **2023**, *380*, 133379. [CrossRef]

95. Zhang, Q.; Wang, C.-F.; Lv, Y.-K. Luminescent switch sensors for the detection of biomolecules based on metal–organic frameworks. *Analyst* **2018**, *143*, 4221–4229. [CrossRef]

96. Aggarwal, V.; Solanki, S.; Malhotra, B.D. Applications of metal–organic framework-based bioelectrodes. *Chem. Sci.* **2022**, *13*, 8727–8743. [CrossRef]

97. Lv, M.; Zhou, W.; Tavakoli, H.; Bautista, C.; Xia, J.; Wang, Z.; Li, X. Aptamer-functionalized metal-organic frameworks (MOFs) for biosensing. *Biosens. Bioelectron.* **2021**, *176*, 112947. [CrossRef]

98. Wu, X.-Q.; Ma, J.-G.; Li, H.; Chen, D.-M.; Gu, W.; Yang, G.-M.; Cheng, P. Metal–organic framework biosensor with high stability and selectivity in a bio-mimic environment. *Chem. Commun.* **2015**, *51*, 9161–9164. [CrossRef]

99. Cai, Y.; Zhu, H.; Zhou, W.; Qiu, Z.; Chen, C.; Qileng, A.; Li, K.; Liu, Y. Capsulation of AuNCs with AIE Effect into Metal–Organic Framework for the Marriage of a Fluorescence and Colorimetric Biosensor to Detect Organophosphorus Pesticides. *Anal. Chem.* **2021**, *93*, 7275–7282. [CrossRef]

100. Liang, W.; Wied, P.; Carraro, F.; Sumby, C.J.; Nidetzky, B.; Tsung, C.-K.; Falcaro, P.; Doonan, C.J. Metal–Organic Framework-Based Enzyme Biocomposites. *Chem. Rev.* **2021**, *121*, 1077–1129. [CrossRef]

101. Huang, S.; Kou, X.; Shen, J.; Chen, G.; Ouyang, G. "Armor-Plating" Enzymes with Metal–Organic Frameworks (MOFs). *Angew. Chem. Int. Ed.* **2020**, *59*, 8786–8798. [CrossRef] [PubMed]

102. Ade, C.; Brodszkij, E.; Thingholm, B.; Gal, N.; Itel, F.; Taipaleenmäki, E.; Hviid, M.J.; Schattling, P.S.; Städler, B. Small Organic Catalase Mimic Encapsulated in Micellar Artificial Organelles as Reactive Oxygen Species Scavengers. *ACS Appl. Polym. Mater.* **2019**, *1*, 1532–1539. [CrossRef]

103. Karim, M.N.; Singh, M.; Weerathunge, P.; Bian, P.; Zheng, R.; Dekiwadia, C.; Ahmed, T.; Walia, S.; Della Gaspera, E.; Singh, S.; et al. Visible-Light-Triggered Reactive-Oxygen-Species-Mediated Antibacterial Activity of Peroxidase-Mimic CuO Nanorods. *ACS Appl. Nano Mater.* **2018**, *1*, 1694–1704. [CrossRef]

104. Wang, M.; Zhou, X.; Wang, S.; Xie, X.; Wang, Y.; Su, X. Fabrication of Bioresource-Derived Porous Carbon-Supported Iron as an Efficient Oxidase Mimic for Dual-Channel Biosensing. *Anal. Chem.* **2021**, *93*, 3130–3137. [CrossRef] [PubMed]

105. Zhang, D.; Du, P.; Chen, J.; Guo, H.; Lu, X. Pyrazolate-based porphyrinic metal-organic frameworks as catechol oxidase mimic enzyme for fluorescent and colorimetric dual-mode detection of dopamine with high sensitivity and specificity. *Sens. Actuators B Chem.* **2021**, *341*, 130000. [CrossRef]

106. Ouyang, Y.; O'Hagan, M.P.; Willner, I. Functional catalytic nanoparticles (nanozymes) for sensing. *Biosens. Bioelectron.* **2022**, *218*, 114768. [CrossRef]

107. Abdel-Lateef, M.A. Utilization of the peroxidase-like activity of silver nanoparticles nanozyme on O-phenylenediamine/H_2O_2 system for fluorescence detection of mercury (II) ions. *Sci. Rep.* **2022**, *12*, 6953. [CrossRef]

108. Li, X.; He, Z.; Li, C.; Li, P. One-step enzyme kinetics measurement in 3D printed microfluidics devices based on a high-performance single vibrating sharp-tip mixer. *Anal. Chim. Acta* **2021**, *1172*, 338677. [CrossRef]

109. Khandan-Nasab, N.; Askarian, S.; Mohammadinejad, A.; Aghaee-Bakhtiari, S.H.; Mohajeri, T.; Kazemi Oskuee, R. Biosensors, microfluidics systems and lateral flow assays for circulating microRNA detection: A review. *Anal. Biochem.* **2021**, *633*, 114406. [CrossRef]

110. Rossini, E.L.; Milani, M.I.; Lima, L.S.; Pezza, H.R. Paper microfluidic device using carbon dots to detect glucose and lactate in saliva samples. *Spectrochim. Acta A Mol. Biomol. Spectrosc.* **2021**, *248*, 119285. [CrossRef]

111. Wu, Y.; Chen, J.-Y.; He, W.-M. Surface-enhanced Raman spectroscopy biosensor based on silver nanopar-ticles@metal-organic frameworks with peroxidase-mimicking activities for ultrasensitive monitoring of blood cholesterol. *Sens. Actuators B Chem.* **2022**, *365*, 131939. [CrossRef]

112. Xia, X.; Weng, Y.; Zhang, L.; Tang, R.; Zhang, X. A facile SERS strategy to detect glucose utilizing tandem enzyme activities of Au@Ag nanoparticles. *Spectrochim. Acta A Mol. Biomol.* **2021**, *259*, 119889. [CrossRef] [PubMed]

113. Zhang, J.; Mai, X.; Hong, X.; Chen, Y.; Li, X. Optical fiber SPR biosensor with a solid-phase enzymatic reaction device for glucose detection. *Sens. Actuators B Chem.* **2022**, *366*, 131984. [CrossRef]

114. Zhu, G.; Wang, Y.; Wang, Z.; Singh, R.; Marques, C.; Wu, Q.; Kaushik, B.K.; Jha, R.; Zhang, B.; Kumar, S. Lo-calized Plasmon-Based Multicore Fiber Biosensor for Acetylcholine Detection. *IEEE Trans. Instrum. Meas.* **2022**, *71*, 1–9. [CrossRef]

115. Avan, A.N.; Demirci-Çekiç, S.; Apak, R. Colorimetric Nanobiosensor Design for Determining Oxidase Enzyme Substrates in Food and Biological Samples. *ACS Omega* **2022**, *7*, 44372–44382. [CrossRef]

116. Zhang, Y.; Xu, Q.; Wang, F.; Gao, T.; Wei, T. Enzyme powered self-assembly of hydrogel biosensor for col-orimetric detection of metabolites. *Sens. Actuators B Chem.* **2023**, *375*, 132942. [CrossRef]

117. You, X.; Pak, J.J. Graphene-based field effect transistor enzymatic glucose biosensor using silk protein for enzyme immobilization and device substrate. *Sens. Actuators B Chem.* **2014**, *202*, 1357–1365. [CrossRef]
118. Alvarado-Ramírez, L.; Rostro-Alanis, M.; Rodríguez-Rodríguez, J.; Sosa-Hernández, J.E.; Melchor-Martínez, E.M.; Iqbal, H.M.N.; Parra-Saldívar, R. Enzyme (Single and Multiple) and Nanozyme Biosensors: Recent Developments and Their Novel Applications in the Water-Food-Health Nexus. *Biosensors* **2021**, *11*, 410. [CrossRef]
119. Pathak, A.; Gupta, B.D. Ultra-selective fiber optic SPR platform for the sensing of dopamine in synthetic cerebrospinal fluid incorporating permselective nafion membrane and surface imprinted MWCNTs-PPy matrix. *Biosens. Bioelectron.* **2019**, *133*, 205–214. [CrossRef]
120. Caldara, M.; Lowdon, J.W.; Rogosic, R.; Arreguin-Campos, R.; Jimenez-Monroy, K.L.; Heidt, B.; Tschulik, K.; Cleij, T.J.; Diliën, H.; Eersels, K.; et al. Thermal Detection of Glucose in Urine Using a Molecularly Imprinted Polymer as a Recognition Element. *ACS Sens.* **2021**, *6*, 4515–4525. [CrossRef]
121. Kim, H.; Seong, W.; Rha, E.; Lee, H.; Kim, S.K.; Kwon, K.K.; Park, K.-H.; Lee, D.-H.; Lee, S.-G. Machine learning linked evolutionary biosensor array for highly sensitive and specific molecular identification. *Biosens. Bioelectron.* **2020**, *170*, 112670. [CrossRef] [PubMed]
122. Mross, S.; Pierrat, S.; Zimmermann, T.; Kraft, M. Microfluidic enzymatic biosensing systems: A review. *Biosens. Bioelectron.* **2015**, *70*, 376–391. [CrossRef] [PubMed]

 **biosensors**

Review

Proteolytic Biosensors with Functional Nanomaterials: Current Approaches and Future Challenges

Jin-Ha Choi

School of Chemical Engineering, Clean Energy Research Center, Jeonbuk National University, Jeonju 54896, Republic of Korea; jhchoi@jbnu.ac.kr; Tel.: +82-63-270-4854

Abstract: Proteolytic enzymes are one of the important biomarkers that enable the early diagnosis of several diseases, such as cancers. A specific proteolytic enzyme selectively degrades a certain sequence of a polypeptide. Therefore, a particular proteolytic enzyme can be selectively quantified by changing detectable signals causing degradation of the peptide chain. In addition, by combining polypeptides with various functional nanomaterials, proteolytic enzymes can be measured more sensitively and rapidly. In this paper, proteolytic enzymes that can be measured using a polypeptide degradation method are reviewed and recently studied functional nanomaterials-based proteolytic biosensors are discussed. We anticipate that the proteolytic nanobiosensors addressed in this review will provide valuable information on physiological changes from a cellular level for individual and early diagnosis.

Keywords: proteolytic biosensors; enzymatic biosensors; matrix metalloproteinase (MMP); caspase family; protease

1. Introduction

Accurate and early diagnosis is essential for various diseases to reduce the fatality rate and increase the recovery rate [1–3]. To this end, there is an urgent need for biosensors that precisely and sensitively measure potential biomarkers such as proteins or nucleic acids. In particular, high-performance biosensors have been actively developed to effectively treat infectious diseases such as SARS-CoV-2 and to help prevent transmission [4–10]. Among various biomarkers, protein could be used as an accurate diagnosis biomarker since proteins are end products produced by the central dogma process, unlike nucleic acid biomarkers, which have inaccuracies due to post-transcriptional and post-translational processes [11,12]. In general, sandwich immunoassay is the most typical protein detection method using an antigen–antibody binding reaction [13–17]. This immunoassay-based measurement method, represented by enzyme-linked immunosorbent assays (ELISAs), has the advantage of being able to selectively detect various proteins, and the results can be intuitively viewed because the results can be confirmed with the naked eye using an enzyme-substrate reaction. However, at least two antibodies are required to measure the target protein, and colorimetric reactions with labeling enzymes are required to confirm the signal. Due to these disadvantages, there are also severe limitations in that it is cost-ineffective, labor-intensive, and time-consuming. In addition, the detection accuracy of the biosensor could be reduced owing to the complicated process.

Among numerous protein biomarkers, there are many enzymatic proteins, including matrix metalloproteinase (MMP) and the caspase family, with proteolytic properties [18–23]. Some of the proteolytic enzymes have the particular property of recognizing and degrading specific peptide sequences. Using this property, many proteolytic biosensors have been developed without using an antigen–antibody reaction [24–26]. Compared to immunoassay-based biosensors, proteolytic biosensors have numerous advantages. For example, since the proteolytic biomarker is measured by a specific peptide degradation

Citation: Choi, J.-H. Proteolytic Biosensors with Functional Nanomaterials: Current Approaches and Future Challenges. *Biosensors* **2023**, *13*, 171. https://doi.org/10.3390/bios13020171

Received: 29 December 2022
Revised: 19 January 2023
Accepted: 19 January 2023
Published: 21 January 2023

reaction, there is an advantage in the reduced cost compared to using multi-antibodies, and the measurement time is shortened because of the simple peptide-cleavage reaction step. In addition, since the peptide degradation reaction is composed of just one step, it is easy to induce a reaction at the intracellular level and in vivo system. In particular, developing nanobiosensors using multifunctional nanomaterials has also been actively studied [27–31]. Nanomaterials have a wide surface area, high reaction rate, and ease of immobilizing various biological materials. Therefore, they have a wide range of applications in the biomedical field, including biosensors. In addition, it is possible to simply check the diagnosis results by improving the measurement sensitivity by using various characteristics (optical, electrical, and mechanical properties, etc.) of nanomaterials. In this review, we will investigate proteolytic biosensors integrated with recently announced nanotechnology. In addition, we will discuss the prospects and complementary points of proteolytic biosensors in the field of disease diagnosis.

2. Proteolytic Enzymes

2.1. Serine Proteases

Nearly one-third of all proteases can be classified as serine proteases. Serine proteases are a large family of protein-degrade enzymes that play a crucial role in processes such as apoptosis, blood coagulation, and inflammation [32–34]. Serine proteases are broadly dispersed in nature and found in all cellular metabolisms. Proteases can be grouped into four major clans, which are the SB clan (subtilisin), the PA clan (chymotrypsin), and SF and SC clans that contain various proteases. Approximately 75% of human serine proteases are part of the PA clan. The PA serine proteases can be divided into three subgroups: the trypsin-like serine proteases, which cleave peptide substrates after positively charged arginine and lysine residues; chymotrypsin-like serine proteases, which cleave after large hydrophobic amino acids, including leucine and alanine; elastase-like enzymes, which cleave substrates after hydrophobic residues. In particular, the neutrophil elastase family, which is related to the immune response, is part of this subgroup [35]. This serine protease family plays an important role in the control of intracellular and extracellular activities. For example, several serine proteases are collected inside granules attached to proteoglycans, avoiding leakage into the cytoplasm and communicating with their cellular objectives [36]. On the other hand, serpins, which were first recognized for serine protease inhibition, inhibit their target protease via a unique suicide mechanism, blocking the protease into an irreversible state [37]. Generally, serine proteases are endoproteases that degrade polypeptide bonds by hydrolysis within a polypeptide chain.

2.2. Cysteine Proteases

Cysteine proteases (cathepsins) have been known to be involved in many physiological processes, such as regulation of proteolytic cascades, cytokine maturation, expression of proteins, and antigen presentation from the cell surface [38–40]. In addition, cysteine proteases have been understood to be involved in numerous pathologies for a long time [41–43]. Cysteine proteases are synthesized as preproenzymes and produced as lysosomes, where they provide their function of protein hydrolysis. After the removal of signal peptides, the molecular mass of these proteases is within the range of 20–35 kDa. Typically, they are involved in precursor protein activation, such as proenzymes and prohormones, bone remodeling, MHC-II-mediated antigen presentation, keratinocytes differentiation, and cell reproduction and apoptosis. In the case of the proteolytic mechanism, Cys25 and His159 form the active site for full catalytic activity. The imidazole group of the histidine polarizes the SH group of the cysteine and enables deprotonation, and a thiolate–imidazolium ion pair is produced. The thiolate anion attacks the carbonyl carbon of the peptide bond to be cleaved, and a tetrahedral intermediate is produced. These cysteine proteases are closely related to several fetal diseases. For example, cathepsin K could be a major target in bone syndromes such as osteoporosis [44,45]. In addition, lots of cancers and arthritis correlate to the expression and activation of cysteine cathepsins [46,47]. Furthermore, the inhibitors of cysteine proteases are highly effective and specific molecules that inhibit cancer progression with fewer adverse effects [48].

2.3. Matrix Metalloproteinase (MMP) Family

Matrix metalloproteases (MMPs) are a zinc-dependent endopeptidases family that has a similar structure and the capability to decompose every part of the extracellular matrix (ECM) [49–51]. The MMP enzyme essentially consists of extracellular matrix remodeling proteases. Using this important physiological role, they have extensive proteolytic activity and contribute to diverse physiological and pathological processes, such as cancer, chronic wounds, cardiovascular diseases, chronic inflammation, and others. MMPs are universal multi-domain enzymes, with 23 MMPs identified in humans (MMPs 1–3, 7–17, 19–21, 23–28) [52–55]. The structure of MMPs composes of a propeptide that keeps the inactive state (pro-MMPs) by blocking the interaction between the active site and the substrate by a cysteine switch. The catalytic domain (active site) is characterized by the existence of a Zn-binding site, where the Zn ion is organized by three histidines and a glutamate, resulting in a linker peptide of varying lengths and a hemopexin-like domain. Besides the extracellular matrix (ECM), MMPs are engaged in a number of inter- and intracellular activities and contribute to functional networks, systematically cooperating with other biomolecules. In addition, MMPs are implicated in specific pathological processes; this could lead to the significant application of MMPs as potential biomarkers for the prognosis of disease and early diagnostic approaches with detailed and useful information for effective therapy.

3. Extracellular Detection of Protease for Diagnosis Using Nanotechnology

3.1. Fluorescence-Based Detection

Proteases released outside the cell can provide important signals that inform the detailed status of the cell. Thus, extracellular proteases could be one of the important biomarkers for disease diagnosis, and it is necessary to measure them precisely and rapidly for effective diagnosis and therapy. Fluorescence-based biosensors using fluorescent materials, including organic dye and fluorescent nanoparticles, have been widely used to measure extracellular proteases by using the phenomenon in which fluorescence signals are converted by specific reactions between proteases and substrates. Renault et al. showed a novel protease-sensitive fluorescent probe based on the covalent-assembly approach [56]. The protease-sensitive fluorescent probe was designed to be degraded by the targeting enzyme (penicillin G acylase (PGA)) to develop a detectable pyronin fluorescent molecule under physiological conditions through the in situ structure of unsymmetrical pyronin AR116. In addition, c-nucleophile (C-Nu) attached to the pyronin precursor can be screened for the optimization of fluorescent signal generation for sensitive protease measurement. Zhang et al. developed an ultrasensitive and rapid thrombin biosensor composed of trifunctional protein (Figure 1a) [57]. This particular trifunctional protein consists of three functional parts: a thrombin cleavage site (TCS), far-red fluorescent protein (smURFP), and hydrophobin (HGFI). HGFI plays a role in attaching the multi-well plate of the trifunctional proteins, and the TCS is a bridge between the plate and the fluorophore. Once the thrombin cleaves the TCS, a red fluorescent signal can be measured with proportional signal intensity to the thrombin concentration. The detection range of this proteolytic biosensor is from 1.02 aM to 0.01 mM, and the limit of detection (LOD) is 0.2 aM within 20 min. Since the above two proteolytic sensors measure biomarkers based on fluorescent emission, the detection results can be known easily and simply, and the sensitivity is also excellent (aM level). However, there is a limitation in that the measured biomarkers must have proteolytic properties.

For proteolytic detection, fluorescence resonance energy transfer (FRET) is a representative analytical method, which is a distance-dependent fluorescent energy-transfer phenomenon between the donor and the acceptor [58]. If the distance between the donor and acceptor is below 10 nm, the FRET will show and the signal from the acceptor is released, whereas the donor fluorophore is recovered if the distance is above 10 nm. Therefore, distance is a major factor in regulating the FRET phenomenon. In this FRET system, the cleavable peptide sequence, which is a degradable site by a specific protease, can be an essential part of determining the distance between the donor and the acceptor. Zhang et al. exhibited graphene oxide (GO)-assisted fluorescence biosensors for the detection of the human immunodeficiency virus (HIV) protease [59]. Fluorescent-labeled peptide molecules are covalently attached to the surface of GO. Fluorescein is effectively quenched by the FRET effect on GO. Once the HIV protease cleaves the peptide substrate, the fluorescence signal is recovered, proportional to the protease level. The authors claimed that HIV-1 protease could be measured at as low as 1.18 ng/mL in a rapid and accurate manner. Brown et al. revealed a FRET-based biosensor for the detection of the 3-chymotrypsin-like cysteine protease, which is highly related to SARS-CoV-2 [60]. For the FRET effect, the authors used an eCFP (Em: 434 nm) and Venus (Em: 528 nm) pair, connected to the cleavable peptide by a cysteine protease. In addition, it was applied to the high-throughput screening platform for several new inhibitors of SARS-CoV-2. The screening recognized 65 inhibitors, with 20 most active inhibitions of SARS-CoV-2. The aforementioned protease biosensors successfully measured the virus-related protease for the diagnosis of viral infection by using the FRET effect in a simple and highly sensitive manner. FRET-based biosensors have the advantage of measuring target proteases simply and sensitively because they induce changes in fluorescence signals with a single peptide degradation reaction. On the other hand, in the case of virus diagnosis, since nucleic acid biosensors can show relatively accurate diagnostic results for viral diseases, protease biosensors are not yet commonly used diagnostic methods.

Except for virus-related protease detection, other particular proteolytic enzymes, including the MMP family and trypsin, have been measured by integrating the FRET phenomenon. Zhang et al. developed a multiplex fluorescence biosensor for the simultaneous detection of multiple protease activities, such as MMPs and a disintegrin and metalloproteinases (ADAMs) (Figure 1b) [61]. Conventional multiplex biosensing platforms consist of an individual array of different biomarkers, whereas the developed multiplexed nGO−peptide sequence biosensor was fabricated by multiple fluorophore-labeled peptides on an nGO sheet. Using this platform, they found the specific combinations of biomarkers for cancer diagnosis (MMP-9, ADAM-10, and ADAM-17) with joint entropy and programming. Li et al. displayed a sensitive and simple trypsin assay by using stable fluorescent polydopamine nanoparticles with the integration of protamine, which induced the quenching effect (Figure 1c) [62]. Due to the proteolytic effect of trypsin on protamine, the aggregated polydopamine nanoparticle and protamine were degraded, and the fluorescent signal was recovered. An increase in the fluorescent signal was exhibited with the concentration of trypsin (0.01 to 0.1 mg/mL). In addition, this biosensing system showed good practicability in human serum and trypsin inhibitor screening. Xu et al. developed a FRET-based turn-on fluorescent biosensor for trypsin detection based on carbon dots as a donor and Au nanoparticles (AuNPs) as an acceptor [63]. Compared with traditional quantum dots (QDs) and fluorophores, carbon dots have advantages for biosensing applications, such as high photostability, water solubility, and low toxicity. Trypsin-specific peptide sequences (Arg-Cys-Phe-Arg-Gly-Gly-Asp-Asp, RCFRGGDD) were covered AuNPs via the Au-SH bond. Due to the negatively charged ASP-covered AuNPs, the AuNPs were dispersed, and the carbon dot could not emit the fluorescent signal. In the trypsin-degraded peptide sequence, the AuNPs were aggregated, and the carbon dots could recover their fluorescent emission. This system could measure trypsin as low as 0.84 ng/mL with high selectivity. Bui et al. presented the protease-to-DNA converting biosensor to provide the protease activity to be transformed to generating specific DNA sequences [64]. Cy3-labeled peptide-DNAs are attached onto QD donors as the input gate. Once trypsin cleaves the peptide sequence, the donor QD emits its fluorescent signal and the DNA-Cy3 complex interacts with a tetrahedral output gate, resulting in Cy5 emission via the FRET effect. As such, in order to improve the sensitivity and simplicity of the proteolytic biosensor, the FRET phenomena are applied as a potential sensing strategy. Since the FRET-based biosensor for measuring proteolytic enzyme can sensitively change the fluorescence signal with a simple degradation reaction, it is possible to measure a small amount of proteolytic enzyme in a time-and cost-effective manner. In contrast, since FRET is only possible when the donor's emission wavelength and the acceptor's excitation wavelength overlap, there is a limitation to the combination of fluorescent materials. In addition, it may not be suitable for developing a multiplex biosensor that needs to measure multiple proteases at once for precise diagnosis.

On the other hand, the combination of fluorescent materials can induce a change in fluorescence property with plasmonic nanomaterials, whose surface plasmon properties can transfer absorbed optical energy into the fluorescence emission of proximal fluorescence molecules. This phenomenon is called metal-enhanced fluorescence (MEF), which substantially boosts the fluorescent signal [65]. This enhancing effect can be applied to proteolytic biosensors with similar strategies to a FRET-based analytical system. Choi et al. developed simple MEF-based proteolytic biosensors composed of DNA, peptide sequence bifunctional AuNPs, and fluorophore (fluorescein isothiocyanate; FITC) levels (Figure 1d) [66]. As the optimal distance of the MEF effect is about 8 nm between the AuNP and FITC, the quenching state is induced as the length of a single-stranded DNA (ssDNA) is 7~8 nm and the peptide sequence is smaller than the ssDNA. When the peptide sequence is degraded by caspase-3, the optimal distance of the ssDNA will show the MEF effect corresponding to the caspase-3 level. Using this biosensing system is possible for the simple (one-step proteolytic reaction) and rapid (<1 h) detection of caspase-3 as low as 10 pg/mL. This simple and sensitive detection system can also be applied to measure intracellular caspase-3. Lucas et al. exhibited MEF-based trypsin biosensors using Ag nanoparticle-modified nano-slivered 96-well plates and FITC-labeled YeBF protein [67]. The authors claimed up to $11,000\times$ signal enhancement for fluorophores due to the effective coupling or enhancement volume region of the silver surface. This biosensing system achieved a detection limit of 1.89 ng of enzymes (2.8 mBAEE activity units). In addition, no washing steps were needed in this MEF system because of the use of the low quantum yield fluorescent label, resulting in a very low background signal of the detached fluorophore. Compared to FRET-based proteolytic biosensors, MEF-based biosensors have the advantage of excellent sensitivity due to the enhanced fluorescent signal. However, it is important to maintain an accurate distance between the donor and the acceptor (around <10 nm); hence, it is difficult to induce the MEF effect in proteolytic enzymatic detection.

3.2. Colorimetric Detection

Besides fluorescence-based biosensors, color change is another optical biosensor where one can observe the target proteolytic protein intuitively and easily with the naked eye. For example, lots of pregnancy tests and SARS-CoV-2 test strips have been widely used for simple diagnosis [68–71]. Change in the absorbance peak is one of the distinguished phenomena for color change. For the induction of a significant change in color or absorbance peak, AuNPs have been utilized because their absorbance property depends on their size and the degree of aggregation. Specific peptide sequences, which can be dissociated by proteolytic enzymes, can be applied to the colorimetric biosensor with AuNPs. Liu et al. developed a colorimetric biosensor for the measurement of protease activity by the growth of AuNPs [72]. Cleavage of the specific peptide against β-secretase induces the exposure of an amino-terminal Cu^{2+}- and Ni^{2+}-binding (ATCUN) motif. This ATCUN peptide can seize Cu^{2+} and, thus, weaken the oxidation effect of ascorbic acid, which induces the reduction of $HAuCl_4$ into AuNPs. Using this strategy, β-secretase was measured as low as 0.1 nM by monitoring the color change of AuNP and ascorbic acid consumption with UV/vis spectroscopy. Creyer et al. showed a particular gold−silver core−shell nanoparticle structure for dramatic color change by the cleavage reaction between the GCGKGCG dithiol peptide and trypsin (Figure 2a) [73]. As a result, the degree of aggregation and the absorbance peak shift can be correlated with trypsin concentration with high linearity ($R^2 = 0.99$). The detection limit of trypsin was 0.47 nM. Liu et al. demonstrated a colorimetric protease assay using two separate biological reactions based on proteolysis-responsive transcription

(Figure 2b) [74]. MMP-2-mediated proteolysis triggers the in vitro transcription of RNAs, which induces the aggregation of DNA-functionalized AuNPs by a complimentary binding reaction. This proteolysis-responsive transcription sensor showed a sensitive detection result, as low as 3.3 pM of MMP-2. Moreover, other protease biomarkers, such as thrombin and the hepatitis C virus, could also be measured with the test strip format for smartphone analysis. Feng et al. developed a label-free peptide-AuNP biosensor for the measurement of the SARS-CoV-2 main protease (Mpro), which is one of the potential targets for the development of drugs [75]. The specific peptide sequence for the SARS-CoV-2 Mpro induces the aggregation of AuNPs by electrostatic interaction. Once the Mpro degrades the peptide, the AuNPs are dissociated and the absorbance peak is blue-shifted, resulting in a clear color change. Moreover, the electrode-modified peptide can be aggregated with the AuNPs, and the proteolytic event can induce a change in the electrochemical signal. In the colorimetric and electrochemical sensing systems, the detection limits were 10 and 0.1 pM, respectively. Ling et al. proposed a nanocellulose-based colorimetric biosensor for the facile detection of human neutrophil elastase [76]. Due to its crystallinity, high surface area, and biocompatible properties, nanocellulose can be utilized as an efficient sensing transducer. Through a deep eutectic solvent treatment, cotton cellulose nanocrystals are formed, and they modify specific peptides for the human neutrophil elastase. The sensitivity of this colorimetric sensor is around 0.005 U/mL. The authors claimed that it could provide a sensitive and convenient sensor platform applicable for point-of-care protease detection.

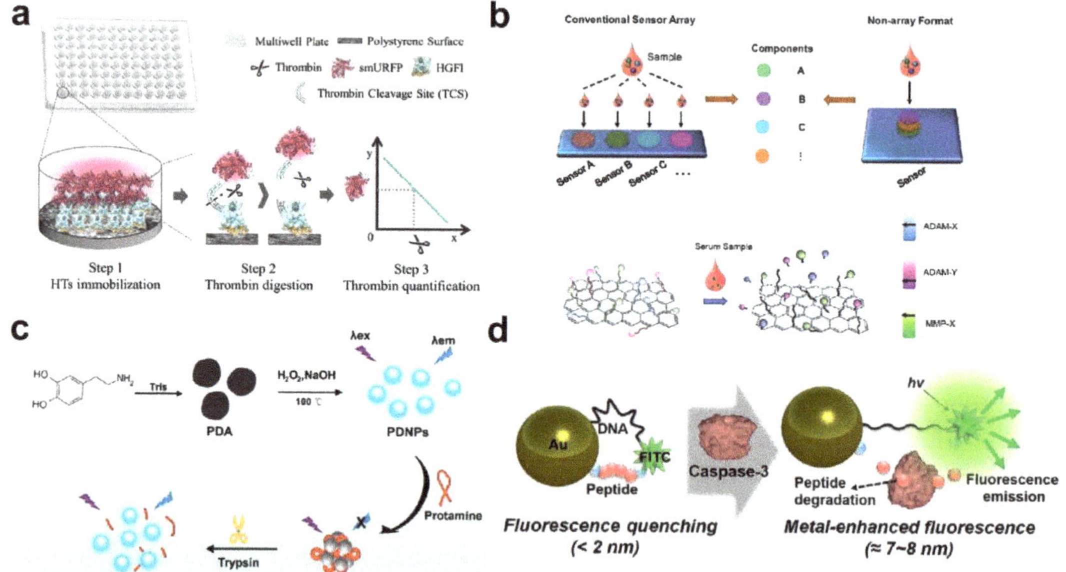

Figure 1. Fluorescence-based biosensors for extracellular protease detection. (**a**) Schematic diagram of the multi-well-plate-based thrombin biosensors composed of trifunctional protein, including a far-red fluorescent protein smURFP, hydrophobin HGFI, and a thrombin cleavage site. This figure is adapted with permission from Ref. [57] (© 2020 WILEY-VCH). (**b**) Schematic diagram of the graphene oxide-based multiplexing biosensor for multiple proteases, ADAM-10, ADAM-17, and MMP-9. This figure is adapted with permission from Ref. [61] (© 2020 American Chemical Society). (**c**) Fluorescent polydopamine nanoparticles for a sensitive and simple trypsin assay via the hydrolysis of protamine by trypsin. This figure is adapted with permission from Ref. [62] (© 2021 Elsevier B.V.). (**d**) MEF-based DNA detection system using a plasmonic Au-assisted MEF effect by CRISPR-Cas12a reaction. This figure is adapted with permission from Ref. [66] (© 2021 American Chemical Society).

3.3. Electrochemical Detection

Besides the optical biosensing approaches for protease biomarker detection, electrochemical measurement has been developed using specific peptides and the corresponding proteases. Xia et al. proposed a novel protease biosensor by altering a homogeneous assay into a surface-bound electrochemical analysis (Figure 2c) [77]. A streptavidin-covered electrode functionalizes the biotin-specific peptide–biotin complex for streptavidin–biotin coupling chemistry. The repeated formation of streptavidin−biotin−GDEVDGK−biotin aggregates provides an insulating layer, thus reducing the electron transfer of ferricyanide. If the peptide sequence is cleaved by caspase-3, the resulting products are distributed into a bulk solution and electron transfer is increased. Therefore, the amplification of the electrochemical signal is achieved with the simple principle of substrate-induced streptavidin assembly. Zhang et al. developed a thiol-sensitive electrochemical probe for the measurement of the SARS-CoV-2 main protease, targeted by a short probe mimicking its substrate [78]. In the probe–protein interaction, a specific fluorescent molecule interacts with the free thiol groups on the target protein, producing a fluorescence and electrochemical signal proportional to the target concentration. The LOD of the SARS-CoV-2 main protease was 1 pM, whereas the clinically required LOD was around 182 pM. The authors claimed that this analytical method could measure the virus protease in clinical SARS-CoV-2-infected patient samples in a simple one-step reaction and in a reagent-less fashion. Shi et al. revealed a label-free electrochemical biosensor for the sensitive detection of MMP-2, with signal amplification using a proteolysis-triggered transcription method (Figure 2d) [79]. The authors utilized RNA polymerase to transduce and amplify the proteolysis reaction by MMP-2 into multiple RNA productions that can be bound by the complementary DNA probes modified on the Au electrode. By integrating the G4/hemin complex, this RNA polymerase-assisted electrochemical biosensor facilitates the highly sensitive detection of MMP-2, with a wide dynamic range from 10 fM to 1.0 nM and 7.1 fM LOD value. Eissa et al. developed a specific peptide sequence-functionalized magnetic bead biosensor for the detection of *Staphylococcus aureus* using dual colorimetric and electrochemical analyses [80]. The peptide-magnetic bead complex is immobilized on the screen printed Au electrode. The protease released from *Staphylococcus aureus* degrades the specific peptide, and the color of the Au electrode changes due to the dissociation of the magnetic bead. In addition, the protease can be detected by monitoring the change in the peak current of square wave voltammetry within 1 min. The LOD value of this electrochemical assay was 3 CFU mL^{-1}, and it was tested with a spiked milk and water sample.

3.4. Others

Because the peptide-cleavage reaction can change various output signals, including fluorescence and electrochemical readouts, other simple and sensitive biosensing approaches have also been researched. Among them, bioluminescence resonance energy transfer (BRET)-based biosensors are similar to the sensing mechanism of FRET. Weihs et al. developed a red-shifted BRET biosensor for the measurement of plasmin activity [81]. Conventional BRET-based proteolytic biosensors use the blue-shifted BRET system, which suffers from background signals due to light absorption and scattering in plasma samples. To overcome this limitation, the authors applied a red-shifted RLuc8.6 luciferase instead of Renilla luciferase Rluc2. As a result, the proposed biosensor exhibited up to a 5-fold increase in sensitivity for plasma samples as low as 11.90 nM within 10 min. In addition, they also presented the real-time-on-chip detection of thrombin activity by integrating the BRET phenomenon [82]. The real-time-on-chip is composed of a compact microreactor and a reusable glass chip with a mixer, an incubation channel, and a detection chamber. This compact chip platform provides a minimum of handling steps, which can reduce the handling error for the precise measurement of thrombin. This sensitive chip can detect thrombin as low as 38 μU/μL in human serum within only 5 min, which is 90% faster than conventional assays.

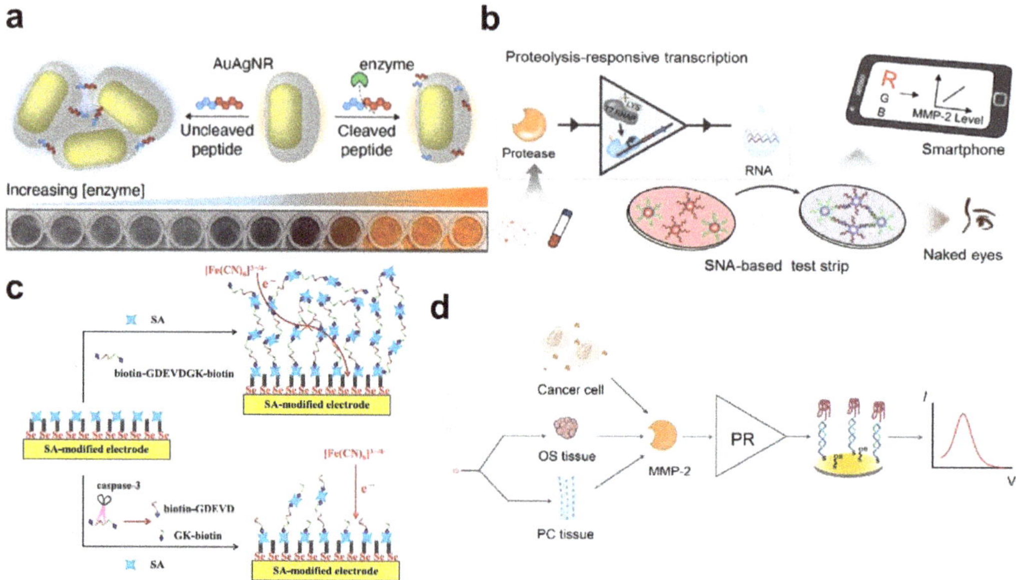

Figure 2. Colorimetric- and electrochemical-based biosensors for extracellular protease detection: (**a**) schematic diagram of the colorimetric biosensors using a plasmonic gold–silver core–gold nanoparticle aggregate for the measurement of trypsin activity. This figure is adapted with permission from Ref. [73] (© 2022 American Chemical Society). (**b**) Schematics of the smartphone-based colorimetric detection of MMP-2 using a modular combination of proteolysis-responsive transcription and nucleic acid-modified AuNPs. This figure is adapted with permission from Ref. [74] (© 2021 American Chemical Society). (**c**) Schematic diagram of the electrochemical protease biosensor composed of a streptavidin-modified Au electrode with a biotin–peptide–biotin complex as a caspase-3 substrate. This figure is adapted with permission from Ref. [77] (© 2021 American Chemical Society). (**d**) Label-free electrochemical detection of MMP-2 by proteolysis-triggered transcription strategy on a Au electrode. This figure is adapted with permission from Ref. [79] (© 2021 Elsevier B.V.).

Surface-enhanced Raman spectroscopy (SERS) could also be a potential sensitive biosensing method for protease detection. Due to the proportional signal intensity to the distance between the Raman dye and SERS-active substrates, the peptide-cleavage event can induce a dramatic change in SERS signals. Choi et al. revealed a SERS-active platform consisting of an Ag-coated hollow polypyrrole (hPPy) nanohorn with a specific peptide sequence [83]. To maximize the SERS signal, Ag is electrodeposited on both sides of the hPPy nanohorn structure. Using this core/double-shell nanohorn substrate, the Raman signal of the peptide-functionalized AuNPs is significantly improved, and caspase-3 can be sensitively measured, with a wide detection range from 10 pg/mL to 10 µg/mL. Wei et al. developed O-GlcNAc transferase (OGT) biosensors by integrating SERS tags and a magnetic bead complex [84]. The SERS tags and magnetic bead complex are linked to the two ends of the peptide by thiolated and biotin residues. In the presence of OGT, peptide glycosylation protects the peptide from the cleavage reaction by proteinase K, resulting in more SERS signals from the left Raman tags on the surface of magnetic beads. This system can sensitively detect OGT as low as 0.1 nM, ranging from 10^{-10} to 10^{-7} M. Besides optical analytical methods, a giant magnetoresistive spin-valve (GMR SV) sensor has been developed for the detection of cysteine protease [85]. Magnetic nanoparticles are immobilized on the surface of GMR SV using a cysteine-protease-specific peptide. This magnetic field measurement approach can detect papain, one of the cysteine proteases, as low as 4 nM in only 3.5 min.

4. Intracellular Detection of Protease for In Situ Monitoring

4.1. In Vitro Proteolytic Analysis at a Cellular Level

As described in Section 3, since extracellularly released proteases can be biomarkers of specific diseases, the development of sensitive and selective biosensors is necessary for the accurate and early diagnosis of several diseases. In addition to measuring extracellular protease in vitro, it is also possible to measure intracellular proteases quantitatively and qualitatively. As illustrated above, as measuring proteases does not involve complex reactions such as antigen–antibody reactions, intracellular proteases can be measured in a simple manner and in real-time. Because intracellular proteases, similar to extracellular proteases, are important biomarkers for determining the state of cells, many researchers have studied intracellular protease biosensors using specific peptide degradation.

For the detection of intracellular proteases, fluorescence-based analytical methods have been mainly utilized using the distance-dependent FRET phenomenon. In particular, there have been virus-infected cells that are enabled to measure virus-related proteases in real-time to diagnose viral infection. Guerruiro et al. exhibited a genetically-engineered turn-on fluorescent biosensor containing a cyclized green fluorescent protein (GFP)-conjugated cVisensor with a specific peptide sequence for adenoviral protease detection [86]. Structural distortion of circular permuted superfolder-GFP is cyclized by *Nostoc punctiforme* DnaE. This in situ live cell biosensor works to restore the deformation by the peptide-cleavage reaction and emits the fluorescent signal of GFP. This label-free sensing system can detect adenovirus 2 days post-infection, faster than conventional plaque-forming assays that require 14 days. Dey-Rao et al. developed a live cell-based fluorescence biosensor to evaluate the intracellular function of the SARS-CoV-2 main protease and its inhibitor [87]. The authors designed it to express a red fluorescence protein (RFP) biosensor unless the SARS-CoV-2 main protease cleaves the peptide and to facilitate the loss of the fluorescence emission of RFP. Inhibition of the main protease function provides the synthesis of working RFP, resulting in the recovery of fluorescence. In addition, GC376, which is the pan-coronavirus's main protease inhibitor, shows effective inhibition of the intracellular CoV2 main protease. The authors claim that this intracellular sensing system can offer an extremely efficient high-throughput screening system for the SARS-CoV-2 main protease. Guerreiro et al. developed GFP-based switch-on split fluorescent biosensors to examine viral infection [88]. Through the protein distortion, which is composed of GFP11 embedment distortion (SNR 6.0) or GFP11 cyclization (SNR 3.5), GFP is split and cannot be assembled for the fluorescent emission. Once adenovirus- and lentivirus-promoted proteolysis cleaves the peptide sequence, this intracellular biosensor shows the fluorescent signal in live cells after 24 h post-infection with a high signal-to-noise ratio of up to 97. Gerber et al. revealed a live cell-based luminescent biosensor for the verification of SARS-CoV-2 infection using recombinant SARS-CoV-2 proteases [89]. This luminescent-based assay applies the cleavage of specific peptide linkers, which are degraded by the main protease of SARS-CoV-2, allowing viral infection to be measured within 24 h in the multi-well plate format for high-throughput analysis. The authors claim that the developed luminescent SARS-CoV-2 reporter live-cell-based biosensor can demonstrate the comparative quantitation of the SARS-CoV-2 virus and the titration of neutralizing antibodies.

Besides virus-related protease detection, other intracellular proteases have also been measured with specific peptide-cleavage reactions in live cells. Luo et al. showed the highly sensitive and precise detection of MMP-2 activity using a rolling circle transcription assay-integrated proteolytic reaction (Figure 3a) [90]. The proteolytic reaction induces the activation of the bacteriophage T7 RNA polymerase (T7 RNAP) and produces long RNA chains, including tandem G-quadruplexes (G4s). As a result, the activity of MMP-2 can be transduced into multiple fluorescent signals from G4s RNAs. For verification of the sensing principle for cell imaging, six distinct cell lines with different MMP-2 expression levels were employed, and they indicated different signals for the MMP-2 expression levels. Braun et al. exhibited a cell-surface-displayable biosensor for MMP-14 detection (Figure 3b) [91]. The cell-surface biosensor consisted of two scFv domains; one was blocked against MG2P fluorescent dye, and another was scFv, a connected specific MMP-14 cleavable sequence. Once the cell-surface MMP-14 degrades the peptide bridge, the blocker is dissociated, and the fluorogen (MG2P) is bound to the scFv domains. Subsequently, the MMP14–fluorogen (MG2P) complex is activated, and the fluorescence signal is generated. This switch-on cell-surface biosensor was enabled to measure the activity of MMP14, as well as location and temporal dynamics. Xu et al. developed a multi-color fluorescent polydopamine nanobiosensor for the multiple sensing of cancer-related proteases in living cells, such as urokinase-type plasminogen activator (uPA), cathepsin B (CTB), matrix metalloproteinase-2, and matrix metalloproteinase-7 (MMP-7) [92]. The polydopamine nanoprobe is functionalized by the four different peptide sequences with different fluorescent dyes. Due to the quenching effect of the polydopamine nanoprobe, the four fluorescent dyes cannot emit a signal. If the polydopamine nanoprobe is intracellularly delivered into the cancer-related cells, the intracellular protease can particularly cleave each peptide, allowing the recovery of the fluorescent signals. Due to the multiplexed intracellular detection, this sensing system can evaluate the stages of tumor progression and provide early and multiple diagnoses of cancers. Peyressatre et al. exhibited a cyclin-dependent kinase 5 (CDK5) kinase activity biosensor for the diagnosis of neurodegenerative pathologies, including glioblastoma and neuroblastoma [93]. The authors developed a CDK5-specific fluorescent peptide biosensor through the selection of peptide sequences from the CDK5 substrate. The quantification of CDK5 kinase activity can be measured on treatment with ATP-competitive inhibitors in both cell extracts and the living cell environment by time-lapse fluorescence microscopy. These kinds of intracellular biosensing systems can facilitate the evaluation of the functional status of intracellular proteases in several cancers and neurodegenerative pathologies by real-time monitoring. The intracellular biosensing method can also be applied to floating cells, such as immune cells, for monitoring endogenous metabolic processes during the immune response. Sun et al. developed a genetically programmed fluorescent itaconate biosensor (BioITA) for the immediate monitoring of itaconate dynamics, which is an anti-inflammatory program of the innate immune response in living macrophages [94]. The authors succeeded in monitoring the itaconate progress by stimulating lipopolysaccharides in living macrophages. In addition, through the BioITA, the response and changing itaconate level in activated macrophages could be observed by the injection of an adeno-associated virus (AAV) with spatiotemporal resolution. Hassanzadeh-Barforoushi et al. utilized microfluidic channels for the measurement of released proteases from single cells using capillary force-assisted separation (Figure 3c) [95]. For the isolation and easy and reliable capture of a small number of cells (~500 single cells), the capillary-stoppage method was applied to the microfluidic channel by shearing with air and sheathing with FC-40 oil. Key parameters, including single-cell encapsulation efficiency (38.8%) and droplet volume evaporation rate (10%) in 48 h, were achieved in the microfluidic platform, and the manipulation of droplet composition through controlled gradient generation were studied. Based on the separation of nanoliters at the single-cell level, the authors employed a simple FRET-based biosensor for the monitoring of secreted MMP-2 activity. The authors claim that this microfluidic platform offers a rapid, simple, and single-cell-level detection method without engineering expertise. Zhong et al. developed an in situ ratiometric SERS nanoprobe for the intracellular imaging of proteases (MMP-2) in different cancer cells for sensitive detection (Figure 3d) [96]. Due to the advantages of the SERS analytical method, such as high sensitivity, resistance to photobleaching and quenching, and trivial autofluorescence, SERS-based visualization using the 2-naphthalenethiol (NT)-labeled Au nanoprobe was applied to measure MMP-2 activity in single living cancer cells. Upon the internalization of the Au nanoprobe into cancer cells, the proteolytic cleavage of the peptide sequence resulted in the dissociation of the 2-NT molecules, and the Raman signal was decreased. Live cell imaging results showed that the MMP-responsive nanoprobe differentiated normal breast cells from breast cancer cells and also differentiated two different breast cancer cell lines with different malignant properties. Cheng et al. developed a selective in vivo imaging and inhibition system against SARS-CoV-2 infection. This amphiphilic assembly system consists of aggregation-induced emission (AIEgen), self-assembly, spacer, and main protease-responsive and cell-penetrating peptide domains. The degradation of this amphiphilic assembly by the main protease from SARS-CoV-2 induces aggregation with increased fluorescence and mitochondrial interference of the infected cells. As a result, this system can perform the selective bioimaging and treatment of SARS-CoV-2-infected cells [97].

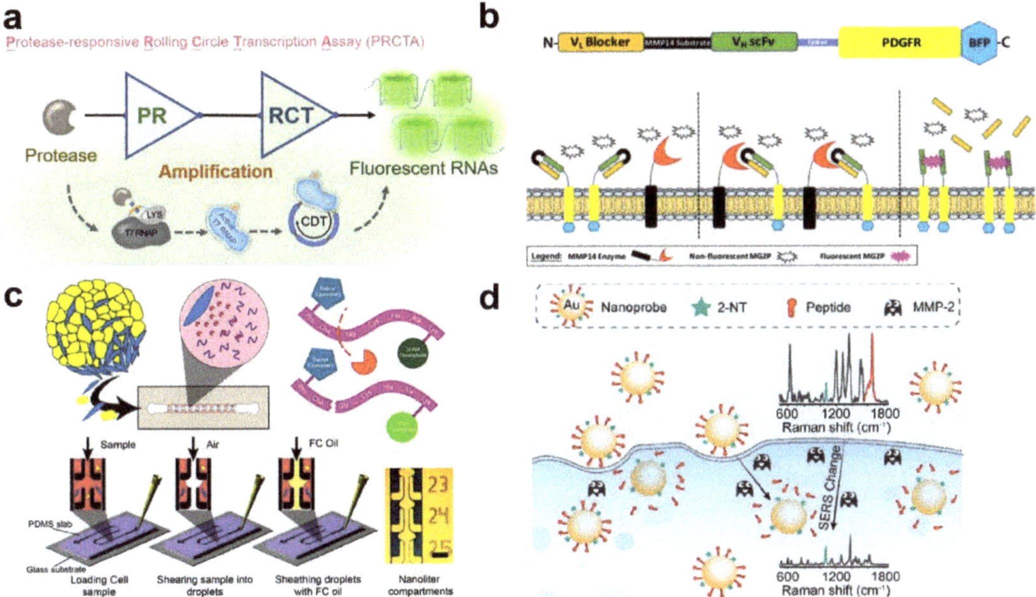

Figure 3. In vitro proteolytic analyzing biosensors: (**a**) Schematic diagram of the proteolysis-responsive rolling circle transcription assay (PRCTA) by integrating protease-responsive RNA polymerases and rolling circle transcription. This figure is adapted with permission from Ref. [90] (© 2020 American Chemical Society). (**b**) Plasmid-based cell-surface fluorescent biosensor for measurement of the location and activity of MMP-14. This figure is adapted with permission from Ref. [91] (© 2018 Springer). (**c**) Microfluidic capillary biosensor for quantifying secreted protease activity at the single cell level. This figure is adapted with permission from Ref. [95] (© 2020 Elsevier B.V.). (**d**) SERS imaging of intracellular MMP-2 activity with the ratiometric SERS Nanoprobe. This figure is adapted with permission from Ref. [96] (© 2020 Elsevier B.V.).

4.2. In Vivo Proteolytic Analysis for Bioimaging

Proteolytic cleavage-based biosensing strategies can be applied to in vivo imaging using optical probes such as fluorescent molecules. Yim et al. developed ear inflammatory biosensors by measuring the short-wave infrared (SWIR) otoscope (Figure 4a) [98]. Cysteine cathepsin proteases, which are upregulated in the inflamed state by immune cells, are released from the otitis media (middle ear infection). The fluorescent probe is linked to the quencher (6QC-ICG) through the cleavable peptide, and this probe is emitted into the lysosome by the latent lysosomotropic effect in acute otitis media. As a result, middle ear infection shows a clear fluorescent emission with a 2.0 signal-to-background ratio. Moore et al. demonstrated a novel detection method for in vivo periodontal inflammation by measuring gingipain proteases released by *P. gingivalis* [99]. For the transduction of a specific peptide cleavage into the signal, the activatable photoacoustic and fluorescent molecular probe, which consisted of a dye-conjugated peptide, [Cy5.5]₂[APRIK], was utilized as an imaging agent for gingipain proteases. The sensitivity of this in vivo biosensor showed 5-fold photoacoustic and >100-fold fluorescence enhancement, with 1.1 nM and

4.4×10^{-4} CFU/mL LOD. In addition, the photoacoustic imaging of the gingipain protease probe was demonstrated in the subgingival pocket of porcine mandibles and the murine brain due to the check on the biological relationship between gingipain and Alzheimer's disease. Xiang et al. exhibited a DNA/peptide/PNA triblock copolymer for in vitro ATP imaging and in vivo tumor imaging (Figure 4b) [100]. The sensing system is composed of a rationally designed peptide nucleic acid (PNA)–peptide–PNA triblock copolymer, which can be divided by tumor-overexpressed cathepsin B. The cathepsin B-activatable probe is in double-stranded form, which is the Cy5-labeled DNA aptamer strand and the quencher (BHQ2)-labeled short complementary DNA strand. The cathepsin B-induced activation of the probe enables tumor cell-specific molecular imaging by the folding of the aptamer with binding ATP, and the fluorescence signal is recovered. This ATP sensing-based fluorescence imaging probe can selectively measure the cathepsin B protease in tumor cells both in vitro and in vivo. Kang et al. developed a chronic wound evaluation platform by integrating a fluorophore–peptide quencher into a maleimide-functionalized polyethylene glycol–diacrylate (PEG-DA) hydrogel [101]. Using the in vivo fluorescence imaging method, a high level of MMP-2 and MMP-9 was clearly demonstrated on the chronic wound site. The authors claim that this simple protease sensor can be applied to indicate the severity of the wound to the surgical doctor and other medical staff by providing the patch format of a fluorescence sensor. Liu et al. reported a noninvasive diagnostic platform of common pulmonary diseases by measuring the activation of human neutrophil elastase (Figure 4c) [102]. This imaging probe consists of a QD–human neutrophil elastase-specific peptide substrate fluorescent dye for the induction of the FRET effect. This system showed both the in vitro and in vivo detection of human neutrophil elastase, with 7.15 pM in aqueous solution, and clear imaging in lung cancer and acute lung injury mouse models. Using this sensor could differentiate pulmonary patients from the healthy with the minimization of environmental interference of the fluorescence.

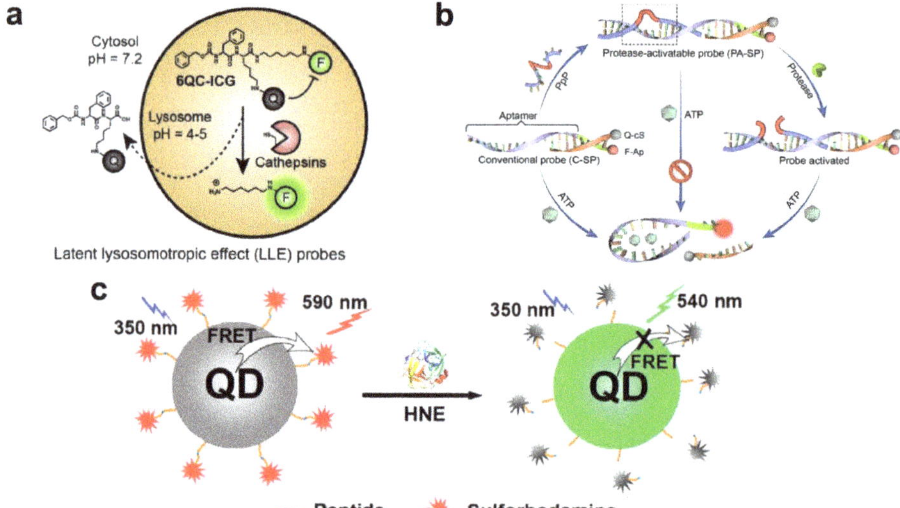

Figure 4. In vivo proteolytic analysis of intracellular proteases for bioimaging: (**a**) Schematics of the fluorescence-guided detection of otitis media (OM) using a protease-cleavable biosensor. This figure is adapted with permission from Ref. [98] (© 2020 American Chemical Society). (**b**) PNA-assisted DNA aptamer sensor with peptide for cathepsin B-activatable ATP detection. This figure is adapted with permission from Ref. [100] (© 2021 Wiley-VCH). (**c**) FRET-based neutrophil elastase biosensor for in vitro detection and in vivo imaging using the peptide substrate, QDs, and organic dyes. This figure is adapted with permission from Ref. [102] (© 2020 American Chemical Society).

5. Outlook and Conclusions

In this review, proteolytic enzymes that can be used as biomarkers for diagnosing various diseases have been briefly demonstrated, with the latest research trends showing the integration of functional nanomaterials. As mentioned above, proteolytic enzymatic biosensors have various advantages over immunoassays represented by ELISA. First, the measurement process of proteolytic biosensors is simpler than ELISA. Compared to multiple antigen–antibody reactions (two or more separated reactions), proteolytic enzymes can be measured using only a peptide degradation reaction. In addition, it is cost-effective and time-saving as the complexity of the measurement can be reduced. Unlike ELISA, one of the advantages is that the additional labeling step of the signaling molecule is not required after the biological reaction (i.e., antigen–antibody reaction) of the target biomarker. Finally, false-negative and false-positive signals caused by antibody instability can also be reduced. However, while most proteins have their specific antibodies, there is a limitation that only proteolytic biomarkers with specific enzymatic properties can be measured. This is a major disadvantage that reduces the versatility of the sensor.

Recently, CRISPR-based DNA and RNA biosensors have been actively studied for sensitive and simple detection. These analytical methods are based on random single nucleic acid degradation reactions if the target DNA/RNA lets the CRISPR complex activate, similar to the strategies of proteolytic enzymatic biosensors. Since the decomposition reactions of peptides and nucleic acids are irreversible, these kinds of sensors cannot be reused. Paradoxically, the non-reusability of the biosensor could be an advantage, namely, as a disposable sensor, because it can improve accuracy by reducing errors due to the reduction of biological reactions due to the regeneration of sensors. In addition, a system for measuring multiple biomarkers is essential for the accurate and early diagnosis of diseases. From this point of view, the proteolytic sensor can easily apply a multiplex system because the peptides that each proteolytic enzyme degrades are different. Furthermore, integration with the CRISPR sensor enables the simultaneous measurement of various types of biomarkers, such as DNA, RNA, and other proteins, on one platform. This will provide not only the accurate diagnosis of the disease mentioned above but also detailed disease information, for example, the progress of the disease and the prognosis after treatment.

Studies measuring the activity of proteolytic enzymes can be applied to various biomedical fields besides in vitro diagnosis. For example, it can be used for real-time in vivo imaging using intracellular and extracellular proteolytic enzymes. In addition, when proteolytic enzymes are used with therapeutic drugs and nanoparticles, they can be used for the development of nanotheragnosis, which can be treated and diagnosed at the same time. For instance, since the MMP family are typically highly expressed in various cancers, it will be possible to develop nanotherapeutics with high potential by inducing drug release through peptide degradation events while detecting them in the body. In this way, it is believed that in combination with various nanotechnology, it will be possible to secure a platform technology that can simultaneously perform various functions while improving measurement sensitivity. In addition, proteolytic biosensors could be applied to toxicological and pharmacological screening systems. For example, sentrin-specific protease 1, which is related to many diseases, including cancers and cardiovascular diseases, is measured for the identification of anti-cancer agents that have toxic effects [103]. Another example is the SARS-CoV-2 main protease inhibitors screened using a proteolytic-based biochemical high-throughput screening (HTS) system. These screening platforms are feasible for simple and sensitive tests for drug screening and pharmacological applications [104]. Proteolytic enzymatic biosensors are expected to be used more actively in the field of early diagnosis of diseases in the future, and it is expected to be widely used for basic biological research, in vitro model systems with sensitive analysis platforms at the cell level, and other biomedical fields such as drug delivery and in vivo imaging.

Funding: This work was supported by a National Research Foundation of Korea (NRF) grant funded by the Korean government (MSIT) (NRF-2022R1C1C1008797) and the National R&D Program through the NRF, as funded by the Ministry of Science and ICT (NRF-2022M3H4A1A01005271).

Institutional Review Board Statement: Not applicable.

Informed Consent Statement: Not applicable.

Data Availability Statement: Not applicable.

Conflicts of Interest: The author declares no conflict of interest. The funders had no role in the design of the study; in the collection, analyses, or interpretation of data; in the writing of the manuscript; or in the decision to publish the results.

References

1. Blandin Knight, S.; Crosbie, P.A.; Balata, H.; Chudziak, J.; Hussell, T.; Dive, C. Progress and prospects of early detection in lung cancer. *Open Biol.* **2017**, *7*, 170070. [CrossRef] [PubMed]
2. Mitsuyoshi, A.; Obama, K.; Shinkura, N.; Ito, T.; Zaima, M. Survival in nonocclusive mesenteric ischemia: Early diagnosis by multidetector row computed tomography and early treatment with continuous intravenous high-dose prostaglandin E(1). *Ann. Surg.* **2007**, *246*, 229–235. [CrossRef] [PubMed]
3. Thomson, C.S.; Forman, D. Cancer survival in England and the influence of early diagnosis: What can we learn from recent EUROCARE results? *Br. J. Cancer* **2009**, *101* (Suppl. S2), S102–T109. [CrossRef] [PubMed]
4. Fani, M.; Zandi, M.; Soltani, S.; Abbasi, S. Future developments in biosensors for field-ready SARS-CoV-2 virus diagnostics. *Biotechnol. Appl. Biochem.* **2021**, *68*, 695–699. [CrossRef]
5. Aziz, A.; Asif, M.; Ashraf, G.; Farooq, U.; Yang, Q.; Wang, S. Trends in biosensing platforms for SARS-CoV-2 detection: A critical appraisal against standard detection tools. *Curr. Opin. Colloid Interface Sci.* **2021**, *52*, 101418. [CrossRef]
6. Orooji, Y.; Sohrabi, H.; Hemmat, N.; Oroojalian, F.; Baradaran, B.; Mokhtarzadeh, A.; Mohaghegh, M.; Karimi-Maleh, H. An Overview on SARS-CoV-2 (COVID-19) and Other Human Coronaviruses and Their Detection Capability via Amplification Assay, Chemical Sensing, Biosensing, Immunosensing, and Clinical Assays. *Nano-Micro Lett.* **2021**, *13*, 18. [CrossRef]
7. Mahshid, S.S.; Flynn, S.E.; Mahshid, S. The potential application of electrochemical biosensors in the COVID-19 pandemic: A perspective on the rapid diagnostics of SARS-CoV-2. *Biosens. Bioelectron.* **2021**, *176*, 112905. [CrossRef]
8. Das Mukhopadhyay, C.; Sharma, P.; Sinha, K.; Rajarshi, K. Recent trends in analytical and digital techniques for the detection of the SARS-CoV-2. *Biophys. Chem.* **2021**, *270*, 106538. [CrossRef]
9. Alafeef, M.; Dighe, K.; Moitra, P.; Pan, D. Rapid, Ultrasensitive, and Quantitative Detection of SARS-CoV-2 Using Antisense Oligonucleotides Directed Electrochemical Biosensor Chip. *ACS Nano* **2020**, *14*, 17028–17045. [CrossRef]
10. Zhao, H.; Liu, F.; Xie, W.; Zhou, T.C.; OuYang, J.; Jin, L.; Li, H.; Zhao, C.Y.; Zhang, L.; Wei, J.; et al. Ultrasensitive supersandwich-type electrochemical sensor for SARS-CoV-2 from the infected COVID-19 patients using a smartphone. *Sens. Actuators B Chem.* **2021**, *327*, 128899. [CrossRef]
11. Xu, L.; Wen, Y.; Pandit, S.; Mokkapati, V.; Mijakovic, I.; Li, Y.; Ding, M.; Ren, S.; Li, W.; Liu, G. Graphene-based biosensors for the detection of prostate cancer protein biomarkers: A review. *BMC Chem.* **2019**, *13*, 112. [CrossRef]
12. Luo, X.; Davis, J.J. Electrical biosensors and the label free detection of protein disease biomarkers. *Chem. Soc. Rev.* **2013**, *42*, 5944–5962. [CrossRef]
13. Pei, X.; Zhang, B.; Tang, J.; Liu, B.; Lai, W.; Tang, D. Sandwich-type immunosensors and immunoassays exploiting nanostructure labels: A review. *Anal. Chim. Acta* **2013**, *758*, 1–18. [CrossRef]
14. Wu, A.H. A selected history and future of immunoassay development and applications in clinical chemistry. *Clin. Chim. Acta* **2006**, *369*, 119–124. [CrossRef]
15. Borrebaeck, C.A. Antibodies in diagnostics—From immunoassays to protein chips. *Immunol. Today* **2000**, *21*, 379–382. [CrossRef]
16. Wang, Z.; Zong, S.; Wu, L.; Zhu, D.; Cui, Y. SERS-Activated Platforms for Immunoassay: Probes, Encoding Methods, and Applications. *Chem. Rev.* **2017**, *117*, 7910–7963. [CrossRef]
17. Wu, J.; Fu, Z.F.; Yan, F.; Ju, H.X. Biomedical and clinical applications of immunoassays and immunosensors for tumor markers. *Trac Trends Anal. Chem.* **2007**, *26*, 679–688. [CrossRef]
18. Halade, G.V.; Jin, Y.F.; Lindsey, M.L. Matrix metalloproteinase (MMP)-9: A proximal biomarker for cardiac remodeling and a distal biomarker for inflammation. *Pharmacol. Ther.* **2013**, *139*, 32–40. [CrossRef]
19. Huang, H. Matrix Metalloproteinase-9 (MMP-9) as a Cancer Biomarker and MMP-9 Biosensors: Recent Advances. *Sensors* **2018**, *18*, 3249. [CrossRef]
20. Roy, R.; Yang, J.; Moses, M.A. Matrix metalloproteinases as novel biomarkers and potential therapeutic targets in human cancer. *J. Clin. Oncol.* **2009**, *27*, 5287–5297. [CrossRef]
21. Ward, T.H.; Cummings, J.; Dean, E.; Greystoke, A.; Hou, J.M.; Backen, A.; Ranson, M.; Dive, C. Biomarkers of apoptosis. *Br. J. Cancer* **2008**, *99*, 841–846. [CrossRef] [PubMed]

22. Silva, F.; Padin-Iruegas, M.E.; Caponio, V.C.A.; Lorenzo-Pouso, A.I.; Saavedra-Nieves, P.; Chamorro-Petronacci, C.M.; Suarez-Penaranda, J.; Perez-Sayans, M. Caspase 3 and Cleaved Caspase 3 Expression in Tumorogenesis and Its Correlations with Prognosis in Head and Neck Cancer: A Systematic Review and Meta-Analysis. *Int. J. Mol. Sci.* **2022**, *23*, 11937. [CrossRef] [PubMed]
23. Yabluchanskiy, A.; Ma, Y.; Iyer, R.P.; Hall, M.E.; Lindsey, M.L. Matrix metalloproteinase-9: Many shades of function in cardiovascular disease. *Physiology* **2013**, *28*, 391–403. [CrossRef] [PubMed]
24. Oliveira-Silva, R.; Sousa-Jeronimo, M.; Botequim, D.; Silva, N.J.O.; Paulo, P.M.R.; Prazeres, D.M.F. Monitoring Proteolytic Activity in Real Time: A New World of Opportunities for Biosensors. *Trends Biochem. Sci.* **2020**, *45*, 604–618. [CrossRef] [PubMed]
25. Welser, K.; Adsley, R.; Moore, B.M.; Chan, W.C.; Aylott, J.W. Protease sensing with nanoparticle based platforms. *Analyst* **2011**, *136*, 29–41. [CrossRef]
26. Ong, I.L.H.; Yang, K.L. Recent developments in protease activity assays and sensors. *Analyst* **2017**, *142*, 1867–1881. [CrossRef]
27. Vanova, V.; Mitrevska, K.; Milosavljevic, V.; Hynek, D.; Richtera, L.; Adam, V. Peptide-based electrochemical biosensors utilized for protein detection. *Biosens. Bioelectron.* **2021**, *180*, 113087. [CrossRef]
28. Weihs, F.; Anderson, A.; Trowell, S.; Caron, K. Resonance Energy Transfer-Based Biosensors for Point-of-Need Diagnosis-Progress and Perspectives. *Sensors* **2021**, *21*, 660. [CrossRef]
29. Karimzadeh, A.; Hasanzadeh, M.; Shadjou, N.; de la Guardia, M. Peptide based biosensors. *TrAC Trends Anal. Chem.* **2018**, *107*, 1–20. [CrossRef]
30. Farzin, L.; Shamsipur, M.; Samandari, L.; Sheibani, S. HIV biosensors for early diagnosis of infection: The intertwine of nanotechnology with sensing strategies. *Talanta* **2020**, *206*, 120201. [CrossRef]
31. Morales-Narvaez, E.; Merkoci, A. Graphene oxide as an optical biosensing platform. *Adv. Mater.* **2012**, *24*, 3298–3308. [CrossRef]
32. Hedstrom, L. Serine protease mechanism and specificity. *Chem. Rev.* **2002**, *102*, 4501–4524. [CrossRef]
33. Di Cera, E. Serine proteases. *IUBMB Life* **2009**, *61*, 510–515. [CrossRef]
34. Patel, S. A critical review on serine protease: Key immune manipulator and pathology mediator. *Allergol. Immunopathol.* **2017**, *45*, 579–591. [CrossRef]
35. Pham, C.T. Neutrophil serine proteases: Specific regulators of inflammation. *Nat. Rev. Immunol.* **2006**, *6*, 541–550. [CrossRef]
36. Bird, P.I. Regulation of pro-apoptotic leucocyte granule serine proteinases by intracellular serpins. *Immunol. Cell Biol.* **1999**, *77*, 47–57. [CrossRef]
37. Janciauskiene, S. Conformational properties of serine proteinase inhibitors (serpins) confer multiple pathophysiological roles. *Biochim. Biophys. Acta* **2001**, *1535*, 221–235. [CrossRef]
38. Otto, H.H.; Schirmeister, T. Cysteine proteases and their inhibitors. *Chem. Rev.* **1997**, *97*, 133–171. [CrossRef]
39. Chapman, H.A.; Riese, R.J.; Shi, G.P. Emerging roles for cysteine proteases in human biology. *Annu. Rev. Physiol.* **1997**, *59*, 63–88. [CrossRef]
40. Que, X.; Reed, S.L. Cysteine proteinases and the pathogenesis of amebiasis. *Clin. Microbiol. Rev.* **2000**, *13*, 196–206. [CrossRef]
41. Lecaille, F.; Kaleta, J.; Bromme, D. Human and parasitic papain-like cysteine proteases: Their role in physiology and pathology and recent developments in inhibitor design. *Chem. Rev.* **2002**, *102*, 4459–4488. [CrossRef] [PubMed]
42. Hasanbasic, S.; Jahic, A.; Karahmet, E.; Sejranic, A.; Prnjavorac, B. The Role of Cysteine Protease in Alzheimer Disease. *Mater. Socio-Med.* **2016**, *28*, 235–238. [CrossRef] [PubMed]
43. Cheng, X.W.; Huang, Z.; Kuzuya, M.; Okumura, K.; Murohara, T. Cysteine protease cathepsins in atherosclerosis-based vascular disease and its complications. *Hypertension* **2011**, *58*, 978–986. [CrossRef] [PubMed]
44. Troen, B.R. The role of cathepsin K in normal bone resorption. *Drug News Perspect.* **2004**, *17*, 19–28. [CrossRef]
45. Dai, R.; Wu, Z.; Chu, H.Y.; Lu, J.; Lyu, A.; Liu, J.; Zhang, G. Cathepsin K: The Action in and Beyond Bone. *Front. Cell Dev. Biol.* **2020**, *8*, 433. [CrossRef]
46. Qian, D.; He, L.S.; Zhang, Q.; Li, W.Q.; Tang, D.D.; Wu, C.J.; Yang, F.; Li, K.; Zhang, H. Cathepsin K: A Versatile Potential Biomarker and Therapeutic Target for Various Cancers. *Curr. Oncol.* **2022**, *29*, 471. [CrossRef]
47. Fonovic, M.; Turk, B. Cysteine cathepsins and their potential in clinical therapy and biomarker discovery. *Proteom. Clin. Appl.* **2014**, *8*, 416–426. [CrossRef]
48. Palermo, C.; Joyce, J.A. Cysteine cathepsin proteases as pharmacological targets in cancer. *Trends Pharm. Sci.* **2008**, *29*, 22–28. [CrossRef]
49. Birkedal-Hansen, H.; Moore, W.G.; Bodden, M.K.; Windsor, L.J.; Birkedal-Hansen, B.; DeCarlo, A.; Engler, J.A. Matrix metalloproteinases: A review. *Crit. Rev. Oral Biol. Med.* **1993**, *4*, 197–250. [CrossRef]
50. Klein, T.; Bischoff, R. Physiology and pathophysiology of matrix metalloproteases. *Amino Acids* **2011**, *41*, 271–290. [CrossRef]
51. Chakraborti, S.; Mandal, M.; Das, S.; Mandal, A.; Chakraborti, T. Regulation of matrix metalloproteinases: An overview. *Mol. Cell Biochem.* **2003**, *253*, 269–285. [CrossRef]
52. Hooper, N.M.; Itoh, Y.; Nagase, H. Matrix metalloproteinases in cancer. *Essays Biochem.* **2002**, *38*, 21–36. [CrossRef]
53. Yager, D.R.; Nwomeh, B.C. The proteolytic environment of chronic wounds. *Wound Repair Regen.* **1999**, *7*, 433–441. [CrossRef]
54. Dollery, C.M.; McEwan, J.R.; Henney, A.M. Matrix metalloproteinases and cardiovascular disease. *Circ. Res.* **1995**, *77*, 863–868. [CrossRef]
55. Nissinen, L.; Kahari, V.M. Matrix metalloproteinases in inflammation. *Biochim. Biophys. Acta* **2014**, *1840*, 2571–2580. [CrossRef]

56. Renault, K.; Debieu, S.; Richard, J.A.; Romieu, A. Deeper insight into protease-sensitive "covalent-assembly" fluorescent probes for practical biosensing applications. *Org. Biomol. Chem.* **2019**, *17*, 8918–8932. [CrossRef]
57. Zhang, H.; Yang, L.; Zhu, X.; Wang, Y.; Yang, H.; Wang, Z. A Rapid and Ultrasensitive Thrombin Biosensor Based on a Rationally Designed Trifunctional Protein. *Adv. Healthc. Mater.* **2020**, *9*, e2000364. [CrossRef]
58. Selvin, P.R. The renaissance of fluorescence resonance energy transfer. *Nat. Struct. Biol.* **2000**, *7*, 730–734. [CrossRef]
59. Zhang, Y.; Chen, X.; Roozbahani, G.M.; Guan, X. Graphene oxide-based biosensing platform for rapid and sensitive detection of HIV-1 protease. *Anal. Bioanal. Chem.* **2018**, *410*, 6177–6185. [CrossRef]
60. Brown, A.S.; Ackerley, D.F.; Calcott, M.J. High-Throughput Screening for Inhibitors of the SARS-CoV-2 Protease Using a FRET-Biosensor. *Molecules* **2020**, *25*, 4666. [CrossRef]
61. Zhang, Y.; Chen, X.; Yuan, S.; Wang, L.; Guan, X. Joint Entropy-Assisted Graphene Oxide-Based Multiplexing Biosensing Platform for Simultaneous Detection of Multiple Proteases. *Anal. Chem.* **2020**, *92*, 15042–15049. [CrossRef] [PubMed]
62. Li, F.; Chen, Y.; Lin, R.; Miao, C.; Ye, J.; Cai, Q.; Huang, Z.; Zheng, Y.; Lin, X.; Zheng, Z.; et al. Integration of fluorescent polydopamine nanoparticles on protamine for simple and sensitive trypsin assay. *Anal. Chim. Acta* **2021**, *1148*, 338201. [CrossRef] [PubMed]
63. Xu, S.; Zhang, F.; Xu, L.; Liu, X.; Ma, P.; Sun, Y.; Wang, X.; Song, D.J.S.; Chemical, A.B. A fluorescence resonance energy transfer biosensor based on carbon dots and gold nanoparticles for the detection of trypsin. *Sens. Actuators B Chem.* **2018**, *273*, 1015–1021. [CrossRef]
64. Bui, H.; Brown, C.W.; Buckhout-White, S.; Diaz, S.A.; Stewart, M.H.; Susumu, K.; Oh, E.; Ancona, M.G.; Goldman, E.R.; Medintz, I.L. Transducing Protease Activity into DNA Output for Developing Smart Bionanosensors. *Small* **2019**, *15*, 1805384. [CrossRef]
65. Geddes, C.D.; Lakowicz, J.R. Metal-enhanced fluorescence. *J. Fluoresc.* **2002**, *12*, 121–129. [CrossRef]
66. Choi, J.H.; Choi, J.W. Metal-Enhanced Fluorescence by Bifunctional Au Nanoparticles for Highly Sensitive and Simple Detection of Proteolytic Enzyme. *Nano Lett.* **2020**, *20*, 7100–7107. [CrossRef]
67. Lucas, E.; Knoblauch, R.; Combs-Bosse, M.; Broedel, S.E., Jr.; Geddes, C.D. Low-concentration trypsin detection from a metal-enhanced fluorescence (MEF) platform: Towards the development of ultra-sensitive and rapid detection of proteolytic enzymes. *Spectrochim. Acta A Mol. Biomol. Spectrosc.* **2020**, *228*, 117739. [CrossRef]
68. Whitman, J.D.; Hiatt, J.; Mowery, C.T.; Shy, B.R.; Yu, R.; Yamamoto, T.N.; Rathore, U.; Goldgof, G.M.; Whitty, C.; Woo, J.M.; et al. Evaluation of SARS-CoV-2 serology assays reveals a range of test performance. *Nat. Biotechnol.* **2020**, *38*, 1174–1183. [CrossRef]
69. Grant, B.D.; Anderson, C.E.; Williford, J.R.; Alonzo, L.F.; Glukhova, V.A.; Boyle, D.S.; Weigl, B.H.; Nichols, K.P. SARS-CoV-2 Coronavirus Nucleocapsid Antigen-Detecting Half-Strip Lateral Flow Assay Toward the Development of Point of Care Tests Using Commercially Available Reagents. *Anal. Chem.* **2020**, *92*, 11305–11309. [CrossRef]
70. Ongaro, A.; Oselladore, E.; Memo, M.; Ribaudo, G.; Gianoncelli, A. Insight into the LFA-1/SARS-CoV-2 Orf7a complex by protein–protein docking, molecular dynamics, and MM-GBSA calculations. *J. Chem. Inf. Model.* **2021**, *61*, 2780–2787. [CrossRef]
71. Velay, A.; Gallais, F.; Benotmane, I.; Wendling, M.J.; Danion, F.; Collange, O.; De Seze, J.; Schmidt-Mutter, C.; Schneider, F.; Bilbault, P.; et al. Evaluation of the performance of SARS-CoV-2 serological tools and their positioning in COVID-19 diagnostic strategies. *Diagn. Microbiol. Infect. Dis.* **2020**, *98*, 115181. [CrossRef]
72. Liu, L.; Deng, D.H.; Wang, Y.R.; Song, K.W.; Shang, Z.L.; Wang, Q.; Xia, N.; Zhang, B. A colorimetric strategy for assay of protease activity based on gold nanoparticle growth controlled by ascorbic acid and Cu(II)-coordinated peptide. *Sens. Actuators B Chem.* **2018**, *266*, 246–254. [CrossRef]
73. Creyer, M.N.; Jin, Z.; Retout, M.; Yim, W.; Zhou, J.; Jokerst, J.V. Gold-Silver Core-Shell Nanoparticle Crosslinking Mediated by Protease Activity for Colorimetric Enzyme Detection. *Langmuir* **2022**, *38*, 14200–14207. [CrossRef]
74. Liu, F.; Chen, R.; Song, W.; Li, L.; Lei, C.; Nie, Z. Modular Combination of Proteolysis-Responsive Transcription and Spherical Nucleic Acids for Smartphone-Based Colorimetric Detection of Protease Biomarkers. *Anal. Chem.* **2021**, *93*, 3517–3525. [CrossRef]
75. Feng, Y.; Liu, G.; La, M.; Liu, L. Colorimetric and Electrochemical Methods for the Detection of SARS-CoV-2 Main Protease by Peptide-Triggered Assembly of Gold Nanoparticles. *Molecules* **2022**, *27*, 615. [CrossRef]
76. Ling, Z.; Xu, F.; Edwards, J.V.; Prevost, N.T.; Nam, S.; Condon, B.D.; French, A.D. Nanocellulose as a colorimetric biosensor for effective and facile detection of human neutrophil elastase. *Carbohydr. Polym.* **2019**, *216*, 360–368. [CrossRef]
77. Xia, N.; Sun, Z.; Ding, F.; Wang, Y.; Sun, W.; Liu, L. Protease Biosensor by Conversion of a Homogeneous Assay into a Surface-Tethered Electrochemical Analysis Based on Streptavidin–Biotin Interactions. *ACS Sens.* **2021**, *6*, 1166–1173. [CrossRef]
78. Zhang, K.; Fan, Z.; Ding, Y.; Li, J.; Li, H. Thiol-sensitive probe enables dynamic electrochemical assembly of serum protein for detecting SARS-CoV-2 marker protease in clinical samples. *Biosens. Bioelectron.* **2021**, *194*, 113579. [CrossRef]
79. Shi, K.; Cao, L.; Liu, F.; Xie, S.; Wang, S.; Huang, Y.; Lei, C.; Nie, Z. Amplified and label-free electrochemical detection of a protease biomarker by integrating proteolysis-triggered transcription. *Biosens. Bioelectron.* **2021**, *190*, 113372. [CrossRef]
80. Eissa, S.; Zourob, M. A dual electrochemical/colorimetric magnetic nanoparticle/peptide-based platform for the detection of Staphylococcus aureus. *Analyst* **2020**, *145*, 4606–4614. [CrossRef]
81. Weihs, F.; Peh, A.; Dacres, H. A red-shifted Bioluminescence Resonance Energy Transfer (BRET) biosensing system for rapid measurement of plasmin activity in human plasma. *Anal. Chim. Acta* **2020**, *1102*, 99–108. [CrossRef] [PubMed]
82. Weihs, F.; Gel, M.; Wang, J.; Anderson, A.; Trowell, S.; Dacres, H. Development and characterisation of a compact device for rapid real-time-on-chip detection of thrombin activity in human serum using bioluminescence resonance energy transfer (BRET). *Biosens. Bioelectron.* **2020**, *158*, 112162. [CrossRef] [PubMed]

83. Choi, J.H.; El-Said, W.A.; Choi, J.W. Highly sensitive surface-enhanced Raman spectroscopy (SERS) platform using core/double shell (Ag/polymer/Ag) nanohorn for proteolytic biosensor. *Appl. Surf. Sci.* **2020**, *506*, 144669. [CrossRef]
84. Wei, C.H.; Sun, R.; Jiang, Y.N.; Guo, X.Y.; Ying, Y.; Wen, Y.; Yang, H.F.; Wu, Y.P. Protease-protection strategy combined with the SERS tags for detection of O-GlcNAc transferase activity. *Sens. Actuators B Chem.* **2021**, *345*, 130410. [CrossRef]
85. Adem, S.; Jain, S.; Sveiven, M.; Zhou, X.; O'Donoghue, A.J.; Hall, D.A. Giant magnetoresistive biosensors for real-time quantitative detection of protease activity. *Sci. Rep.* **2020**, *10*, 7941. [CrossRef]
86. Guerreiro, M.R.; Freitas, D.F.; Alves, P.M.; Coroadinha, A.S. Detection and Quantification of Label-Free Infectious Adenovirus Using a Switch-On Cell-Based Fluorescent Biosensor. *ACS Sens.* **2019**, *4*, 1654–1661. [CrossRef]
87. Dey-Rao, R.; Smith, G.R.; Timilsina, U.; Falls, Z.; Samudrala, R.; Stavrou, S.; Melendy, T. A fluorescence-based, gain-of-signal, live cell system to evaluate SARS-CoV-2 main protease inhibition. *Antivir. Res.* **2021**, *195*, 105183. [CrossRef]
88. Guerreiro, M.R.; Fernandes, A.R.; Coroadinha, A.S. Evaluation of Structurally Distorted Split GFP Fluorescent Sensors for Cell-Based Detection of Viral Proteolytic Activity. *Sensors* **2020**, *21*, 24. [CrossRef]
89. Gerber, P.P.; Duncan, L.M.; Greenwood, E.J.; Marelli, S.; Naamati, A.; Teixeira-Silva, A.; Crozier, T.W.; Gabaev, I.; Zhan, J.R.; Mulroney, T.E.; et al. A protease-activatable luminescent biosensor and reporter cell line for authentic SARS-CoV-2 infection. *PLoS Pathog.* **2022**, *18*, e1010265. [CrossRef]
90. Luo, X.; Zhao, J.; Xie, X.; Liu, F.; Zeng, P.; Lei, C.; Nie, Z. Proteolysis-Responsive Rolling Circle Transcription Assay Enabling Femtomolar Sensitivity Detection of a Target Protease Biomarker. *Anal. Chem.* **2020**, *92*, 16314–16321. [CrossRef]
91. Braun, A.; Farber, M.J.; Klase, Z.A.; Berget, P.B.; Myers, K.A. A cell surface display fluorescent biosensor for measuring MMP14 activity in real-time. *Sci. Rep.* **2018**, *8*, 5916. [CrossRef]
92. Xu, J.; Fang, L.; Shi, M.; Huang, Y.; Yao, L.; Zhao, S.; Zhang, L.; Liang, H. A peptide-based four-color fluorescent polydopamine nanoprobe for multiplexed sensing and imaging of proteases in living cells. *Chem. Commun.* **2019**, *55*, 1651–1654. [CrossRef]
93. Peyressatre, M.; Laure, A.; Pellerano, M.; Boukhaddaoui, H.; Soussi, I.; Morris, M.C. Fluorescent Biosensor of CDK5 Kinase Activity in Glioblastoma Cell Extracts and Living Cells. *Biotechnol. J.* **2020**, *15*, e1900474. [CrossRef]
94. Sun, P.; Zhang, Z.; Wang, B.; Liu, C.; Chen, C.; Liu, P.; Li, X. A genetically encoded fluorescent biosensor for detecting itaconate with subcellular resolution in living macrophages. *Nat. Commun.* **2022**, *13*, 6562. [CrossRef]
95. Hassanzadeh-Barforoushi, A.; Warkiani, M.E.; Gallego-Ortega, D.; Liu, G.; Barber, T. Capillary-assisted microfluidic biosensing platform captures single cell secretion dynamics in nanoliter compartments. *Biosens. Bioelectron.* **2020**, *155*, 112113. [CrossRef]
96. Zhong, Q.; Zhang, K.; Huang, X.; Lu, Y.; Zhao, J.; He, Y.; Liu, B. In situ ratiometric SERS imaging of intracellular protease activity for subtype discrimination of human breast cancer. *Biosens. Bioelectron.* **2022**, *207*, 114194. [CrossRef]
97. Cheng, Y.; Clark, A.E.; Zhou, J.; He, T.; Li, Y.; Borum, R.M.; Creyer, M.N.; Xu, M.; Jin, Z.; Zhou, J.; et al. Protease-Responsive Peptide-Conjugated Mitochondrial-Targeting AIEgens for Selective Imaging and Inhibition of SARS-CoV-2-Infected Cells. *ACS Nano* **2022**, *16*, 12305–12317. [CrossRef]
98. Yim, J.J.; Singh, S.P.; Xia, A.; Kashfi-Sadabad, R.; Tholen, M.; Huland, D.M.; Zarabanda, D.; Cao, Z.; Solis-Pazmino, P.; Bogyo, M.; et al. Short-Wave Infrared Fluorescence Chemical Sensor for Detection of Otitis Media. *ACS Sens.* **2020**, *5*, 3411–3419. [CrossRef]
99. Moore, C.; Cheng, Y.; Tjokro, N.; Zhang, B.; Kerr, M.; Hayati, M.; Chang, K.C.J.; Shah, N.; Chen, C.; Jokerst, J.V. A Photoacoustic-Fluorescent Imaging Probe for Proteolytic Gingipains Expressed by Porphyromonas gingivalis. *Angew. Chem. Int. Ed. Engl.* **2022**, *61*, e202201843. [CrossRef]
100. Xiang, Z.; Zhao, J.; Yi, D.; Di, Z.; Li, L. Peptide Nucleic Acid (PNA)-Guided Peptide Engineering of an Aptamer Sensor for Protease-Triggered Molecular Imaging. *Angew. Chem. Int. Ed. Engl.* **2021**, *60*, 22659–22663. [CrossRef]
101. Kang, S.M.; Cho, H.; Jeon, D.; Park, S.H.; Shin, D.S.; Heo, C.Y. A Matrix Metalloproteinase Sensing Biosensor for the Evaluation of Chronic Wounds. *BioChip J.* **2019**, *13*, 323–332. [CrossRef]
102. Liu, S.Y.; Yan, A.M.; Guo, W.Y.; Fang, Y.Y.; Dong, Q.J.; Li, R.R.; Ni, S.N.; Sun, Y.; Yang, W.C.; Yang, G.F. Human Neutrophil Elastase Activated Fluorescent Probe for Pulmonary Diseases Based on Fluorescence Resonance Energy Transfer Using CdSe/ZnS Quantum Dots. *ACS Nano* **2020**, *14*, 4244–4254. [CrossRef] [PubMed]
103. Taghvaei, S.; Sabouni, F.; Minuchehr, Z.; Taghvaei, A. Identification of novel anti-cancer agents, applying in silico method for SENP1 protease inhibition. *J. Biomol. Struct. Dyn.* **2022**, *40*, 6228–6242. [CrossRef] [PubMed]
104. Coelho, C.; Gallo, G.; Campos, C.B.; Hardy, L.; Wurtele, M. Biochemical screening for SARS-CoV-2 main protease inhibitors. *PLoS ONE* **2020**, *15*, e0240079. [CrossRef]

 *biosensors*

MDPI

Review

Recent Advances of Point-of-Care Devices Integrated with Molecularly Imprinted Polymers-Based Biosensors: From Biomolecule Sensing Design to Intraoral Fluid Testing

Rowoon Park [1,†], Sangheon Jeon [1,†], Jeonghwa Jeong [1,†], Shin-Young Park [2], Dong-Wook Han [1,3] and Suck Won Hong [1,3,*]

1   Department of Cogno-Mechatronics Engineering, Pusan National University, Busan 46241, Korea; rowoon.p153@gmail.com (R.P.); jsfhse@pusan.ac.kr (S.J.); j_purification@pusan.ac.kr (J.J.); nanohan@pusan.ac.kr (D.-W.H.)
2   Department of Dental Education and Dental Research Institute, School of Dentistry, Seoul National University, Seoul 03080, Korea; nalby99@snu.ac.kr
3   Department of Optics and Mechatronics Engineering, Pusan National University, Busan 46241, Korea
*   Correspondence: swhong@pusan.ac.kr
†   These authors contributed equally to this work.

Abstract: Recent developments of point-of-care testing (POCT) and in vitro diagnostic medical devices have provided analytical capabilities and reliable diagnostic results for rapid access at or near the patient's location. Nevertheless, the challenges of reliable diagnosis still remain an important factor in actual clinical trials before on-site medical treatment and making clinical decisions. New classes of POCT devices depict precise diagnostic technologies that can detect biomarkers in biofluids such as sweat, tears, saliva or urine. The introduction of a novel molecularly imprinted polymer (MIP) system as an artificial bioreceptor for the POCT devices could be one of the emerging candidates to improve the analytical performance along with physicochemical stability when used in harsh environments. Here, we review the potential availability of MIP-based biorecognition systems as custom artificial receptors with high selectivity and chemical affinity for specific molecules. Further developments to the progress of advanced MIP technology for biomolecule recognition are introduced. Finally, to improve the POCT-based diagnostic system, we summarized the perspectives for high expandability to MIP-based periodontal diagnosis and the future directions of MIP-based biosensors as a wearable format.

Keywords: molecularly imprinted polymer; point-of-care test; biomolecule; oral disease; wearable device

Citation: Park, R.; Jeon, S.; Jeong, J.; Park, S.-Y.; Han, D.-W.; Hong, S.W. Recent Advances of Point-of-Care Devices Integrated with Molecularly Imprinted Polymers-Based Biosensors: From Biomolecule Sensing Design to Intraoral Fluid Testing. *Biosensors* **2022**, *12*, 136. https://doi.org/10.3390/bios12030136

Received: 8 January 2022
Accepted: 21 February 2022
Published: 22 February 2022

1. Introduction

Molecular diagnostics point-of-care (POC) is a technology belonging to the field of personalized healthcare and refers to clinical pathology tests for the diagnosis of disease [1]. Generally, it has been used to enhance the therapeutic effect by enabling an immediate test next to the patient, which was tested in the field of immunology and clinical chemistry [2]. POC devices are a type of in vitro diagnostic (IVD) medical device, designed for the purpose of diagnosing various diseases to determine prognosis, evaluating health status by medical treatment effect and even preventing disease [3]. The market growth for IVD devices can be attributed to the increasing proportion of the geriatric population and technological advancement in diagnostics [4]. Recently, during the COVID-19 pandemic, growing interest in public healthcare has rapidly boosted up rapid testing kits, anticipating the development of various types of devices with market expansion. Although the expensive cost of product development may defer research demand for POC testing (POCT), some progressive technologies are continuously introduced by merging existing portable sensing platforms in line with recently developed bioelectronic devices with suitable configurations in medical diagnosis applications [5–7]. One aspect of bioelectronics is the application of physicochemical signal transducers to detect substances at the molecular level and recognize interactions through signal processing. Biosensors used in POCT encompass a wide range of topics for the detection of analytes, various types of receptors, such as enzymes, antibodies, antigens, proteins at the interface of biological molecules and sensors. Therefore, since the POCT can be performed in close proximity to the location where the patient is being treated, emerging technologies as a potential alternative may replace the conventionally used laboratory-based diagnostic testing, including different combinations of components for sample handling and recognition elements. Thus far, the recent trend in the integration of diagnostic devices has moved to cost-effective programable tools for rapid and sensitive detection of biomarkers in biofluids, such as sweat, tear, saliva and urine [8–10]. However, many biomarkers in biological samples (i.e., biological fluids) are often present at very limited concentrations, coexisting with unwanted interfering species. Therefore, the detection of biomarkers usually requires highly qualified antibodies for sensitivity detection techniques together with sampling purification. To analyze one type of biomarker, enzyme-linked immunosorbent assay (ELISA) is a widely used immunological assay in diagnostic research [11], which provides quantitative data on specific proteins in serum samples. Despite its high specificity and low limit of detection (LOD), some drawbacks still arise from relatively long procedures with moderate reliability or expensive bioassay kits' specified protocols, depending on the primarily designed binding affinity for different targets [12].

In this context, as a rational synthetic strategy and biomimetic design in the field of biotechnology, molecularly imprinted polymers (MIPs) have been revisited in response to the continuous demand for rapid, accurate and cost-effective analytical platforms. MIPs, crosslinked polymer matrices with molecular recognition sites formed by synthesizing in the presence of a target template, have received immense attention to guarantee affordable detection modules for target molecules for decades [13–15]. Historically, although viewed as an old material system, the MIP technology has progressed with a renewed field of research and expanded the area by combining emerging nanomaterials and advanced detection techniques with new applications [16–18]. Specifically, MIPs can be considered synthetic chemocavities or antibodies, as tailor-made artificial receptors that recognize and bind target molecules with high selectivity and chemical affinity [19–21]. The MIP matrices can be synthesized simply by polymerization of monomers forming a complex with target molecules, in which a relatively weak bonding was set between the template molecules and crosslinked monomers. In detail, starting with the prepolymer/template mixture, the spatial arrangement of MIPs can be determined by several well-known interactions, such as hydrogen bonds, Van der Waals forces, hydrophobic interactions and electrostatic interactions [22–24]. Subsequently, after the removal of template molecules from the crosslinked polymeric matrices, copious cavities with specified chemical end groups can be easily crafted, depending on complemental templates defined by size, shape and chemical functionality. Indeed, a large number of results on MIP techniques have reported newly developed molecular imprinting strategies with small molecules, such as sugars, steroids, pesticides, epitopes and amino acid derivatives [25–27]. These previous elaborated works have demonstrated the reliable capabilities of MIPs in highly targetable recognition systems on specific molecules, used in chemical sensors, analytic separations, solid-phase extractions, drug delivery systems, catalysts and library screening methods [28–30].

Having been progressively specialized in the field of biotechnology, MIPs were successfully commercialized for the solid phase of drugs and pesticides for an extraction toolbox to rebind small molecules toward improved sample refinement of chromatographic analysis [31,32]. In addition, molecular imprinting for other larger-scale substances in particular is also considered a candidate for expandible technology and remains under development with plenty of potential. Biomacromolecules, including antibodies, viruses, proteins, enzymes, nucleic acids and even living cells, can readily be imprinted in precisely designed polymer matrices with the help of other interfacial molecules or additives. However, the biomacromolecule approaches in the MIP system will confront serial problems with less reliability in the sophisticated capturing process because the classical bulk methodologies for target templates may fail to precisely recognize the protein target molecules. The lack of accessibility lies primarily in the intrinsic properties of protein molecules themselves. Complete rebinding may be difficult due to imprinted sites that differ from the original conformation by irreversible structural reconfiguration during polymerization [33]. In other words, templated proteins embedded in the polymerized matrix can be partially immobilized with strong physical bonds in the crosslinked polymeric network during the template removal step, which provides fewer rebinding sites [34]. Thus, improved techniques are needed, especially in protein imprinting processes, to prevent irreversible entrapment in 3D polymeric networks. Hence, the large number of uncontrolled interaction sites by the imprinted proteins may lead to cross-reactivity on the originally provided cavities along with non-specific adsorption. Biomacromolecule recognition systems have now become a new growing research area in MIP approaches, and the control of the biological environment associated with an appropriated function (i.e., artificial recognizable antibody) requires further development in practical use and biomedical diagnostic applications [35].

In view of the above, the MIP-based pseudo-immunoassays may be developed and may strengthen the biosensing capability and related POCT to measure the concentration of small and macromolecules with the help of specialized 'artificial antibodies' to antigen counterparts [36]. Thus, an enhanced accuracy of MIP-based POCT will accelerate the molecular diagnostic and has critical potential to play an important role in analytical tests in various fields with different macromolecular targets for more accurate results. For now, commonly used biosensors are mostly based on the detection of antigen–antibody interactions, which are evaluated and quantified with respect to each proposed transducing mechanism [37–39]. Moreover, most antibodies used are proteins, which are physically, chemically and biochemically unstable for use in a medical-grade immunoassay. Therefore, healthcare devices based on MIP-technology-based POCT may provide suitable access for specific patients by providing reliable results from artificial antibody-integrated biosensors (i.e., MIP-based testing). The MIP-based biosensing platform can be robust, sensitive and accurate to enable label-free detection of biomolecular analytes. Such beneficial embodiments will meet endless supply demands in the segmental market to develop MIP technology, leading to an integration of cost-effective portable POCT with an expansion of the overall industrial progression.

In this review, we will highlight the recent progress of MIP technology, particularly with a wide range of examples of POCT-based biomarker detection. Because the biomolecules used in developing diagnostic POCT are generally complex in nature, integrating the test kits (i.e., cartridges) with appropriate antibodies requires a highly sophisticated coupling between the two components [40]. We believe that the existing problems of POCT can be solved by utilizing novel MIP technology through the sensitive detection of biomarkers in the sampling of biological fluids extracted from sweat, tears and urine, and directly allowing the detection of biomolecules [41–43]. However, to be commercialized for POCT, the MIP-based biosensors must overcome several challenges related to non-specific rebinding, cross-reactivity, acceptable analytical performance and practical use models. Moreover, due to continuous exposure to the fluid of interest simultaneously with other biological components, the biosensors targeting analytes may need recalibration to correct for signal responses over time. On this issue, at the end of this review, we will also suggest wearable POCT devices and advanced MIP-based technology to determine periodontal diagnostics by identifying oral fluid-based biomarkers for precision oral medicine. Since periodontitis is a complex and multifactorial disease, prognostic progression is hardly detected, subject even to chronic conditions [44]. The discovered biomarkers that cause oral disease can be molecularly imprinted on conductive polymer surfaces. With easy accessibility from saliva sampling, the MIP-based periodontic diagnosis to improve POCT-based diagnostic systems is still under investigation for clinical application, which will be an important field of study and highly beneficial for overall public healthcare.

2. MIPs for Biomolecule Recognition: Concepts of POCT and Synthetic Approaches

2.1. Concepts of the MIP-Technology-Based Portable POCT Devices

The most important feature for MIP-enabled biosensors is the comparability that can be integrated with the existing systems, which provides high recognition ability [45]. The persistent durability allows MIPs to be used in various types of POCT applications, depending on the types of samples to be tested, as presented in Figure 1. Benchtop-scale POCT devices, incorporating MIP-based biosensors, are poised to transform the healthcare device platforms. Conventional biosensors combining biological elements have been produced in a form of chips, disposable strips, cartridges or electrodes in the application of POC devices [46,47]. However, in certain situations, the diagnostic devices have some limitations in new classes of POC devices [48]. For example, the short shelf life of biomolecular immobilized biosensors is less cost effective for manufactured products because they should be refrigerated in transport and storage [49]. Another potential issue might cause the low activity of biomolecular functions under harsh chemical conditions in some cases of biofluids or sampling, such as extreme changes in pH, saline or highly reactive solvents in certain treatment controls for the POCT devices [50], which critically affect the performance of biosensors by fundamental degradation of the disease-specific biomarkers [51]. Besides, although conventionally used bioreceptors are suitable for achieving selectivity, multiple processing with technical difficulties and delicate interface control are required in the implementation onto biosensors [52]. Since the immobilization of recognition sites is essential to transduce the signals for the operation of biosensors, advantageous materials and alternative strategies will be needed to resolve the shortcomings linked with conventional biomarker integration while achieving selectivity. By the aforementioned motivations, MIP techniques have been progressed for biosensor applications with carefully contrived design. The MIP-based biosensor can be considered an artificial antibody-integrated polymeric active layer that readily sustains stability in challenging testing chemical environments, such as high-temperature limits up to ~300 °C [53,54]. Since general proteins are usually denatured in irreversible forms higher than ~80 °C [55,56], MIP-based biosensors are more stable in storage and even suitable for applications requiring a high-temperature range. In the scene of biomolecule imprinting with low-cost materials, by taking advantage of the MIP technique to mimic biological sensing elements, such as antibodies and bioreceptors [57], a variety of single-target biosensors can be developed for diagnostic biosensors and assays for POCT devices [58]. Because the MIP-enabled biosensors extracted out the biological antibodies or other elements, the desired receptor surface can be tailored for relatively high selectivity and specificity. Compared to the biosensors integrated with natural antibodies, MIP-based biosensors have exhibited a comparable or even decreased limit of detection (LOD) with signal-to-noise enhancement and improved stability resulting in potential use for biosensing platforms [59]. The continued progress of MIP technology holds great potential with innovative key aspects of inexpensive, rapid and sensitive detection for desired POCT systems, providing other opportunities in demanding medical environments.

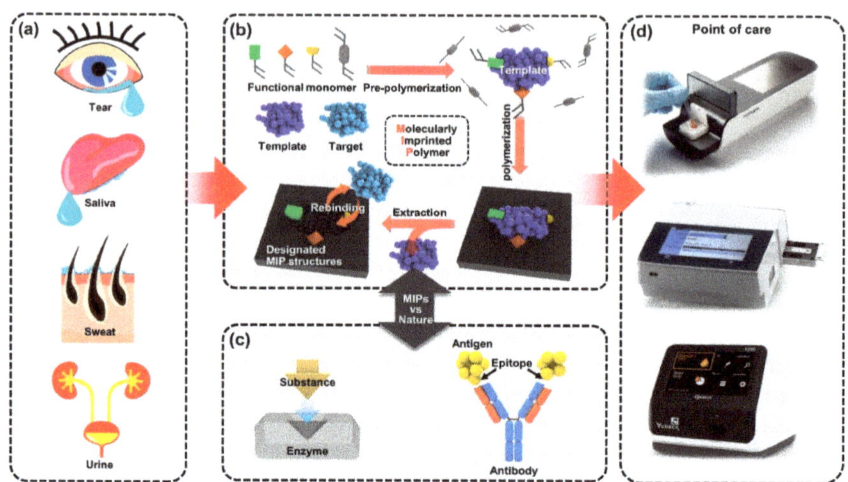

Figure 1. A new class of benchtop-scale POCT devices utilized by MIP-based biosensors for precision diagnostic technology to detect biomarkers in biofluids. (**a**) The four types of representative human biofluid reflecting health conditions; tears, saliva, sweat and urine. (**b**) Schematic illustration for fabricating the molecular imprinting system that contains the biorecognition sites and (**c**) the example of natural biorecognition system; enzyme–substrate complex (**left**) and antigen–antibody reaction (**right**); the biomimetic functional similarity of the MIP biosensing system is comparable to natural antibodies. (**d**) Various types of immunoassay-based benchtop-scale POCT devices.

2.2. Biomolecule Imprinted Polymers Based on Bulk Imprinting Techniques

At the beginning of the development of synthetic process for MIPs, target-oriented techniques, capable of recognizing and binding biomacromolecules (e.g., proteins), have been introduced along with practical use in numerous applications, including clinical diagnostics [60], drug delivery systems [61], proteomics and environmental analysis [62]. Despite the widespread research work on MIPs, there have been some limitations in the design of the MIP material system in protein detection by the intrinsic conditions of templated proteins, such as size, complexity and structural instability. Recently, however, the synthetic strategies for protein-based MIPs have been extensively developed to improve the selective recognition capability, named by bulk, suspension, emulsion and epitope imprinting, depending on the materials mainly featured [63–65]. As one synthetic process for protein-based MIPs, the so-called bulk imprinting polymerization is the most commonly accepted method with apparent advantage of the simplicity of the processing scheme. In designing bulk imprinting, by using the crosslinker and the functional monomer, protein molecules can be imprinted entirely on the growing polymer matrix with randomly distributed configuration, and subsequent extraction of the templates from the MIPs complete the process with high yield performance. Finally, for the collected MIP powders, the mechanical grinding process separates the bulk-imprinted polymers into the form of micron-scale particles or beads. This sequential process suggests a viable route to produce a large number of bulk MIP particles that can be used in several commercial applications [66]. However, for a typical bulk MIP system, a random free diffusion of the templates (i.e., small molecules) was subjected to the formation of microcavities in the densely networked MIP structures [67]. Therefore, bulk imprinting is adaptable for imprinting for small molecules because the adsorption/release of templated molecules is easily expected and represents relatively fast and reversible interactions, which add value to the small-molecule imprinted matrix as multiple-time reusable assay in cost-effective benefits [68]. On the other hand, a limited synthetic condition for the bulk-imprinted biomacromolecules has been reported due to partially trapped templated molecules in the polymer chains, commonly featured with a complex distribution in the mixed state prior to the polymerization. Such drawback lies in the limited diffusion rate of biomacromolecules from the nature of bulk MIP manufacturing system. Consequently, low diffusion rates attenuate the quality of MIPs with lacking accessibility on the rebinding sites. Moreover, the mechanical grinding process as a final stage is strictly controlled to maintain the original integrity of the prepared samples, that is, damages of the recognition sites reduce the adsorption capacity of the bulk MIP system. Although crushing films into smaller dimensional microparticles notably expands binding recognition sites, the irregular shapes and sizes of the resulting bulk MIP particles lead to less reliable signal detection for accurate biosensing of biomacromolecules with high precision [69]. Thus, to avoid the instability of protein-imprinted bulk MIPs, one key parameter can be the homogeneous combination between template sizes (i.e., the large size of proteins) of the incorporated monomers by careful design to provide protein recognition sites with high reproducibility. The nanoscale protein-imprinting polymer in uniform 3D bulk scale is inevitable for an improved binding site accessibility, meeting quality requirements for biosensor application on the appropriate analytical performance [70]. The advances of protein-imprinting technique have also expanded to direct construction of micro/nanoscale surface-imprinted MIP systems on planar surfaces with the development of combinatorial materials composition, which has suggested novel sensing mechanisms in various types of signal transducing systems [71]. In competition with the bulk MIP technique, other strategic templating processes have been introduced as alternative methods to resolve the problems with diffusion limitations, uniform features and improved selectivity.

2.3. Biomolecule-Imprinted Polymers Based on Surface Imprinting Techniques

As an effective way of integrating biomolecule-imprinted polymer into biosensor systems, newly developed surface imprinting techniques have been directly applied to transducing elements, such as chemically derived electric signals [72–74]. By the advantageous feature of the surface MIP system, the increased specific binding sites are exposed only on the surface of the polymer matrix for effective recognition, which thus accelerates mass transfer and accurate rebinding capacity (i.e., adsorption/desorption efficiency). To generate a protein-imprinted polymeric surface, a suitable monomer selection for the templates is a crucial factor in the rational design through high affinity of chemical composition for the advanced surface MIP system. Similar to the bulk MIP system, since the binding strength and stability between monomer and template depend on non-covalent weak forces, such as hydrogen/hydrophobic or electrostatic interactions, designing a template/monomer complex on the active surface area to effectively recognize the rebinding biomolecules is necessary. As shown in Figure 2, the interactions between proteins and monomers for constructing protein-imprinted polymers (i.e., artificial receptor formations) on the electrically conductive surface can be classified into several types: (i) formation of a prepolymerized complex on the electrode; (ii) sequential electropolymerization of functional monomers after template physisorption; and (iii) immobilization of the target protein using a complemental linker. As an easily accessible process, protein–monomer mixtures were introduced onto electrode surfaces using drop casting [75], spin casting [76] and spray coating [77]. After that, the prepolymerized complex was crosslinked under specific electrosynthesis conditions, and the templated protein could be extracted by physical or chemical methods to form steric cavities, constructing a surface-MIP recognition system (Figure 2a). On the other hand, Figure 2b represents another developed surface imprinting method that induces spontaneous adsorption of proteins to the electrode surface to increase templated cavities. The templates (i.e., proteins) built on the electrode surface can be imprinted by physicochemical interactions following the electrosynthetic process, in which the configured specific cavities are mostly on the surface of the MIP matrix. During this electropolymerization approach, the isoelectric point (pI) of the protein can be considered an important factor in designing sophisticated MIP–protein complexes. As an inherent property of proteins (i.e., amphiphilicity), the pI is usually defined by the pH value of a solution, at which the net charge is zero [78]. Thus, the electrostatic behavior of proteins with pH is subtle in the process because when the pH of a solution is higher than the pI value, the surface of the protein becomes predominantly negatively charged, resulting in a repulsive force on the same charged molecules. In contrast, in the case of lower pH of the solution than the pI value, the protein surface can be positively charged. However, under conditions with a pH value close to the pI, the attractive force prevails between the proteins by balancing the negative and positive charges, leading to the aggregation or precipitation of the protein [79,80]. To resolve this problem, the surface-MIP approach using a sacrificial carrier was demonstrated. By the fact that the aggregated form of proteins can lead to inadequate sensing properties in MIP-based biosensor systems, the formation of covalently immobilized proteins prior to electrodeposition of functional monomers on the electrode surface could obviously enhance the sensing performance for target proteins (Figure 2c). As one clear demonstration, the sequential molecular imprinting process of specific protein binding sites for selective recognition system in the surface-MIP structure is as follows [81]. First, the electrode surface could be chemically modified via a 4-ATP/DTSSP linker system and immobilized with a target protein (CDNF). After electrochemical polymerization with functional monomers, selective molecular cavities could be formed simply by the S–S bond cleavage process. This experimental approach implies that target protein immobilization can be facilitated by a simple combination of conventional chemistries (i.e., linkers such as self-assembled monolayers, SAMs) to create more uniform specific binding sites with finely tuned affinity [82–84], instead of random electropolymerization from the protein/monomer mixture. Notably, in this case, the optimized thickness of the surface-MIP film was precisely controlled during the electropolymerization process not to exceed the height of the immobilized target protein, which is the most important factor in avoiding irreversible entrapment of proteins in the imprinted polymer matrix [85].

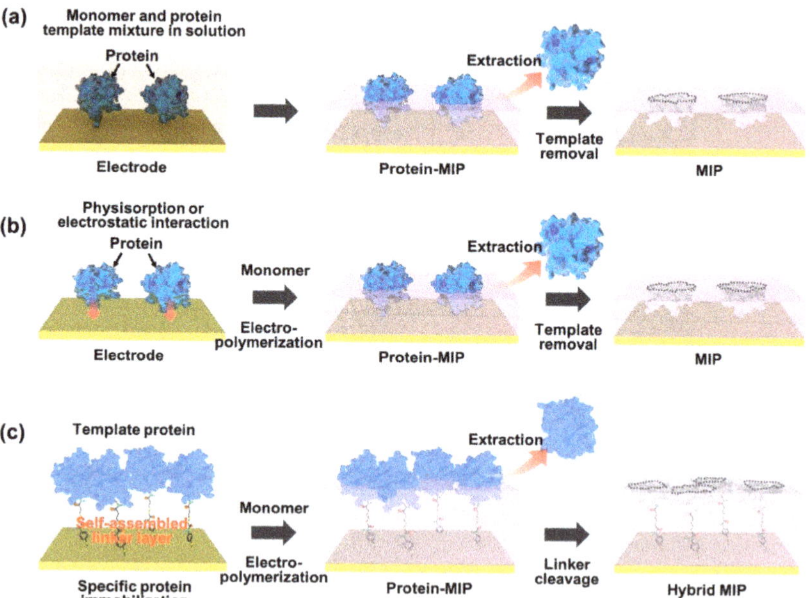

Figure 2. Representative strategies on the surface imprinting process to construct specific protein recognition cavities. In an appropriate design concept, the selective rebinding site can be generated by using a functional monomer for electropolymerization on a prepared electrode surface, which includes the formation of the pre-polymerization complex (**a**), the template physisorption (**b**) and the immobilization of the target protein (**c**).

2.4. Electrosynthetic Strategies for Biomolecule-Imprinted Polymers

As described in the previous section, direct electropolymerization has proved to be an efficient technique to construct surface-imprinted MIPs by the adequate combination of monomers and templates [86–88]. More importantly, the selection of biomolecular templates has not been limited within an allowed experimental condition and is ready to be applicable to state-of-the-art electropolymerization strategies on the electrically conductive electrode surfaces, such as RNAs, DNAs, peptides, aptamers, antibodies, proteins, viruses, bacteria and even living cells [89–96]. A suitable choice of monomers and templates plays a key role in the molecular design of electrodeposition techniques with delicate modulation of parameters for the MIP-enabled conducting polymer matrix. Thus, synergistic influences on surface MIPs have been evaluated as a result of highly specified analytical performance in the biosensing platforms [97,98]. As reported earlier, the main parameters for electrosynthetic process can be summarized as follows [99]: (i) voltage or current of applied potential; (ii) potential scan rate and periodic potential pulses during deposition cycles; and (iii) the restriction of electrical density on the electrode. Therefore, the electrosynthesis of conductive polymers in the surface-MIP system highly depends on the series of optimization by a control of the surface morphology, density and film thickness to tune the capability of charge transfer passing through the electrode [100]. Figure 3a represents a typical process of electropolymerization for protein-imprinted polymers. In the protein-imprinted polymerization step of the electrosynthesis process, the stacking monomer layers and boundaries define the shape of the complemental recognition sites according to the size of biomolecules, such as proteins with embedded functional groups. Thus, sequentially designed processing steps with tailored compositions can be important to define the exposed areas of the cavities and controlled distributions of biomolecules prior to the electrodeposition of monomers. Such electroactive monomers on a prepared electrode surface to be grown as an electrically conductive polymeric matrix should be carefully selected according to the electrosynthesis conditions because there are many combinatorial options with other binding assistant chemicals, including phenol, o-aminophenol, o-phenylenediamine, aminophenyl boronic acid, scopoletin, aniline, pyrrole, 3,4-ethylene dioxythiophene, 2,2′-bithiophene-5-carboxylic acid and dopamine, as previously reported [101–104]. For example, as illustrated in Figure 3b, a variety of combinations has been demonstrated by using o-phenylenediamine (o-PD) on template proteins for the surface-MIP integration [105,106]. Moreover, based on a computational approach, Raziq et al. recently demonstrated that o-PD could be a reasonable choice for biologically active MIPs compared to a set of molecules, such 3,4-ethylenedioxythiophen (EDOT) and dopamine (DA), in binding to the SARS-CoV-2 viral protein [107]. Looking into the detailed performance, the Glide empirical scoring function (GScore) values for the scoring binding pose of other monomers (i.e., o-PD, dopamine and EDOT) docked to SARS-CoV-2 nucleoprotein (ncovNP) were similar, in the ranges of -25.2 and -29.5 kJ mol^{-1}, confirming that they could form stable pre-polymerized complexes with ncovNP. As a result of the quantum chemical calculation, the H-bond interactions on the ncovNP molecules adjunct with NH_2 groups were determined decisively; the o-PD monomer was found to be a superior option compared to other monomers. This computational modeling approach can be highly useful and expanded to the advanced designing of MIPs, especially for the electrosynthetic process, because the molecular reaction and energy startup guidance based on the predominant parametric assumptions might be derived without repetitive control experiments [108]. In the biomolecular imprinting field, this high-end computational approach may be advantageous in rapid development in choosing correct parameters between the template and functional monomer to realize MIP-based biosensors, yielding highly selective recognition sites by scrutinizing the critical interaction energies [109]. By doing this, the electrosynthetic strategies for viral-protein-based MIPs (i.e., SARS-CoV-2) can contribute to producing a new concept of POCT devices to fully utilize the electrically operational transducers [110], which is under development with the popularly introduced small-scale device integrated with microchips for the wearable or skin-attachable format [111]. We will discuss this in more detail in the following sections.

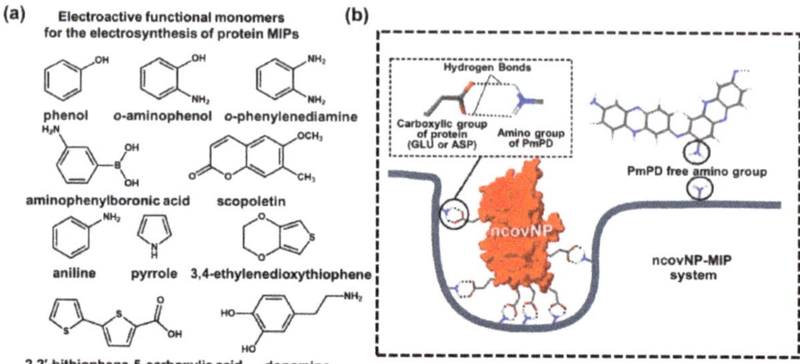

Figure 3. (a) Representative electroactive monomers used for implementation of the MIP-based biorecognition system; chemical structures of each functional monomer are listed as follows: phenol, o-aminophenol, o-phenylenediamine, aminophenylboronic acid, scopoletin, aniline, pyrrole, 3,4-ethylenedioxythiophene, 2,2'-bithiophene-5-carboxylic acid and dopamine. (b) Schematic illustration of the ncovNP–MIP system to form multiple non-covalent bonds between ncovNP and cavity. Reproduced with permission from [80]. Copyright Elsevier, 2020.

3. Transducing Systems and Practical Approaches for MIP-Based Biosensors

3.1. Mass Sensing Approaches

To date, MIP-based quartz crystal microbalance (QCM) sensors have been regarded as one of the most promising sensing techniques and have been developed as an analytical method to detect various biomolecules, such as DNA, peptide and viral protein [112–114]. In a typical configuration of a QCM-based sensor, an AT-cut quartz wafer with metal electrodes on either side is coupled with an oscillator circuit that drives the QCM to resonate at an its intrinsic frequency. According to the basic principle of QCM operation, an analyte adsorbed on the electrode surface causes a change in the fundamental resonant frequency of the crystal, which precisely determines the mass of the adsorbed analyte, that is, the resonant frequency is decreased depending on loaded mass [115]. Because MIP membranes contain specific molecular recognition sites with size, shape and chemical features complementary to the desired target molecule, the combination of a QCM sensor and MIP membrane with artificial receptors can selectively recognize target molecules in complex biological samples and minimize cross-sensitivity with sufficient susceptibility (Figure 4a). Therefore, as biomolecules are adsorbed into the cavity of the MIP membrane, the fundamental resonant frequency gradually decreases in the MIP-based QCM sensor system (Figure 4b). The technical parameter for imprinting small molecules or metastable biologicals lies in the selection of functional monomers and appropriate solvents for MIP synthesis. For example, small molecules are stable in organic solvents, whereas biomolecules are unstable and can be denatured under similar conditions [116]. Therefore, many efforts have been made to construct water-soluble polymer systems, such as hydrogels for imprinting biomolecules [117,118].

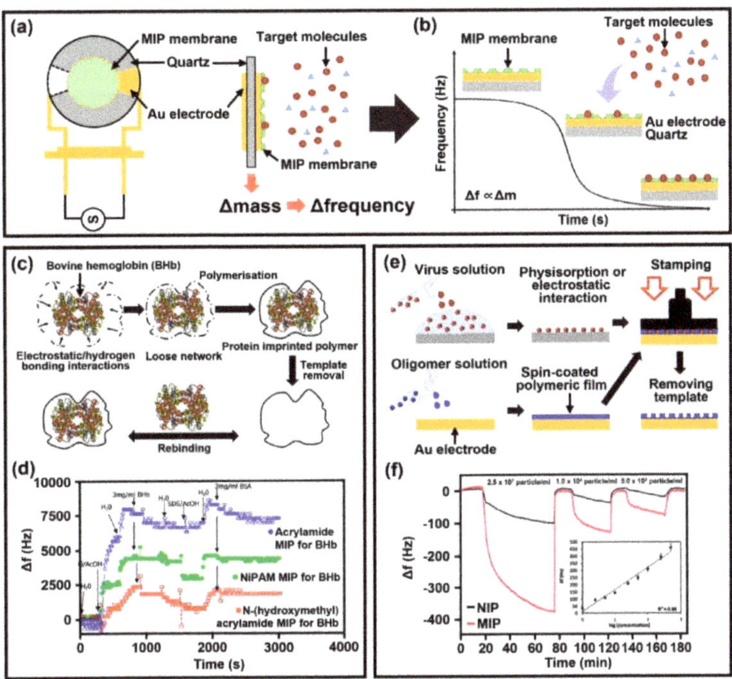

Figure 4. (**a**) A conceptual design for an MIP-based QCM sensor capable of selective adsorption of target molecules. (**b**) Frequency change in QCM transducer according to mass change due to binding between the target molecules and artificial receptors formed on the MIP membrane. (**c**) A synthetic strategy to prepare hydrogel-based MIPs for BHb protein adsorption. Reproduced with permission from [119]. Copyright The royal society of chemistry, 2014. (**d**) Frequency response of polyacrylamide-based MIPs to BHb and BSA protein. Reproduced with permission from [119]. Copyright The royal society of chemistry, 2014. (**e**) Schematic illustration of a molecular imprinting process using template stamp coated with influenza virus. (**f**) Frequency changes of MIP- and NIP-based QCM sensor at different concentrations of H1N3 influenza A virus; the inset shows a change in the frequency according to virus concentration ($r^2 = 0.98$). Reproduced with permission from [114]. Copyright The royal society of chemistry, 2013.

As illustrated in Figure 4c, a conceptual demonstration was performed as a QCM-based diagnostic system using molecularly imprinted hydrogels to detect bovine hemoglobin (BHb) [119]. By the optimized synthesis condition for the hydrogel-based MIP membrane with specific binding capacity, three distinct types of the acrylamide functional monomer were utilized, such as acrylamide (AA), N-hydroxymethylacrylamide (NHMA) and N-isopropylacrylamide (NiPAm). The graph in Figure 4d describes the gradual QCM-based frequency shifts through the sequential immersion process in BHb, SDS/AcOH and bovine serum albumin (BSA) solutions; the molecular weight of BSA and BHb corresponds to 66.5 kDa and 64.5 kDa, respectively. The degree of cross-selectivity for non-target proteins (i.e., BSA) revealed the recognition site formation (i.e., cavities) with protein complemental to the target (i.e., BHB) in a hydrogel-based MIP system. In the BHb-imprinted MIPs formed by AA, NHMA and NIPAM-based matrices, an apparent decrease in resonant frequency by the BHb adsorption process was confirmed without changes in additional BSA proteins. In particular, as a hydrogel-based MIP system, the detection performance for BHb was correlated with the degree of hydrophilicity and increased in the order of polyNHMA, polyacrylamide and polyNIPAM. Consequently, the high selectivity of NHMA–MIP for the BHb protein was observed by the presence of a hydroxyl group in the cavity architecture.

Although AA-based MIP was equally selective for both BHb and BSA proteins, the absence of hydroxyl groups in the cavities derived a relatively weak ability to distinguish between cognate and non-cognate proteins.

As another outstanding example of QCM-based biomolecules detection system, Figure 4e shows the selective capture of infectious influenza A (i.e., H1N3, H1N1, H5N3, H5N1 and H6N1 viruses) by combining MIPs and QCM-based gravimetric transducer [120]. To prepare the MIP-based QCM sensor, a delicate sequential process was used utilizing an MIP precursor solution containing functional monomer, cross-linking agent and a photoinitiator. The MIP solution was spin-casted on an Au electrode, and then a stamp coated with the template (i.e., virus stamp) was pressed onto the spin-coated MIP prepolymer film. In this experimental scheme, acrylamide (AMM), methacrylic acid (MAA), methylmethacrylate (MMA) and N-vinylpyrrolidone (VP) were used as functional monomers. In the following, MIP prepolymer film incorporating virus particles was polymerized under a UV light source with 254 nm wavelength overnight, and the template (i.e., virus) was extracted by denaturing the virus from the polymeric network by using 10% hydrochloric acid. As a result, the limitation of selective recognition of virus subtypes was dramatically enhanced due to the addition of the VP monomer. The specificity and sensitivity for the template were further improved by controlling the ratio between the different monomers and the cross-linking agent.

Figure 4f shows a response curve for the QCM-based MIP sensor capable of detecting the H1N3 influenza A virus over time. The calibration curve showed that the logarithmic relationship between frequency and virus concentrations responds to mass changes in a linear form ($r^2 = 0.98$). When equilibrium is reached between the MIP membrane and the surrounding solution, the MIP signals increase at least 5-fold compared to non-imprinted polymer (NIPs) signals. In the case of NIPs, it is desirable that specific binding to the target molecule be restricted, but this is not achievable in practice due to unpredictable and non-specific adsorption or physical influences between the NIP membrane and the target molecule. Accordingly, the QCM sensing system integrated with an MIP membrane that contains a memory effect from the template molecules (i.e., recognition sites) provides high affinity to the target virus in a reproducible manner for more reasonable molecular detecting tools. In this way, the MIP-coated QCM sensor can be expected in the next generation of molecular sensor platform to support diagnostic POCT to evaluate small proteins, artificial enzymes or viruses.

3.2. Electrochemical Sensing Approaches

For the MIP-based biosensors, to evaluate the complemental interactions between the analytes and cavities (i.e., receptors) formed on the electrode surface, the analytical signals should be transduced and converted into a quantitative range of values [120]. On this, we first focus on the electrochemical sensing approach, which is quite suitable for MIP-based biosensors with various types of electrode configurations to validate a rebinding of the analytes [121]. Due to ease of access, biomolecule analytes can be measured quantitatively, combined with external redox materials in the solution-type tests. The measured value changes are originated from the Faraday current, corresponding to the redox reactions on the MIP-based electrode in cyclic voltammetry (CV) or differential pulse voltammetry (DPV) [122]. By well-known basic fundamental ideas, the mechanism, caused by the diffusion of the redox probe, is usually understood in specific operating conditions for the MIP-based electrochemical sensors [123,124]. For example, the physical docking of analytes (e.g., viruses, small molecules and proteins) into the surface cavities can generally block the diffusion of the redox probe on the MIP electrode surface in amenable chemical reactions [125,126].

Figure 5 illustrates the basic concept of electrochemical transduction on MIP-based sensors. The sensing procedure requires sequential steps, including rebinding of the target analytes on the MIP electrode and washing to remove the non-specific binding elements. Using voltammetry test, the detection of analytes on MIP-based electrochemical methods can be performed to evaluate the current density according to the potential range for waveform techniques (e.g., differential pulse, square wave, linear sweep or staircase) [127]. On the basis of this technical support, the advantages of the voltammetry test for MIP-based biosensors are prompt analyses to determine the selectivity and sensitivity of the engaged analytes by electrochemical reacting on the electrode within a given concentration range. Among the voltammetry test, the DPV technique is the most acceptable technique for MIP-based biosensors, with relatively simple implementation and a low level of noise by the capacitive current. Amperometric sensing approach by measuring the generated current at the sensing electrode surface with respect to a fixed time interval can be an effective method at a constant single-potential step, known as the chronoamperometric technique. This amperometric method directly reflected the measured current with the concentration of the analytes at a constant applied potential; therefore, the analytes are detected by the facilitation of the built-in MIP electrode. The signal-changing value collected from the amperometric device is of an apparent reading gauge, interfaced with the MIP-based electrode, since the mass transfer rate was delivered from the signal-changing value by electrochemically active analytes. Electrochemical impedance spectroscopy (EIS) is a sensitive characterization method of collecting the electrical signals from chemical responses, in which the time response with the chemical systems was susceptible based on low-amplitude alternating current and voltages over a range of frequencies. Therefore, the quantitative values can be obtained by this technique, but the chemical mechanisms for the localized conditions at the electrode surface and electrolyte solution would be more complicated depending on the analytes [123,127]. However, the EIS system covers a wide range of MIP-based sensing matters as an effective analytical tool to find analytic characterization for biosensor transductions. The following Table 1 summarizes recently presented works on electrochemical sensors that detect various molecules.

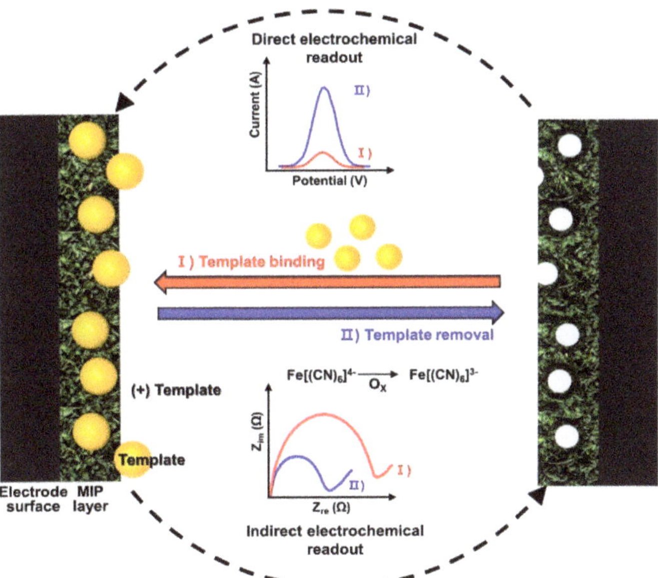

Figure 5. The basic concept of the electrochemical transduction on the MIP-based sensors using voltammetry test; the sensing procedure requires typical sequential steps, including the removal of the templates and rebinding of the target analytes on the MIP electrode. An electrochemical analytical method can be selected depending on the electroactive property of the analytes, which can be categorized either through direct detection (i.e., direct electrochemical readout) of the analyte or indirect detection (i.e., indirect electrochemical readout) using a redox probe, such as $[Fe(CN)6]^{3/4-}$.

Table 1. Summary of versatile approaches for detecting biomolecules using MIP-based biosensors.

	Template	Monomer	Form of MIPs	Electrochemical Techniques	Template Removal	Selectivity	LOD	Ref.
Protein	Human interleukin-1β (IL-1β)	EriochromeBlack T (EBT)	Carbon	EIS	0.1 M PBS/CV	IgG, Myo	1.5 pM	[86]
	Human interleukin-2 (IL-2)	MAA/MBA	CdTe QDs	Fluorescence measurements	.	.	5.91 fg mL^{-1}	[128]
	C-reactive proteins	O-4-nitrophenylphosphorylcholine (O-4NPPC)	Gold	Circular dichroism (CD) measurements	10% sodium dodecyl sulfate	.	.	[129]
	BSA	Raethyleneglycol diacrylate (TEGMPA), diacryloyl urea (DAU), ammonium persulphate (APS)	MWCNT	DPV	Methanol	.	.	[130]
	Thrombin	Acrylamide (AM), Methylenebisacrylamide (MBAA)	Hydrogels	Shrinking measurements	4.3 M GuCl/1.4 M NaCl.	BSA	1000 fM	[131]
Small molecule	2.4-dichlorophenoxyacetic acid (2.4-D)	PMMA (Poly(methyl methacrylate)	Gold	QCM	Acetic acid	Atrazine, Ametryn, BA	7.73 µg mL^{-1}	[62]
	Formaldehyde (HCHO)	TFMAA	Gold	QCM	.	HCl, HF	24.2 and 8.0 ppm	[132]
	Chloramphenicol	EriochromeBlack T (EBT)	Laser-induced graphene (LIG)	EIS	ACN solution	Amoxicillin/clavulanic acid (AMC), oxytetracycline (OTC), sodium sulfadiazine	0.62 nM	[133]
	Cortisol	Polypyrrole (PPy)-Prussian blue (PB)	Carbon	CA	0.1 M PBS/CV	Glucose, lactate, urea, ascorbic acid, acetaminophen, uric acid	0.9 and 0.2 nM	[134]
	Caffeine	Pyrrole (PPy)	Gold	EIS	FBS/resonance frequency	Theophylline	.	[135]
Virus	SARS-CoV-2 nucleoprotein	Poly-m-phenylenediamine (PmPD)	Gold-TFE	DPV	10% acetic acid solution	S1, E2 HCV CD48 and BSA	15 fM, −50 fM	[107]
	SARS-CoV-2 nucleoprotein	Dithiothreitol (DTT)	Gold	SWV, CV	10% acetic acid	IgG, E2, HSA	15 fM −64 fM	[136]

3.3. Practical Uses of MIP-Based Biosensors: Urgent Demand and Immediate Contribution

Disposable POC diagnostic devices have supported the patient-centered healthcare system. With rapid signal acquisition, some device platforms are convenient and cost effective in assisting clinicians and patients with reasonable diagnostic coverage. However, in some cases, these diagnostic devices have inevitable limitations, such as short shelf life to keep the sensors refrigerated in transport and storage because of the antibodies and receptors embedded in the disposable kits. Another potential issue can concern the chemical conditions when performing analyte detection by changing the conditions of pH, saltwater or highly reactive organic solvents, as previously described. Notably, as an innovative artificial 'plastic' receptor, the MIP-based POC diagnostic platform offers high stability and can be stored at room temperature, operating in challenging physical and chemical environments to expand its applications.

As one example of the disposable POC diagnostic platform, Figure 6a illustrates recently reported MIP-based biosensor by utilizing a typical three-terminal electrode, in which the MIP working electrode was prepared by an electrochemical synthesis and plated with a porous membrane (PVA hydrogel) to impregnate PB (Prussian blue) redox probe. In this configuration, endogenous cortisol levels were detected from the sweat sampling. In this approach, a cheap screen-printed electrode (SPE) was covered with a porous MIP membrane containing a PB redox probe. This biosensor can readily detect the exposed cortisol analytes in sweat upon fingertip palpation through a specific rebinding on a cortisol-imprinted polymeric membrane (i.e., PVA hydrogel). The MIP membrane was found to be sensitive to the amperometric method by providing engraved cavities from the PB probe. Compared to the previous methods, such 'touch/incubate/detect' protocol is innovative in the development of the POC device, highlighted by rapid detection time (~3.5 min) in quantifying cortisol levels simply from a fingertip based on the current changes (Figure 6b). Since cortisol is linked to mental health, monitoring cortisol levels will be an important indicator of early detection of psychological conditions. The highly permeable liquid-absorbing porous membrane, according to a capillary reaction of sweat, can be extended to other examples of MIP-based biosensors to instantaneously collect other secreted biofluids from the human body [134,137,138]. Besides, the authors also demonstrated the expandability of MIP-based biosensors to a conformal epidermal-integrated patch by tracking changes in cortisol levels during on-body exercise, showing excellent performance after repeated use (60 cycles) for real-time cortisol monitoring (Figure 6c). The recent development of healthcare systems already moves to skin-attachable devices or even implantable electronic circuits [139–141], and thus, MIP-based biosensors will be able to be securely integrated with other platforms for potential application to meet the required high selectivity.

The worldwide pandemic situation by an infectious coronavirus disease (COVID-19), caused by the SARS-CoV-2 virus, accelerated the realistic MIP-based biosensing system [107,136,142]. Since rapid detection by using an MIP-based electronic device provides an easily accessible and stress-free approach, a portable electrochemical biosensor can obviously be useful for POC tests when molecularly imprinted with a SARS-CoV-2 nucleoprotein (i.e., ncovNP-MIP). As an example, a synthetic recognition element integrated biosensor for selective detection of ncovNP was recently reported, as presented in Figure 6d [143]. In their demonstration, the MIP-based COVID-19 sensor showed linear responses to the cavities from ncovNP in the apparent concentration range of 2.22–111 fM in the lysis buffer; within the measured values, LOD and LOQ were valued as 15 and 50 fM, respectively. In this straightforward approach, the ncovNP-imprinted biosensor could transduce the signals from the specific rebinding of the ncovNP in nasopharyngeal swab samples collected from COVID-19-positive patients. This demonstration implies that the MIP-based virus sensors have a great potential to extend their application to other infectious mutant viruses for rapid diagnostic tools as POCT kits [144]. Referring to the recently reported COVID-19 POCT with a detection performance within ~2 min (LOD: ~5 fg mL^{-1}), it may be an efficient assessment to simply measure current changes through optimized settings in the MIP-integrated electrode [136].

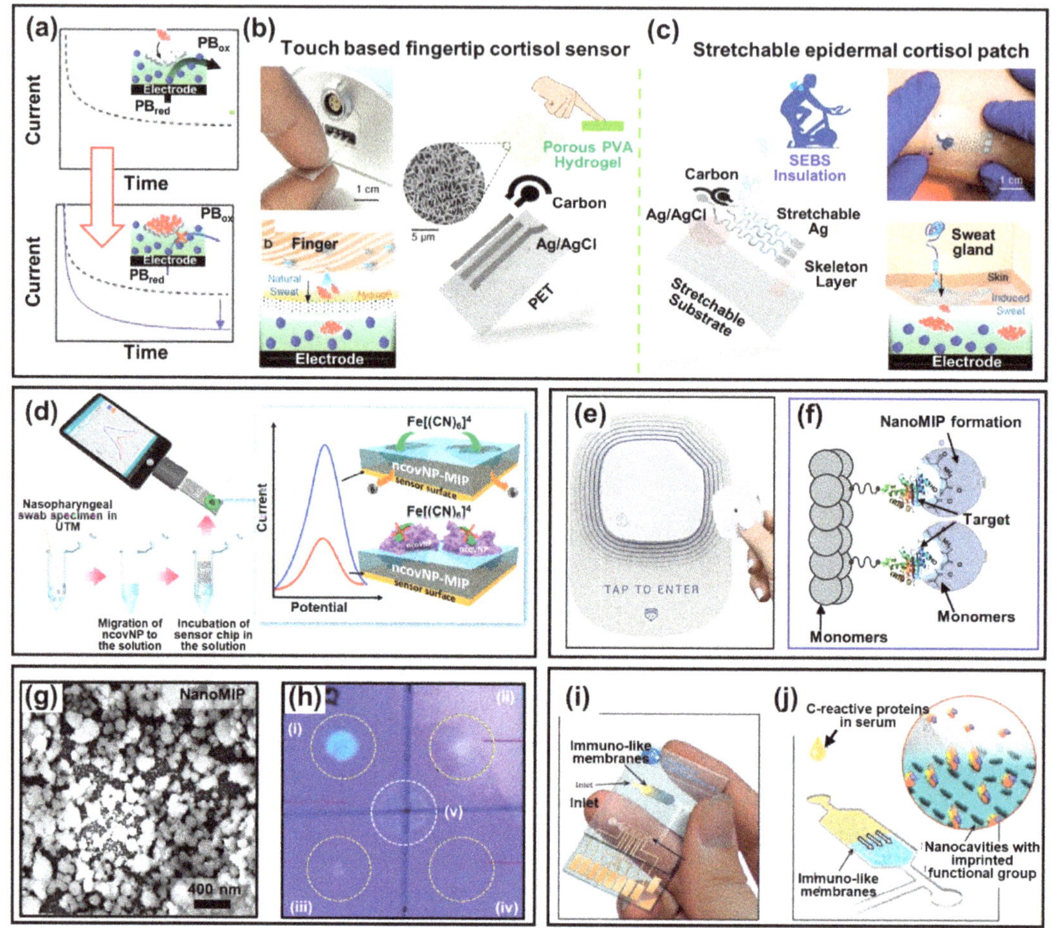

Figure 6. Disposable POC diagnostic biosensors. (**a**) The stress-free sensing platform to detect endogenous cortisol levels from the sweat sampling in an electrode structure with a porous MIP membrane; the reduced current can be measured by oxidation of the imprinted PB by cortisol rebinding process. Reproduced with permission from [133]. Copyright Wiley–VCH, 2021. (**b**) Schematic diagram of MIP-based cortisol sensor based on 'touch/incubate/detect' protocol. Reproduced with permission from [133]. Copyright Wiley–VCH, 2021. (**c**) A demonstration of the MIP-based biosensor for a conformal epidermal-integrated patch that tracks changes in cortisol levels during on-body exercise. Reproduced with permission from [133]. Copyright Wiley–VCH, 2021. (**d**) A mobile COVID-19 diagnostic system equipped with a disposable MIP-based biosensor. Reproduced with permission from [106]. Copyright Elsevier, 2021. (**e**,**f**) An image and concept of commercialized COVID-19 biosensor produced by solid-phase polymerization for nanoMIP to capture SARS-CoV-2. (**g**) SEM image of the nanoMIP film surface. (**h**) Optical micrographs captured from the dot blot assay for SARS-CoV-2 detection; (i) SARS-CoV-2 full-length spike protein trimer (positive control), (ii,iii) test position (2×10^4 PFU), (vi) viral culture media only (negative control), (v) CPN 510B (reference signal). (**i**,**j**) Scheme of the MIP-based POCT device based on immuno-polymeric membranes with a confined fluidic flow and defined electrode array to isolate CRPs in human serum samples. Reproduced with permission from [128]. Copyright Elsevier, 2013.

As a similar MIP approach in this categorized application, MIP-based POCT outperforms commercially used antibodies for a biosensor platform by revealing an improved detection capability with lower viral loads during an extended infection period (Figure 6e) [138]. By a designed polymerization process with SARS-CoV-2, precise molecular imprinting could be successfully performed for the receptor-binding domain (RBD) region by mimicking the spike glycoprotein (i.e., target), as described in Figure 6f. A SEM image in Figure 6g shows the surface morphology of the crafted *nanoMIPs* (i.e., nanoparticle-featured surface imprinting). In this experiment, a solid-phase imprinting process was used to promote intimate chemical interactions between the template and functional monomers with stoichiometric chemical moieties during the immobilization process (Figure 6h). For detecting SARS-CoV-2, the nanoMIP was combined with fluorescent polymeric nanoparticles (FPNs) to visualize the virus recognition capability simply by using a dot blot assay, as shown in Figure 6i; these FPN-integrated MIPs yielded significantly brighter signals (i.e., 10,000 times higher level) than other samples [145]. The measured areas coated with nanoMIP film are shown as follows: (i) positive controls for SARS-CoV-2 spike protein; (ii) and (iii) a SARS-CoV-2 capture region; (iv) negative control with virus culture medium only; v) reference control. The imaged dot blot arrays were able to selectively detect SARS-CoV-2 and reported a low LOD value of 5 fg mL^{-2}. By further selectivity evaluation, the SARS-CoV-2-imprinted biosensing platform only recognized SARS-CoV-2 spike glycoprotein in the dot blot assay, whereas no responses with the human coronavirus spike glycoprotein (299E, HKU1, OC43) were detected. Supported by a reliable scale-up manufacturing process, this manipulated nanoMIP platform may give rise to an impact on the regular diagnosis for quick check-up of COVID-19 in hospitals, drive-through sites or at home, as an effective POCT kit. Indeed, the progressive type of nanoparticle-based MIPs (i.e., nanoMIP for a single species) could extend their POCT applications to other target molecules, such as enzymes or proteins, because the system provides more selective and specific rebinding sites for high accuracy in diagnostic testing.

Figure 6i displays a novel MIP-based POCT device for protein recognition based on an immune-polymeric membrane used to isolate C-reactive proteins (CRPs) from serum samples. In their approach, the cavities structured in the MIP-integrated membrane were combined with a confined fluidic flow, interlocked on a defined electrode array. In particular, the biosensing performance was evaluated by the separation principle in a critically aligned configuration of CRPs on the working electrode, as drawn in Figure 6j. By this setting, the impedance changes were detected directly on the applied current, responding to the CRP rebinding reaction in the MIP-integrated membrane. Rapid detection of CRPs was evaluated within 2 min, starting with incubation of serum samples. Their biomimetic immuno-membrane manifests several advantages in the MIP-based biosensor technology by rendering receptors as biological sensing elements. Therefore, the electrochemical detection method is compatible with the structured MIP membrane that is addressed in the defined sensing area. With regard to its high compatibility with microfabrication processes, it is possible that other advanced techniques can be applied to 3D nanoporous vertical channels to engineer high specificity.

4. Concept of Oral POCT to Detect Diseases: Novel Detection in Salivary Biomarkers

The advantage of the user-friendly POCT as a wearable form is perfectly fit for new diagnostic concepts by detecting small molecules from the collected biofluid sampling, since that process does not require specialists or complicated treatment with medical equipment [134]. As is well known, saliva includes tremendous biomarkers, including substances secreted from salivary glands, external substances, microorganisms and blood-derived compounds, reflecting oral diseases or systemic diseases [146–148]. However, given the low concentration of biomarkers in saliva, effective detection can easily lead to false signals by contamination of external factors [149]. However, the continuous interest in molecule sensing from saliva has extended the research area in wearable device applications, from which in situ saliva analysis has been rapidly developed. Thus, several intuitive ideas have been suggested to minimize the contamination of saliva sample, divided mainly into a mouthguard platform for direct measurement of biomarkers from saliva in the oral cavity or external sensing with a microfluidic system as IVD devices. In this final section, we summarize the recent development of wearable oral biosensing devices for detecting a set of biomarkers in saliva and conclude with the proposal of MIP-integrated biosensing platform as a promising approach in the same categorized study.

Biosensors mounted on mouthguards are straightforward as one good example of the POCT approach. Recently, as shown in Figure 7a, Kim et al. presented an integrated wireless mouthguard to sense salivary metabolites based on an amperometric sensing platform to detect uric acid (UA) in diluted saliva for smart healthcare monitoring [150]. The amperometric enzyme electrode is the oldest platform since the first introduction of the glucose biosensor by Clark in 1962. Briefly, the detection of ion presence on solution based on electric current or changes in electric current has been called 'amperometry'. As addressed in Section 3.1, an amperometric biosensor induces a current proportional to the concentration of the substance to be detected. In line with this electrochemical approach for a practical wearable device, the wireless mouthguard biosensor was integrated with a Bluetooth-enabled circuit board, built on flexible PET (polyethylene terephthalate) substrate. A biosensor embedded as a wearable mouthguard was firstly demonstrated by utilizing an SPE transducer in a flexible format and mounted in mouthguard preform [151]. In detail, they used a simple chemical modification on the commonly used screen-printed working electrode by electropolymerized oPD and simultaneous crosslinking of the uricase enzyme. Therefore, a soft mouthguard was facilitated for continuous monitoring of UA in saliva. The wearable device was configured with an amperometric transducer and coupled by wireless communication systems (i.e., Bluetooth), which were readily integrated into a system on chip as a singular product. The saliva sample could be collected from the mouth for real-time sensing, and the current signals were extracted in the continuous operation for 4 h in the monitoring process with 10 min intervals with a stability of the electrochemical response of 300 mM UA. Within the optimized experimental condition, they reported a current response of every 0.5 s at a 2 Hz frequency with a sensitivity of 2.45 mA mM^{-1}. Accordingly, the result from the wireless mouthguard type salivary UA sensor enabled the transfer of data measured from real-time detection. This new concept of biosensors offered an attractive electrochemical sensing platform with high sensitivity and selectivity. The 'in mouth' mounting in the human body still requires a critical assessment of biocompatibility with less toxic materials that are essential for the realization of wearable electronic devices [152].

As a similar approach presented in Figure 7b, a customized mouthguard-type biosensor was also introduced by Arakawa et al. [153]. This wearable oral POCT, so-called 'cavity sensor', was produced on a plastic substrate (i.e., polyethylene terephthalate glycol, PETG) to be a mountable oral cavity for non-invasive saliva analysis. The presented biosensor consisted of Pt, Ag/AgCl and glucose oxidase (GOD) electrodes, immobilized by poly (MPC-co-EHMA) (PMEH) for glucose monitoring. With these configured electrodes, the output current produced by glucose oxidation at GOD was measured by the amperometric method as a function of the concentration of H$_2$O$_2$. The characteristic sensing performance in artificial saliva was set for 1.0 wt% PMEH matrix with an electrode surface area of 16.8 mm^2 to measure the glucose, showing high selectivity in current output only for glucose, compared to other saliva analytes, 100 mM L^{-1} of galactose, sorbitol, fructose, mannitol and xylitol solution. Moreover, this GOD-based biosensor exhibited high sensitivity in PBS and artificial saliva ranging from 0.05 to 1.0 mM L^{-1} under a wireless condition in continuous real-time data collection. Therefore, advanced oral biosensors will be helpful in non-invasive monitoring systems for diabetic patients. Although there may be differences in glucose levels between blood and saliva samples, a personalized POCT device for glucose sensing can be a promising approach for further development into a nontoxic and safe mouthguard-type platform [154].

Another alternative to in vivo oral monitoring devices has been introduced to estimate sodium intake. As illustrated in Figure 7c, Lee et al. presented a novel biosensor with a customized dental brace composed of a biocompatible elastomer that can measure sodium ion concentration in direct contact with the oral cavity [155]. The active electrode for the biosensor was fully embedded in a microporous structured elastomer (i.e., Soma Foama 15, SF15) to package an integrated circuit system equipped with stretchable interconnects. Thus, the hybrid form of bioelectronic device enabled conformal contact with intraoral tissues. Owing to the high permeability based on the breathable SF15 membrane, the sodium ion sensor is perfectly suited for detecting sodium intake in the oral cavity with high selectivity and sensitivity. By optimizing impedance matching to the circuits incorporated with the porous membrane, real-time quantification of sodium intake was realized for wireless data transfer. This sodium sensor in the oral cavity showed clear changes of electrical signals as a function of the sodium concentration ranges, such as 10^{-4}, 10^{-3}, 10^{-2}, 10^{-1} and 1 M, under in vitro experimental condition, demonstrating high sensitivity and selectivity; the level of output voltage was also compatible with the result from in vitro experimental detection capability. In particular, soft electronics with user-friendly interface can take advantage of favorable physical properties to actively collect information about not only sodium intake but also dietary habits and health management [156].

In general, the geometry of the oral cavity varies from person to person. Thus, an externally configured sensing element may be preferred when direct contact in the mouth is limited due to the lack of teeth or the discomfort that would permit using biosensors [157]. As proof of concept, a biosensor-embedded pacifier was designed for wearable oral POCT devices to monitor the glucose concentration levels in saliva, as shown in Figure 7d [158]. In this demonstration, the pacifier made from nontoxic silicone served a useful function as a fluidic saliva collector, integrated with an SPE-based biosensor for electrochemical detection, in which an amperometric circuit was connected with a miniaturized wireless data transfer module. Since the presented POCT device contained a saliva-fluidic channel (i.e., biofluid reservoir), this unique design helped the biosensor operation by soaking the exposed working electrode in unidirectional saliva flow without any suction pressure, allowing the sensor to detect glucose based on the glucose oxidase modified PB transducer. The electrochemical method with functional electrode offered good selectivity for glucose with no responses to 200 µM of UA and 20 µM of ascorbic acid (AA) and without any interfering crosstalk from substances left in the mouth. When the sensitivity was scanned, the current changes in glucose concentration in artificial saliva exhibited a well-defined ranged concentration between 0.1 and 1.4 mM. The PB-oxidase electrode proved to be suitable for monitoring glucose concentration by monitoring the correlation between blood and saliva glucose before and after food intake. Aside from the instability of the enzyme-based electrode and incomplete sterilization with each use, the only minor limit of the pump-free pacifier-based sensing platform was the delayed duration of saliva collection time to reach the exposed working electrode with signal stabilization. However, assisted by this progressive work, additional study on the choices of effective design and more efficient instrumentation of microfluidic channels will pave the way for a viable route for the development of wearable oral POCT.

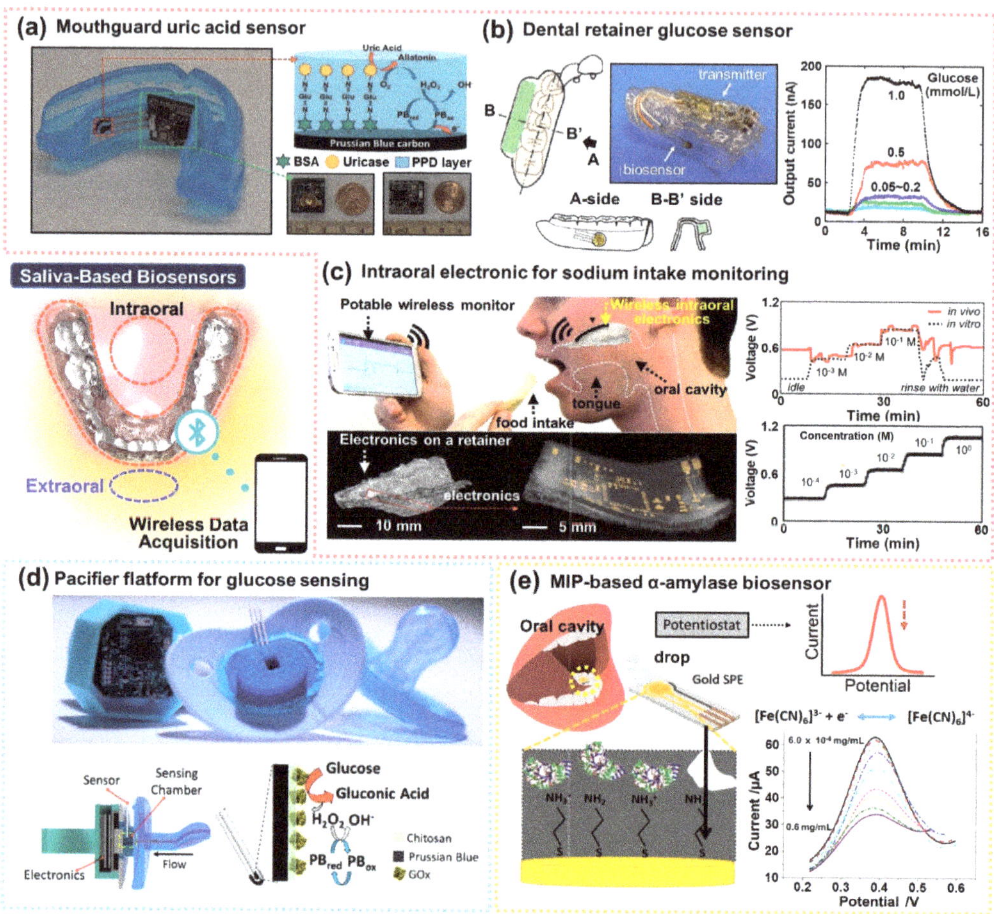

Figure 7. Concept of non-invasive biosensing platforms mounted in the oral cavity for real-time monitoring by wireless communication to detect molecules (middle left image). (**a**) A mouthguard integrated with a wireless circuit board for monitoring salivary UA and the chemically modified SPE sensor for electrochemical detection. Reproduced with permission from [150]. Copyright Elsevier, 2015. (**b**) Customized glucose biosensor mounted on a plastic retainer with a detection range of 0.05–1.0 mM L^{-1} in artificial saliva. Reproduced with permission from [153]. Copyright Elsevier, 2016. (**c**) A wireless electronic device mounted on a retainer with a permeable porous membrane that detects sodium ions. The graphs show clear changes in the electrical signals according to sodium concentration based on in vitro and in vivo experiments. Reproduced with permission from [155]. Copyright National academy of sciences, 2018. (**d**) A pacifier-type glucose biosensor for external monitoring of saliva collected without any pump through the fluidic channel. Reproduced with permission from [158]. Copyright American Chemical Society, 2019. (**e**) Scheme of α-amylase imprinted biosensing platform on Au-SPE electrode to quantify the concentration range of 6.0×10^{-6}– 0.6 mg mL^{-1} through the electrochemical method. Reproduced with permission from [83].

In this section, we have discussed various approaches for salivary biomarker detecting biosensors and PCOT devices in a wearable format. As a similar but slightly different approach, MIP technology as artificial receptors is ready to be combined together with the intraoral POCT devices and amperometric molecular recognition system. Although it has not yet been reported in this field of research, when the MIP system can be directly applied to a biosensing system, especially in an oral disease monitoring system, it could have far-reaching implications and clearly stand out by resolving some drawbacks with improved performance. Since some approaches on the protein-based MIP on the SPE surface have been reported previously for detecting salivary protein as a combinatorial result of the electrochemical method [159–161], it is easily possible to transform the strategies on the bioactive electrode preparation and the selection of suitable materials for nontoxic packaging. A recent α-amylase imprinted biosensor is a representative example of stress-related healthcare monitoring for potential POCT (Figure 7e). In this case, a typical amperometric transducer was used for the ranged quantification of the α-amylase concentration by the utilization of Au-SPE electrodes, at which the surface of the working electrode was electropolymerized with conductive pyrrole and α-amylase. With an accurately controlled Au surface using a cysteamine self-assembled monolayer, the α-amylase template was effectively immobilized. In this MIP process, the biomarker (i.e., α-amylase) can leave behind highly specific cavities after the removal of the templates in the polypyrrole matrix on the electrode (refer to Section 2.4). Next, a typical electroanalysis was conducted using one of the pulse techniques, that is, square wave voltammetry (SWV) [162]. In this discriminated sensing, the MIP biosensor to capture a specific molecule from oral biofluid represented outstanding performance in the α-amylase concentration range from 6.0×10^{-6} to 0.60 mg mL^{-1} in buffer solution with high sensitivity (LOC $< 3.0 \times 10^{-4}$ mg mL^{-1}). Due to the nature of the MIP-based sensing system [163], it showed high sensitivity and selectivity through the rebinding process only on target biomolecules in human saliva and in a buffer solution containing other biomarkers. Conclusively, this MIP biosensor exhibited analytical capabilities as a promising candidate for diagnostic POCT devices when integrated with sophisticated electrodes and a real-time wireless system [83].

5. Conclusions and Outlook

Due to the outbreak of COVID-19, there has been an increasing interest in technology and product developments in the field of molecular diagnostic POCT with the highest accuracy and rapid identification. From a long-term perspective, significant technological progress for on-site medical treatment will persist, and a new format of POC devices and related techniques will be introduced to the market due to the change in work methods caused by the COVID-19 and the generalization of telecommuting. In this context, this review discussed and summarized the engineering innovations of POCT devices equipped with MIP-based biosensing systems. In particular, the integration of MIP-based electrochemical biosensors with POCT-based diagnostic systems can be expected to be adaptable in IVD devices for an accurate and selective detection of biomarkers caused by various diseases, not only for viral proteins.

Up to date, there has been a large set of advances in developing various types of MIP-based POC devices in biomedical diagnostics for emergency assessment. The detection of biomarkers at trace levels in biofluids using the MIP-based biosensing device may provide highly qualified artificial receptors for biologically specific and selective recognition together with an easy manufacturing process. Therefore, with the strong demand for non-invasive and rapid evaluation, the use of MIP-based biosensors offers new opportunities in disease identification and risk-resolving assessment by managing the fundamental preservation of disease-specific biomarker-imprinted cavities in POC devices, free from shelf life, cost effectiveness and storage issues. In addition, in the categorized research field, the detection of biomarkers in periodontology remains limited overall. Artificial antibody-based approaches highlighted in this review imply great potential in dental care and medicine. Because chronic immune-inflammatory responses to microbial biofilms formed in the oral tissue or dental implanted surface are in systemic conditions, such as cardiovascular disease, diabetes or gastrointestinal diseases, the detection of the secreted biomarkers will be important guidelines for oral and other systemic diseases, including identifying SARS-CoV-2 in saliva [164]. Advances in electrochemical biosensing platforms with appropriate transducing systems have recently accomplished remarkable innovations in interfaced wearable electronics with biomarker assessment [110]. As a future direction, our suggested examples highlight MIP-integrated devices in oral biomarker detection and their validation for clinical POCT to advance precision medicine. This suggests that the potential of collecting clinical outcomes from wearable biometric interfaces relevant to chronic prognosis will benefit public health.

Author Contributions: R.P., S.J., J.J. and S.W.H. developed the idea and structure of the review article. R.P., S.J. and J.J. wrote the original manuscript using the materials supplied by S.W.H., S.-Y.P., D.-W.H. and S.W.H. who revised and improved the manuscript. D.-W.H. and S.W.H. supervised the manuscript. All authors have read and agreed to the published version of the manuscript.

Funding: This work was supported by the Korea Medical Device Development Fund grant funded by the Korean government (the Ministry of Science and ICT, the Ministry of Trade, Industry and Energy, the Ministry of Health and Welfare, the Ministry of Food and Drug Safety) (NTIS Number: 9991006781, KMDF_PR_20200901_0108). This work was also supported by the National Research Foundation of Korea (NRF) Grant funded by the Korean government (MSIT) (NRF-2021R1A5A1032937).

Institutional Review Board Statement: Not applicable.

Informed Consent Statement: Not applicable.

Data Availability Statement: Not applicable.

Conflicts of Interest: The authors declare no conflict of interest.

References

1. Gubala, V.; Harris, L.F.; Ricco, A.J.; Tan, M.X.; Williams, D.E. Point of Care Diagnostics: Status and Future. *Anal. Chem.* **2012**, *84*, 487–515. [CrossRef]
2. Gervais, L.; De Rooij, N.; Delamarche, E. Microfluidic Chips for Point-of-Care Immunodiagnostics. *Adv. Mater.* **2011**, *23*, H151–H176. [CrossRef]
3. Nayak, S.; Blumenfeld, N.R.; Laksanasopin, T.; Sia, S.K. Point-of-Care Diagnostics: Recent Developments in a Connected Age. *Anal. Chem.* **2017**, *89*, 102–123. [CrossRef]
4. Rohr, U.P.; Binder, C.; Dieterle, T.; Giusti, F.; Messina, C.G.M.; Toerien, E.; Moch, H.; Schäfer, H.H. The Value of in Vitro Diagnostic Testing in Medical Practice: A Status Report. *PLoS ONE* **2016**, *11*, e0149856. [CrossRef]
5. Vashist, S.K. In Vitro Diagnostic Assays for COVID-19: Recent Advances and Emerging Trends. *Diagnostics* **2020**, *10*, 202. [CrossRef]
6. Oyewole, A.O.; Barrass, L.; Robertson, E.G.; Woltmann, J.; O'Keefe, H.; Sarpal, H.; Dangova, K.; Richmond, C.; Craig, D. COVID-19 Impact on Diagnostic Innovations: Emerging Trends and Implications. *Diagnostics* **2021**, *11*, 182. [CrossRef]
7. St John, A.; Price, C.P. Existing and Emerging Technologies for Point-of-Care Testing. *Clin. Biochem. Rev.* **2014**, *35*, 155.
8. Yang, Y.; Song, Y.; Bo, X.; Min, J.; Pak, O.S.; Zhu, L.; Wang, M.; Tu, J.; Kogan, A.; Zhang, H.; et al. A Laser-Engraved Wearable Sensor for Sensitive Detection of Uric Acid and Tyrosine in Sweat. *Nat. Biotechnol.* **2020**, *38*, 217–224. [CrossRef]
9. Chen, J.; Zhu, X.; Ju, Y.; Ma, B.; Zhao, C.; Liu, H. Electrocatalytic Oxidation of Glucose on Bronze for Monitoring of Saliva Glucose using a Smart Toothbrush. *Sens. Actuators B Chem.* **2019**, *285*, 56–61. [CrossRef]
10. Michael, I.; Kim, D.; Gulenko, O.; Kumar, S.; Clara, J.; Ki, D.Y.; Park, J.; Jeong, H.Y.; Kim, T.S.; Kwon, S.; et al. A Fidget Spinner for the Point-of-Care Diagnosis of Urinary Tract Infection. *Nat. Biomed. Eng.* **2020**, *4*, 591–600. [CrossRef]
11. Gan, S.D.; Patel, K.R. Enzyme Immunoassay and Enzyme-Linked Immunosorbent Assay. *J. Invest. Dermatol.* **2013**, *133*, e12. [CrossRef] [PubMed]
12. Sakamoto, S.; Putalun, W.; Vimolmangkang, S.; Phoolcharoen, W.; Shoyama, Y.; Tanaka, H.; Morimoto, S. Enzyme-Linked Immunosorbent Assay for the Quantitative/Qualitative Analysis of Plant Secondary Metabolites. *J. Nat. Med.* **2018**, *72*, 32–42. [CrossRef] [PubMed]
13. Lavigne, J.J.; Anslyn, E.V. Sensing a Paradigm Shift in the Field of Molecular Recognition: From Selective to Differential Receptors. *Angew. Chem. Int. Ed.* **2001**, *40*, 3118–3130. [CrossRef]
14. Omar, S.A.; Thomas, S.B.; Cem, E.; Alvaro, G.-C.; Sergey, A.P. Molecularly Imprinted Polymers in Electrochemical and Optical Sensors. *Sens. Actuators B Chem.* **2019**, *37*, 294–309.
15. Cai, D.; Ren, L.; Zhao, H.; Xu, C.; Zhang, L.; Yu, Y.; Wang, H.; Lan, Y.; Roberts, M.F.; Chuang, J.H.; et al. A Molecular-Imprint Nanosensor for Ultrasensitive Detection of Proteins. *Nat. Nanotechnol.* **2010**, *5*, 597–601. [CrossRef]
16. Kajisa, T.; Sakata, T. Molecularly Imprinted Artificial Biointerface for an Enzyme-Free Glucose Transistor. *ACS Appl. Mater. Interfaces* **2018**, *10*, 34983–34990. [CrossRef]
17. Deng, J.; Chen, S.; Chen, J.; Ding, H.; Deng, D.; Xie, Z. Self-Reporting Colorimetric Analysis of Drug Release by Molecular Imprinted Structural Color Contact Lens. *ACS Appl. Mater. Interfaces* **2018**, *10*, 34611–34617. [CrossRef]

18. Mugo, S.M.; Alberkant, J. Flexible Molecularly Imprinted Electrochemical Sensor for Cortisol Monitoring in Sweat. *Anal. Bioanal. Chem.* **2020**, *412*, 1825–1833. [CrossRef]
19. Vlatakis, G.; Andersson, L.I.; Müller, R.; Mosbach, K. Drug Assay Using Antibody Mimics Made by Molecular Imprinting. *Nature* **1993**, *361*, 645–647. [CrossRef]
20. Liu, J.; Deng, Q.; Tao, D.; Yang, K.; Zhang, L.; Liang, Z.; Zhang, Y. Preparation of Protein Imprinted Materials by Hierarchical Imprinting Techniques and Application in Selective Depletion of Albumin from Human Serum. *Sci. Rep.* **2014**, *4*, 1–6. [CrossRef]
21. Culver, H.R.; Peppas, N.A. Protein-Imprinted Polymers: The Shape of Things to Come? *Chem. Mater.* **2017**, *29*, 5753–5761. [CrossRef] [PubMed]
22. Zhang, Y.; Qu, X.; Yu, J.; Xu, L.; Zhang, Z.; Hong, H.; Liu, C. ^{13}C NMR Aided Design of Molecularly Imprinted Adsorbents for Selectively Preparative Separation of Erythromycin. *J. Mater. Chem. B* **2014**, *2*, 1390–1399. [CrossRef] [PubMed]
23. Zarycz, M.N.C.; Guerra, C.F. NMR ^{1}H-Shielding Constants of Hydrogen-Bond Donor Reflect Manifestation of the Pauli Principle. *J. Phys. Chem. Lett.* **2018**, *9*, 3720–3724. [CrossRef] [PubMed]
24. Piletska, E.V.; Guerreiro, A.R.; Romero-Guerra, M.; Chianella, I.; Turner, A.P.F.; Piletsky, S.A. Design of Molecular Imprinted Polymers Compatible with Aqueous Environment. *Anal. Chim. Acta* **2008**, *607*, 54–60. [CrossRef] [PubMed]
25. Yang, J.C.; Shin, H.-K.; Hong, S.W.; Park, J.Y. Lithographically Patterned Molecularly Imprinted Polymer for Gravimetric Detection of Trace Atrazine. *Sens. Actuators B Chem.* **2015**, *216*, 476–481. [CrossRef]
26. Doué, M.; Bichon, E.; Dervilly-Pinel, G.; Pichon, V.; Chapuis-Hugon, F.; Lesellier, E.; West, C.; Monteau, F.; Le Bizec, B. Molecularly Imprinted Polymer Applied to the Selective Isolation of Urinary Steroid Hormones: An Efficient Tool in the Control of Natural Steroid Hormones Abuse in Cattle. *J. Chromatogr. A* **2012**, *1270*, 51–61. [CrossRef]
27. Rachkov, A.; Minoura, N. Towards Molecularly Imprinted Polymers Selective to Peptides and Proteins. The Epitope Aproach. *Biochim. Biophys. Acta (BBA)-Protein Struct. Mol. Enzymol.* **2001**, *1544*, 255–266. [CrossRef]
28. Arabi, M.; Ostovan, A.; Bagheri, A.R.; Guo, X.; Wang, L.; Li, J.; Chen, L. Strategies of Molecular Imprinting-Based Solid-Phase Extraction Prior to Chromatographic Analysis. *Trends Analyt. Chem.* **2020**, *128*, 115923. [CrossRef]
29. Wulff, G. Enzyme-Like Catalysis by Molecularly Imprinted Polymers. *Chem. Rev.* **2002**, *102*, 1–28. [CrossRef]
30. Zaidi, S.A. Latest Trends in Molecular Imprinted Polymer Based Drug Delivery Systems. *RSC Adv.* **2016**, *6*, 88807–88819. [CrossRef]
31. Sánchez-González, J.; Odoardi, S.; Bermejo, A.M.; Bermejo–Barrera, P.; Romolo, F.S.; Moreda–Piñeiro, A.; Strano-Rossi, S. HPLC-MS/MS Combined with Membrane-Protected Molecularly Imprinted Polymer Micro-Solid-Phase Extraction for Synthetic Cathinones Monitoring in Urine. *Drug Test. Anal.* **2019**, *11*, 33–44. [CrossRef] [PubMed]
32. Regal, P.; Díaz-Bao, M.; Barreiro, R.; Cepeda, A.; Fente, C. Application of Molecularly Imprinted Polymers in Food Analysis: Clean-Up and Chromatographic Improvements. *Open Chem.* **2012**, *10*, 766–784. [CrossRef]
33. Turner, N.W.; Liu, X.; Piletsky, S.A.; Hlady, V.; Britt, D.W. Recognition of Conformational Changes in β-lactoglobulin by Molecularly Imprinted Thin Films. *Biomacromolecules* **2007**, *8*, 2781–2787. [CrossRef] [PubMed]
34. Sullivan, M.V.; Dennison, S.R.; Archontis, G.; Reddy, S.M.; Hayes, J.M. Toward Rational Design of Selective Molecularly Imprinted Polymers (MIPs) for Proteins: Computational and Experimental Studies of Acrylamide Based Polymers for Myoglobin. *J. Phys. Chem.* **2019**, *123*, 5432–5443. [CrossRef]
35. Schirhagl, R. Bioapplications for molecularly imprinted polymers. *Anal. Chem.* **2014**, *86*, 250–261. [CrossRef]
36. Cieplak, M.; Kutner, W. Artificial Biosensors: How Can Molecular Imprinting Mimic Biorecognition? *Trends Biotechnol.* **2016**, *34*, 922–941. [CrossRef]
37. Turner, A.P.F.; Karube, I.; Wilson, G.S. *Biosensors: Fundamentals and Applications*; Oxford University Press: Oxford, UK, 1987.
38. Aćimović, S.S.; Šípová-Jungová, H.; Emilsson, G.; Shao, L.; Dahlin, A.B.; Käll, M.; Antosiewicz, T.J. Antibody–Antigen Interaction Dynamics Revealed by Analysis of Single-Molecule Equilibrium Fluctuations on Individual Plasmonic Nanoparticle Biosensors. *ACS Nano* **2018**, *12*, 9958–9965. [CrossRef]
39. Sharma, S.; Byrne, H.; O'Kennedy, R.J. Antibodies and Antibody-Derived Analytical Biosensors. *Essays Biochem.* **2016**, *60*, 9–18.
40. Qin, Z.; Peng, R.; Baravik, I.K.; Liu, X. Fighting COVID-19: Integrated Micro-and Nanosystems for Viral Infection Diagnostics. *Matter* **2020**, *3*, 628–651. [CrossRef]
41. Bandodkar, A.J.; Gutruf, P.; Choi, J.; Lee, K.; Sekine, Y.; Reeder, J.T.; Rogers, J.A. Battery-Free, Skin-Interfaced Microfluidic/Electronic Systems for Simultaneous Electrochemical, Colorimetric, and Volumetric Analysis of Sweat. *Sci. Adv.* **2019**, *5*, eaav3294. [CrossRef]
42. Kim, J.; Kim, M.; Lee, M.S.; Kim, K.; Ji, S.; Kim, Y.T.; Park, J.U. Wearable Smart Sensor Systems Integrated on Soft Contact Lenses for Wireless Ocular Diagnostics. *Nat. Commun.* **2017**, *8*, 1–8. [CrossRef] [PubMed]
43. Warren, A.D.; Kwong, G.A.; Wood, D.K.; Lin, K.Y.; Bhatia, S.N. Point-of-Care Diagnostics for Noncommunicable Diseases using Synthetic Urinary Biomarkers and Paper Microfluidics. *Proc. Natl. Acad. Sci. USA* **2014**, *111*, 3671–3676. [CrossRef] [PubMed]
44. Steigmann, L.; Maekawa, S.; Sima, C.; Travan, S.; Wang, C.W.; Giannobile, W.V. Biosensor and Lab-on-a-Chip Biomarker-Identifying Technologies for Oral and Periodontal Diseases. *Front. Pharmacol.* **2020**, *11*, 1663. [CrossRef] [PubMed]
45. Selvolini, G.; Marrazza, G. MIP-Based Sensors: Promising New Tools for Cancer Biomarker Determination. *Sensors* **2017**, *17*, 718. [CrossRef] [PubMed]
46. Ge, S.; Ge, L.; Yan, M.; Song, X.; Yu, J.; Huang, J. A Disposable Paper-Based Electrochemical Sensor with an Addressable Electrode Array for Cancer Screening. *Chem. Commun.* **2012**, *48*, 9397–9399. [CrossRef]

47. Martín-Yerga, D.; Álvarez-Martos, I.; Blanco-López, M.C.; Henry, C.S.; Fernández-Abedul, M.T. Point-of-Need Simultaneous Electrochemical Detection of Lead and Cadmium Using Low-Cost Stencil-Printed Transparency Electrodes. *Anal. Chim. Acta* **2017**, *981*, 24–33. [CrossRef]
48. Kumar, S.; Nehra, M.; Khurana, S.; Dilbaghi, N.; Kumar, V.; Kaushik, A.; Kim, K.-H. Aspects of Point-of-Care Diagnostics for Personalized Health Wellness. *Int. J. Nanomed.* **2021**, *16*, 383–402. [CrossRef]
49. Gouda, M.D.; Kumar, M.A.; Thakur, M.S.; Karanth, N.G. Enhancement of Operational Stability of an Enzyme Biosensor for Glucose and Sucrose Using Protein Based Stabilizing Agents. *Biosens. Bioelectron.* **2002**, *17*, 503–507. [CrossRef]
50. Yarman, A.; Kurbanoglu, S.; Zebger, I.; Scheller, F.W. Simple and Robust: The Claims of Protein Sensing by Molecularly Imprinted Polymers. *Sens. Actuators B Chem.* **2021**, *330*, 129369. [CrossRef]
51. Xu, J.; Miao, H.; Wang, J.; Pan, G. Molecularly Imprinted Synthetic Antibodies: From Chemical Design to Biomedical Applications. *Small* **2020**, *16*, 1906644. [CrossRef]
52. Matharu, Z.; Bandodkar, A.J.; Gupta, V.; Malhotra, B.D. Fundamentals and Application of Ordered Molecular Assemblies to Affinity Biosensing. *Chem. Soc. Rev.* **2012**, *41*, 1363–1402. [CrossRef]
53. Svenson, J.; Nicholls, I.A. On the Thermal and Chemical Stability of Molecularly Imprinted Polymers. *Anal. Chim. Acta* **2001**, *435*, 19–24. [CrossRef]
54. Yan, H.; Sun, N.; Han, Y.; Yang, C.; Wang, M.; Wu, R. Ionic Liquid-mediated Molecularly Imprinted Solid-phase Extraction Coupled with Gas Chromatography-electron Capture Detector for Rapid Screening of Dicofol in Vegetables. *J. Chromatogr. A* **2013**, *1307*, 21–26. [CrossRef]
55. Cavagnero, S.; Debe, D.A.; Zhou, Z.H.; Adams, M.W.; Chan, S.I. Kinetic role of electrostatic interactions in the unfolding of hyperthermophilic and mesophilic rubredoxins. *Biochemistry* **1998**, *37*, 3369. [CrossRef]
56. Sanchez-Ruiz, J.M.; Lopez-Lacomba, J.L.; Cortijo, M.; Mateo, P.L. Differential scanning calorimetry of the irreversible thermal denaturation of thermolysin. *Biochemistry* **1988**, *27*, 1648. [CrossRef]
57. Boonsriwong, W.; Chunta, S.; Thepsimanon, N.; Singsanan, S.; Lieberzeit, P.A. Thin Film Plastic Antibody-Based Microplate Assay for Human Serum Albumin Determination. *Polymers* **2021**, *13*, 1763. [CrossRef]
58. Yeo, C.; Kaushal, S.; Yeo, D. Enteric Involvement of Coronaviruses: Is Faecal–Oral Transmission of SARS-CoV-2 Possible? *Lancet Gastroenterol. Hepatol.* **2020**, *5*, 335–337. [CrossRef]
59. Smolinska-Kempisty, K.; Guerreiro, A.; Canfarotta, F.; Cáceres, C.; Whitcombe, M.J.; Piletsky, S.A. Comparison of the Performance of Molecularly Imprinted Polymer Nanoparticles for Small Molecule Targets and Antibodies in the ELISA Format. *Sci. Rep.* **2016**, *6*, 1–7. [CrossRef]
60. Kartal, F.; Çimen, D.; Bereli, N.; Denizli, A. Molecularly Imprinted Polymer Based Quartz Crystal Microbalance Sensor for the Clinical Detection of Insulin. *Mater. Sci. Eng. C* **2019**, *97*, 730–737. [CrossRef]
61. Cunliffe, D.; Kirby, A.; Alexander, C. Molecularly Imprinted Drug Delivery Systems. *Adv. Drug Deliv. Rev.* **2005**, *57*, 1836–1853. [CrossRef]
62. Lee, J.; Yang, J.C.; Lone, S.; Park, W.I.; Lin, Z.; Park, J.; Hong, S.W. Enabling the Selective Detection of Endocrine-Disrupting Chemicals via Molecularly Surface-Imprinted Coffee Rings. *Biomacromolecules* **2021**, *22*, 1523. [CrossRef]
63. Boitard, C.; Rollet, A.L.; Ménager, C.; Griffete, N. Surface-Initiated Synthesis of Bulk-Imprinted Magnetic Polymers for Protein Recognition. *Chem. Commun.* **2017**, *53*, 8846–8849. [CrossRef]
64. Sun, Y.; Chen, J.; Li, Y.; Li, H.; Zhu, X.; Hu, Y.; Huang, S.; Li, J.; Zhong, S. Bio-Inspired Magnetic Molecularly Imprinted Polymers Based on Pickering Emulsions for Selective Protein Recognition. *New J. Chem.* **2016**, *40*, 8745–8752. [CrossRef]
65. Yang, Y.Q.; He, X.W.; Wang, Y.Z.; Li, W.Y.; Zhang, Y.K. Epitope Imprinted Polymer Coating CdTe Quantum Dots for Specific Recognition and Direct Fluorescent Quantification of the Target Protein Bovine Serum. *Biosens. Bioelectron.* **2014**, *54*, 266–272. [CrossRef]
66. Henthorn, D.B.; Peppas, N.A. Molecular Simulations of Recognitive Behavior of Molecularly Imprinted Intelligent Polymeric Networks. *Ind. Eng. Chem. Res.* **2007**, *46*, 6084–6091. [CrossRef]
67. Refaat, D.; Aggour, M.G.; Farghali, A.A.; Mahajan, R.; Wiklander, J.G.; Nicholls, I.A.; Piletsky, S.A. Strategies for Molecular Imprinting and the Evolution of MIP Nanoparticles as Plastic Antibodies—Synthesis and Applications. *Int. J. Mol. Sci.* **2019**, *20*, 6304. [CrossRef]
68. Gao, B.; Fu, H.; Li, Y.; Du, R. Preparation of Surface Molecularly Imprinted Polymeric Microspheres and Their Recognition Property for Basic Protein Lysozyme. *J. Chromatogr. B* **2010**, *21*, 1731–1738. [CrossRef]
69. Turner, N.W.; Jeans, C.W.; Brain, K.R.; Allender, C.J.; Hlady, V.; Britt, D.W. From 3D to 2D: A Review of the Molecular Imprinting of Proteins. *Biotechnol. Prog.* **2006**, *22*, 1474–1489. [CrossRef]
70. Kalecki, J.; Iskierko, Z.; Cieplak, M.; Sharma, P.S. Oriented Immobilization of Protein Templates: A New Trend in Surface Imprinting. *ACS Sens.* **2020**, *5*, 3710–3720. [CrossRef]
71. Aya, G.A.; Yang, J.C.; Hong, S.W.; Park, J.Y. Replicated Pattern Formation and Recognition Properties of 2,4-Dichlorophenoxyacetic Acid-Imprinted Polymers using Colloidal Silica Array Molds. *Polymers* **2019**, *11*, 1332. [CrossRef]
72. Devkota, L.; Nguyen, L.T.; Vu, T.T.; Piro, B. Electrochemical Determination of Tetracycline using AuNP-Coated Molecularly Imprinted Overoxidized Polypyrrole Sensing Interface. *Electrochim. Acta* **2018**, *270*, 535–542. [CrossRef]
73. Si, B.; Song, E. Molecularly Imprinted Polymers for the Selective Detection of Multi-Analyte Neurotransmitters. *Microelectron. Eng.* **2018**, *187*, 58–65. [CrossRef]

74. Mazzotta, E.; Turco, A.; Chianella, I.; Guerreiro, A.; Piletsky, S.A.; Malitesta, C. Solid-Phase Synthesis of Electroactive Nanoparticles of Molecularly Imprinted Polymers. A Novel Platform for Indirect Electrochemical Sensing Applications. *Sens. Actuators B Chem.* **2016**, *229*, 174–180. [CrossRef]
75. Shi, H.; Tsai, W.B.; Garrison, M.D.; Ferrari, S.; Ratner, B.D. Template-Imprinted Nanostructured Surfaces for Protein Recognition. *Nature* **1999**, *398*, 593–597. [CrossRef]
76. Sun, Y.; Lan, Y.; Yang, L.; Kong, F.; Du, H.; Feng, C. Preparation of Hemoglobin Imprinted Polymers Based on Graphene and Protein Removal Assisted by Electric Potential. *RSC Adv.* **2016**, *6*, 61897–61905. [CrossRef]
77. Tlili, A.; Attia, G.; Khaoulani, S.; Mazouz, Z.; Zerrouki, C.; Yaakoubi, N.; Othmane, A.; Fourati, N. Contribution to the Understanding of the Interaction between a Polydopamine Molecular Imprint and a Protein Model: Ionic Strength and pH Effect Investigation. *Sensors* **2021**, *21*, 619. [CrossRef]
78. Pergande, M.R.; Cologna, S.M. Isoelectric Point Separations of Peptides and Proteins. *Proteomes* **2017**, *5*, 4. [CrossRef]
79. Łapińska, U.; Saar, K.L.; Yates, E.V.; Herling, T.W.; Müller, T.; Challa, P.K.; Christopher, M.D.; Knowles, T.P.J. Gradient-Free Determination of Isoelectric Points of Proteins on chip. *Phys. Chem. Chem. Phys.* **2017**, *19*, 23060–23067. [CrossRef]
80. Kidakova, A.; Boroznjak, R.; Reut, J.; Öpik, A.; Saarma, M.; Syritski, V. Molecularly Imprinted Polymer-Based SAW Sensor for Label-Free Detection of Cerebral Dopamine Neurotrophic Factor Protein. *Sens. Actuators B Chem.* **2020**, *308*, 127708. [CrossRef]
81. Dechtrirat, D.; Gajovic-Eichelmann, N.; Bier, F.F.; Scheller, F.W. Hybrid Material for Protein Sensing Based on Electrosynthesized MIP on a Mannose Terminated Self-Assembled Monolayer. *Adv. Func. Mater.* **2014**, *24*, 2233–2239. [CrossRef]
82. Tretjakov, A.; Syritski, V.; Reut, J.; Boroznjak, R.; Volobujeva, O.; Öpik, A. Surface Molecularly Imprinted Polydopamine Films for Recognition of Immunoglobulin G. *Mikrochim. Acta* **2013**, *180*, 1433–1442. [CrossRef]
83. Rebelo, T.S.; Miranda, I.M.; Brandão, A.T.; Sousa, L.I.; Ribeiro, J.A.; Silva, A.F.; Pereira, C.M. A Disposable Saliva Electrochemical MIP-Based Biosensor for Detection of the Stress Biomarker α-Amylase in Point-of-Care Applications. *Electrochem* **2021**, *2*, 427–438. [CrossRef]
84. Tretjakov, A.; Syritski, V.; Reut, J.; Boroznjak, R.; Öpik, A. Molecularly Imprinted Polymer Film Interfaced with Surface Acoustic Wave Technology as a Sensing Platform for Label-Free Protein Detection. *Anal. Chim. Acta* **2016**, *902*, 182–188. [CrossRef]
85. Cardoso, A.R.; De Sá, M.H.; Sales, M.G.F. An Impedimetric Molecularly-Imprinted Biosensor for Interleukin-1β Determination, Prepared by In-Situ Electropolymerization on Carbon Screen-Printed Electrodes. *Bioelectrochemistry* **2019**, *130*, 107287. [CrossRef]
86. Tavares, A.P.; Sales, M.G.F. Novel Electro-Polymerized Protein-Imprinted Materials using Eriochrome Black T: Application to BSA Sensing. *Electrochim. Acta* **2018**, *262*, 214–225. [CrossRef]
87. Dechtrirat, D.; Sookcharoenpinyo, B.; Prajongtat, P.; Sriprachuabwong, C.; Sanguankiat, A.; Tuantranont, A.; Hannongbua, S. An Electrochemical MIP Sensor for Selective detection of Salbutamol Based on a Graphene/PEDOT: PSS Modified Screen Printed Carbon Electrode. *RSC Adv.* **2018**, *8*, 206–212. [CrossRef]
88. Zhang, L.; Luo, K.; Li, D.; Zhang, Y.; Zeng, Y.; Li, J. Chiral Molecular Imprinted Sensor for Highly Selective Determination of D-carnitine in Enantiomers via dsDNA-Assisted Conformation Immobilization. *Anal. Chim. Acta* **2020**, *1136*, 82–90. [CrossRef]
89. Palladino, P.; Minunni, M.; Scarano, S. Cardiac Troponin T Capture and Detection in Real-Time via Epitope-Imprinted Polymer and Optical Biosensing. *Biosens. Bioelectron.* **2018**, *106*, 93–98. [CrossRef]
90. Jolly, P.; Tamboli, V.; Harniman, R.L.; Estrela, P.; Allender, C.J.; Bowen, J.L. Aptamer–MIP Hybrid Receptor for Highly Sensitive Electrochemical Detection of Prostate Specific Antigen. *Biosens. Bioelectron.* **2016**, *75*, 188–195. [CrossRef]
91. Karami, P.; Bagheri, H.; Johari-Ahar, M.; Khoshsafar, H.; Arduini, F.; Afkhami, A. Dual-Modality Impedimetric Immunosensor for Early Detection of Prostate-Specific Antigen and Myoglobin Markers Based on Antibody-Molecularly Imprinted Polymer. *Talanta* **2019**, *202*, 111–122. [CrossRef]
92. Karimian, N.; Vagin, M.; Zavar, M.H.A.; Chamsaz, M.; Turner, A.P.; Tiwari, A. An Ultrasensitive Molecularly-Imprinted Human Cardiac Troponin Sensor. *Biosens. Bioelectron.* **2013**, *50*, 492–498. [CrossRef] [PubMed]
93. Lu, C.H.; Zhang, Y.; Tang, S.F.; Fang, Z.B.; Yang, H.H.; Chen, X.; Chen, G.N. Sensing HIV Related Protein using Epitope Imprinted Hydrophilic Polymer Coated Quartz Crystal Microbalance. *Biosens. Bioelectron.* **2012**, *31*, 439–444. [CrossRef] [PubMed]
94. Tokonami, S.; Nakadoi, Y.; Nakata, H.; Takami, S.; Kadoma, T.; Shiigi, H.; Nagaoka, T. Recognition of Gram-Negative and Gram-Positive Bacteria with a Functionalized Conducting Polymer Film. *Res. Chem. Intermed.* **2014**, *40*, 2327–2335. [CrossRef]
95. Yongabi, D.; Khorshid, M.; Losada-Pérez, P.; Eersels, K.; Deschaume, O.; D'Haen, J.; Bartic, C.; Hooyberghs, J.; Thoelen, R.; Wübbenhorst, M. Cell Detection by Surface Imprinted Polymers SIPs: A Study to Unravel the Recognition Mechanisms. *Sens. Actuators B Chem.* **2018**, *255*, 907–917. [CrossRef]
96. Cui, F.; Zhou, Z.; Zhou, H.S. Molecularly Imprinted Polymers and Surface Imprinted Polymers Based Electrochemical Biosensor for Infectious Diseases. *Sensors* **2020**, *20*, 996. [CrossRef]
97. Crapnell, R.D.; Hudson, A.; Foster, C.W.; Eersels, K.; Grinsven, B.V.; Cleij, T.J.; Banks, C.E.; Peeters, M. Recent Advances in Electrosynthesized Molecularly Imprinted Polymer Sensing Platforms for Bioanalyte Detection. *Sensors* **2019**, *19*, 1204. [CrossRef]
98. Ji, Z.; Chen, W.; Wang, E.; Deng, R. Electropolymerized Molecular Imprinting & Graphene Modified Electrode for Detection of Melamine. *Int. J. Electrochem. Sci.* **2017**, *12*, 11942–11954.
99. Kan, X.; Xing, Z.; Zhu, A.; Zhao, Z.; Xu, G.; Li, C.; Zhou, H. Molecularly Imprinted Polymers Based Electrochemical Sensor for Bovine Hemoglobin Recognition. *Sens. Actuators B Chem.* **2012**, *168*, 395–401. [CrossRef]
100. Ribeiro, J.A.; Pereira, C.M.; Silva, A.F.; Sales, M.G.F. Electrochemical Detection of Cardiac Biomarker Myoglobin using Polyphenol as Imprinted Polymer Receptor. *Anal. Chim. Acta* **2017**, *981*, 41–52. [CrossRef]

101. Essousi, H.; Barhoumi, H. Electroanalytical Application of Molecular Imprinted Polyaniline Matrix for Dapsone Determination in Real Pharmaceutical Samples. *J. Electroanal. Chem.* **2018**, *818*, 131–139. [CrossRef]

102. Li, H.H.; Wang, H.H.; Li, W.T.; Fang, X.X.; Guo, X.C.; Zhou, W.H.; Cao, X.; Kou, D.X.; Zhou, Z.J.; Wu, S.X. A Novel Electrochemical Sensor for Epinephrine Based on Three Dimensional Molecularly Imprinted Polymer Arrays. *Sens. Actuators B Chem.* **2016**, *222*, 1127–1133. [CrossRef]

103. Stojanovic, Z.; Erdőssy, J.; Keltai, K.; Scheller, F.W.; Gyurcsányi, R.E. Electrosynthesized Molecularly Imprinted Polyscopoletin Nanofilms for Human Serum Albumin Detection. *Anal. Chim. Acta* **2017**, *977*, 1–9. [CrossRef] [PubMed]

104. Choi, D.Y.; Yang, J.C.; Park, J. Optimization and Characterization of Electrochemical Protein Imprinting on Hemispherical Porous Gold Patterns for the Detection of Trypsin. *Sens. Actuators B Chem.* **2022**, *350*, 130855. [CrossRef]

105. Shumyantseva, V.V.; Bulko, T.V.; Sigolaeva, L.V.; Kuzikov, A.V.; Archakov, A.I. Electrosynthesis and Binding Properties of Molecularly Imprinted Poly-o-Phenylenediamine for Selective Recognition and Direct Electrochemical Detection of Myoglobin. *Biosens. Bioelectron.* **2016**, *86*, 330–336. [CrossRef] [PubMed]

106. Raziq, A.; Kidakova, A.; Boroznjak, R.; Reut, J.; Öpik, A.; Syritski, V. Development of a Portable MIP-Based Electrochemical Sensor for Detection of SARS-CoV-2 Antigen. *Biosens. Bioelectron.* **2021**, *178*, 113029. [CrossRef] [PubMed]

107. Boroznjak, R.; Reut, J.; Tretjakov, A.; Lomaka, A.; Öpik, A.; Syritski, V. A Computational Approach to Study Functional Monomer-Protein Molecular Interactions to Optimize Protein Molecular Imprinting. *J. Mol. Recognit.* **2017**, *30*, e2635. [CrossRef]

108. Mazouz, Z.; Mokni, M.; Fourati, N.; Zerrouki, C.; Barbault, F.; Seydou, M.; Kalfat, R.; Yaakoubi, N.; Omezzine, A.; Bouslema, A. Computational Approach and Electrochemical Measurements for Protein Detection with MIP-Based Sensor. *Biosens. Bioelectron.* **2020**, *151*, 111978. [CrossRef]

109. Campuzano, S.; Pedrero, M.; Yáñez-Sedeño, P.; Pingarrón, J.M. New Challenges in Point of Care Electrochemical Detection of Clinical Biomarkers. *Sens. Actuators B Chem.* **2021**, *345*, 130349. [CrossRef]

110. Tu, J.; Torrente-Rodríguez, R.M.; Wang, M.; Gao, W. The Era of Digital Health: A Review of Portable and Wearable Affinity Biosensors. *Adv. Func. Mater.* **2020**, *30*, 1906713. [CrossRef]

111. Diltemiz, S.E.; Hür, D.; Ersöz, A.; Denizli, A.; Say, R. Designing of MIP based QCM Sensor Having Thymine Recognition Sites based on Biomimicking DNA Approach. *Biosens. Bioelectron.* **2009**, *25*, 599–603. [CrossRef]

112. Wasilewski, T.; Szulczyński, B.; Kamysz, W.; Gębicki, J.; Namieśnik, J. Evaluation of Three Peptide Immobilization Techniques on a QCM Surface Related to Acetaldehyde Responses in the Gas Phase. *Sensors* **2018**, *18*, 3942. [CrossRef] [PubMed]

113. Wangchareansak, T.; Thitithanyanont, A.; Chuakheaw, D.; Gleeson, M.P.; Lieberzeit, P.A.; Sangma, C. Influenza A Virus Molecularly Imprinted Polymers and Their Application in Virus Sub-type Classification. *J. Mater. Chem. B* **2013**, *1*, 2190–2197. [CrossRef] [PubMed]

114. Afzal, A.; Mujahid, A.; Schirhagl, R.; Bajwa, S.Z.; Latif, U.; Feroz, S. Gravimetric Viral Diagnostics: QCM based Biosensors for Early Detection of Viruses. *Chemosensors* **2017**, *5*, 7. [CrossRef]

115. Hawkins, D.M.; Stevenson, D.; Reddy, S.M. Investigation of Protein Imprinting in Hydrogel-based Molecularly Imprinted Polymers (HydroMIPs). *Anal. Chim. Acta* **2005**, *542*, 61–65. [CrossRef]

116. Kubo, T.; Arimura, S.; Tominaga, Y.; Naito, T.; Hosoya, K.; Otsuka, K. Molecularly Imprinted Polymers for Selective Adsorption of Lysozyme and Cytochrome c using a PEG-based Hydrogel: Selective Recognition for Different Conformations due to pH Conditions. *Macromolecules* **2015**, *48*, 4081–4087. [CrossRef]

117. Bossi, A.; Bonini, F.; Turner, A.P.F.; Piletsky, S.A. Molecularly Imprinted Polymers for the Recognition of Proteins: The State of the Art. *Biosens. Bioelectron.* **2007**, *22*, 1131–1137. [CrossRef]

118. Reddy, S.M.; Phan, Q.T.; El-Sharif, H.; Govada, L.; Stevenson, D.; Chayen, N.E. Protein Crystallization and Biosensor Applications of Hydrogel-based Molecularly Imprinted Polymers. *Biomacromolecules* **2012**, *13*, 3959–3965. [CrossRef]

119. Ozcelikay, G.; Kaya, S.I.; Ozkan, E.; Cetinkaya, A.; Nemutlu, E.; Kır, S.; Ozkan, S.A. Sensor-Based MIP Technologies for Targeted Metabolomics Analysis. *Trends Analyt. Chem.* **2022**, *146*, 116487. [CrossRef]

120. Costa-Rama, E.; Fernández-Abedul, M.T. Based Screen-Printed Electrodes: A New Generation of Low-Cost Electroanalytical Platforms. *Biosensors* **2021**, *11*, 51. [CrossRef]

121. Lach, P.; Cieplak, M.; Majewska, M.; Noworyta, K.R.; Sharma, P.S.; Kutner, W. Gate Effect" in p-Synephrine Electrochemical Sensing with a Molecularly Imprinted Polymer and Redox Probes. *Anal. Chem.* **2019**, *91*, 7546–7553. [CrossRef] [PubMed]

122. Sharma, P.S.; Garcia-Cruz, A.; Cieplak, M.; Noworyta, K.R.; Kutner, W. 'Gate effect' in Molecularly Imprinted Polymers: The Current State of Understanding. *Curr. Opin. Electrochem.* **2019**, *16*, 50–56. [CrossRef]

123. Iskierko, Z.; Sharma, P.S.; Bartold, K.; Pietrzyk-Le, A.; Noworyta, K.; Kutner, W. Molecularly Imprinted Polymers for Separating and Sensing of Macromolecular Compounds and Microorganisms. *Biotechnol. Adv.* **2016**, *34*, 30–46. [CrossRef] [PubMed]

124. Gui, R.; Guo, H.; Jin, H. Preparation and Applications of Electrochemical Chemosensors Based on Carbon-Nanomaterial-Modified Molecularly Imprinted Polymers. *Nanoscale Adv.* **2019**, *1*, 3325–3363. [CrossRef]

125. Kumar Prusty, A.; Bhand, S. Molecularly Imprinted Polyresorcinol Based Capacitive Sensor for Sulphanilamide Detection. *Electroanalysis* **2019**, *31*, 1797–1808. [CrossRef]

126. Dincer, C.; Bruch, R.; Costa-Rama, E.; Fernández-Abedul, M.T.; Merkoçi, A.; Manz, A.; Güder, F. Disposable Sensors in Diagnostics, Food, and Environmental Monitoring. *Adv. Mater.* **2019**, *31*, 1806739. [CrossRef]

127. Piloto, A.M.L.; Ribeiro, D.S.; Rodrigues, S.S.M.; Santos, J.L.; Sales, M.G.F. Label-Free Quantum Dot Conjugates for Human Protein IL-2 Based on Molecularly Imprinted Polymers. *Sens. Actuators B Chem.* **2020**, *304*, 127343. [CrossRef]

128. Hong, C.C.; Chen, C.P.; Horng, J.C.; Chen, S.Y. Point-of-Care Protein Sensing Platform Based on Immuno-like Membrane with Molecularly-Aligned Nanocavities. *Biosens. Bioelectron.* **2013**, *50*, 425–430. [CrossRef]
129. Prasad, B.B.; Prasad, A.; Tiwari, M.P. Multiwalled Carbon Nanotubes-Ceramic Electrode Modified with Substrate-Selective Imprinted Polymer for Ultra-Trace Detection of Bovine Serum Albumin. *Biosens. Bioelectron.* **2013**, *39*, 236–243. [CrossRef]
130. Bai, W.; Gariano, N.A.; Spivak, D.A. Macromolecular Amplification of Binding Response in Superaptamer Hydrogels. *J. Am. Chem. Soc.* **2013**, *135*, 6977–6984. [CrossRef]
131. Yang, J.C.; Hong, S.W.; Jeon, S.; Park, W.I.; Byun, M.; Park, J. Molecular Imprinting of Hemispherical Pore-Structured Thin Films via Colloidal Lithography for Gaseous Formaldehyde Gravimetric Sensing. *Appl. Surf. Sci.* **2021**, *570*, 151161. [CrossRef]
132. Cardoso, A.R.; Marques, A.C.; Santos, L.; Carvalho, A.F.; Costa, F.M.; Martins, R.; Fortunato, E. Molecularly-Imprinted Chloramphenicol Sensor with Laser-Induced Graphene Electrodes. *Biosens. Bioelectron.* **2019**, *124*, 167–175. [CrossRef] [PubMed]
133. Tang, W.; Yin, L.; Sempionatto, J.R.; Moon, J.M.; Teymourian, H.; Wang, J. Touch-Based Stressless Cortisol Sensing. *Adv. Mater.* **2021**, *33*, 2008465. [CrossRef] [PubMed]
134. Ratautaite, V.; Plausinaitis, D.; Baleviciute, I.; Mikoliunaite, L.; Ramanaviciene, A.; Ramanavicius, A. Characterization of Caffeine-Imprinted Polypyrrole by a Quartz Crystal Microbalance and Electrochemical Impedance Spectroscopy. *Sens. Actuators B Chem.* **2015**, *212*, 63–71. [CrossRef]
135. Ayankojo, A.G.; Boroznjak, R.; Reut, J.; Öpik, A.; Syritski, V. Molecularly Imprinted Polymer Based Electrochemical Sensor for Quantitative Detection of SARS-CoV-2 Spike Protein. *Sens. Actuators B Chem.* **2022**, *353*, 131160. [CrossRef] [PubMed]
136. Lin, Y.; Bariya, M.; Nyein, H.Y.Y.; Kivimäki, L.; Uusitalo, S.; Jansson, E.; Javey, A. Porous Enzymatic Membrane for Nanotextured Glucose Sweat Sensors with High stability Toward Reliable Noninvasive Health Monitoring. *Adv. Funct. Mater.* **2019**, *29*, 1902521. [CrossRef]
137. Parlak, O.; Keene, S.T.; Marais, A.; Curto, V.F.; Salleo, A. Molecularly Selective Nanoporous Membrane-Based Wearable Organic Electrochemical Device for Noninvasive Cortisol Sensing. *Sci. Adv.* **2018**, *4*, eaar2904. [CrossRef]
138. Kim, J.; Sempionatto, J.R.; Imani, S.; Hartel, M.C.; Barfidokht, A.; Tang, G.; Wang, J. Simultaneous Monitoring of Sweat and Interstitial Fluid Using a Single Wearable Biosensor Platform. *Adv. Sci.* **2018**, *5*, 1800880. [CrossRef]
139. De la Paz, E.; Barfidokht, A.; Rios, S.; Brown, C.; Chao, E.; Wang, J. Extended Noninvasive Glucose Monitoring in the Interstitial Fluid Using an Epidermal Biosensing Patch. *Anal. Chem.* **2021**, *93*, 12767–12775. [CrossRef]
140. Manjakkal, L.; Yin, L.; Nathan, A.; Wang, J.; Dahiya, R. Energy Autonomous Sweat-Based Wearable Systems. *Adv. Mater.* **2021**, *33*, 2100899. [CrossRef]
141. Cennamo, N.; D'Agostino, G.; Perri, C.; Arcadio, F.; Chiaretti, G.; Parisio, E.M.; Zeni, L. Proof of Concept for a Quick and Highly Sensitive On-Site Detection of SARS-CoV-2 by Plasmonic Optical Fibers and Molecularly Imprinted Polymers. *Sensors* **2021**, *21*, 1681. [CrossRef]
142. COVID-19 nanoMIPs-Synthetic SARS-CoV-2 Antibodies-MIP Diagnostics Home Page. Available online: https://www.mip-dx.com/covid19-nanomip (accessed on 8 November 2021).
143. Min, J.; Nothing, M.; Coble, B.; Zheng, H.; Park, J.; Im, H.; Lee, H. Integrated Biosensor for Rapid and Point-of-Care Sepsis Diagnosis. *ACS Nano* **2018**, *12*, 3378–3384. [CrossRef] [PubMed]
144. cpn™-Conjugated Polymer Nanoparticles-Stream Bio Home Page. 2020. Available online: https://www.streambio.co.uk/our-technology (accessed on 8 November 2021).
145. Park, J.; Dong, H.H.; Park, J.-K. Towards Practical Sample Preparation in Point-of-Care Testing: User-Friendly Microfluidic Devices. *Lab Chip* **2020**, *20*, 1191–1203. [CrossRef] [PubMed]
146. Malon, R.S.P.; Sadir, S.; Balakrishnan, M.; Córcoles, E.P. Saliva-Based Biosensors: Noninvasive Monitoring Tool for Clinical Diagnostics. *BioMed Res. Int.* **2014**, *2014*, 962903. [CrossRef] [PubMed]
147. Sorsa, T.; Tervahartiala, T.; Leppilahti, J.; Hernandez, M.; Gamonal, J.; Tuomainen, A.M.; Lauhio, A.; Pussinen, P.J.; Mantyla, P. Collagenase-2 (MMP-8) as a Point-of-Care Biomarker in Periodontitis and Cardiovascular Diseases. Therapeutic Response to Non-Antimicrobial Properties of Tetracyclines. *Pharm. Res.* **2011**, *63*, 108–113. [CrossRef]
148. Bellagambi, F.G.; Lomonaco, T.; Salvo, P.; Vivaldi, F.; Hangouët, M.; Ghimenti, S.; Biagini, D.; Francesco, F.D.; Fuoco, R.; Errachid, A. Saliva Sampling: Methods and Devices. An Overview. *TrAC Trends Anal. Chem.* **2020**, *124*, 115781. [CrossRef]
149. Miočević, O.; Cole, C.R.; Laughlin, M.J.; Buck, R.L.; Slowey, P.D.; Shirtcliff, E.A. Quantitative Lateral Flow Assays for Salivary Biomarker Assessment: A Review. *Front. Public Health* **2017**, *5*, 133–145. [CrossRef]
150. Kim, J.; Imani, S.; De Araujo, W.R.; Warchall, J.; Valdés-Ramírez, G.; Paixão, T.R.L.C.; Mercier, P.P.; Wang, J. Wearable Salivary Uric Acid Mouthguard Biosensor with Integrated Wireless Electronics. *Biosens. Bioelectron.* **2015**, *74*, 1061–1068. [CrossRef]
151. Kim, J.; Valdes-Ramirez, G.; Bandodkar, A.J.; Jia, W.; Martinez, A.G.; Ramirez, J.; Mercier, P.; Wang, J. Non-Invasive Mouthguard Biosensor for Continuous Salivary Monitoring of Metabolites. *Analyst* **2014**, *139*, 1632–1636. [CrossRef]
152. Sharma, A.; Badea, M.; Tiwari, S.; Marty, J.L. Wearable Biosensors: An Alternative and Practical Approach in Healthcare and Disease Monitoring. *Molecules* **2021**, *26*, 748. [CrossRef]
153. Arakawa, T.; Kuroki, Y.; Nitta, H.; Chouhan, P.; Toma, K.; Sawada, S.; Takeuchi, S.; Sekita, T.; Akiyoshi, K.; Minakuchi, S. Mouthguard Biosensor with Telemetry System for Monitoring of Saliva Glucose: A Novel Cavitas Sensor. *Biosens. Bioelectron.* **2016**, *84*, 106–111. [CrossRef]
154. Mascarenhas, P.; Fatela, B.; Barahona, I. Effect of Diabetes Mellitus Type 2 on Salivary Glucose—A Systematic Review and Meta-Analysis of Observational Studies. *PLoS ONE* **2014**, *9*, e101706.

155. Lee, Y.; Howe, C.; Mishra, S.; Lee, D.S.; Mahmood, M.; Piper, M.; Kim, Y.; Tieu, K.; Byun, H.-S.; Coffey, J.P. Wireless, Intraoral Hybrid Electronics for Real-Time Quantification of Sodium Intake Toward Hypertension Management. *Proc. Natl. Acad. Sci. USA* **2018**, *115*, 5377–5382. [CrossRef] [PubMed]

156. Vu, T.; Lin, F.; Alshurafa, N.; Xu, W. Wearable Food Intake Monitoring Technologies: A Comprehensive Review. *Computers* **2017**, *6*, 4. [CrossRef]

157. Fălămaș, A.; Rotaru, H.; Hedeșiu, M. Surface-Enhanced Raman Spectroscopy (SERS) Investigations of Saliva for Oral Cancer Diagnosis. *Lasers Med. Sci.* **2020**, *35*, 1393–1401. [CrossRef]

158. García-Carmona, L.; Martín, A.; Sempionatto, J.R.; Moreto, J.R.; González, M.C.; Wang, J.; Escarpa, A. Pacifier Biosensor: Toward Noninvasive Saliva Biomarker Monitoring. *Anal. Chem.* **2019**, *91*, 13883–13891. [CrossRef]

159. Canfarotta, F.; Czulak, J.; Betlem, K.; Sachdeva, A.; Eersels, K.; Van Grinsven, B.; Cleij, T.J.; Peeters, M. A Novel Thermal Detection Method Based on Molecularly Imprinted Nanoparticles as Recognition Elements. *Nanoscale* **2018**, *10*, 2081–2089. [CrossRef]

160. Guerreiro, J.R.L.; Bochenkov, V.E.; Runager, K.; Aslan, H.; Dong, M.; Enghild, J.J.; De Freitas, V.; Ferreira Sales, M.G.; Sutherland, D.S. Molecular Imprinting of Complex Matrices at Localized Surface Plasmon Resonance Biosensors for Screening of Global Interactions of Polyphenols and Proteins. *ACS Sens.* **2016**, *1*, 258–264. [CrossRef]

161. Tabrizi, M.A.; Fernández-Blázquez, J.P.; Medina, D.M.; Acedo, P. An Ultrasensitive Molecularly Imprinted Polymer-Based Electrochemical Sensor for the Determination of SARS-CoV-2-RBD by Using Macroporous Gold Screen-Printed Electrode. *Biosens. Bioelectron.* **2022**, *196*, 113729. [CrossRef]

162. Liu, Y.; Tuleouva, N.; Ramanculov, E.; Revzin, A. Aptamer-Based Electrochemical Biosensor for Interferon Gamma Detection. *Anal. Chem.* **2010**, *82*, 8131–8136. [CrossRef]

163. Belbruno, J.J. Molecularly Imprinted Polymers. *Chem. Rev.* **2019**, *119*, 94–119. [CrossRef]

164. Beck, J.D.; Offenbacher, S. Systemic Effects of Periodontitis: Epidemiology of Periodontal Disease and Cardiovascular Disease. *J. Periodontol.* **2005**, *76*, 2089–2100. [CrossRef] [PubMed]

Review

Micro-/Nano-Structured Biodegradable Pressure Sensors for Biomedical Applications

Yoo-Kyum Shin [1,†], Yujin Shin [2,†], Jung Woo Lee [2,*] and Min-Ho Seo [1,3,*]

[1] Department of Information Convergence Engineering, Pusan National University, 49 Busandaehak-ro, Mulgeum-eup, Yangsan-si 50612, Gyeongsangnam-do, Korea

[2] Department of Materials Science and Engineering, Pusan National University, 2 Busandaehak-ro 63beon-gil, Geumjeong-gu, Busan 46241, Korea

[3] School of Biomedical Convergence Engineering, Pusan National University, 49 Busandaehak-ro, Mulgeum-eup, Yangsan-si 50612, Gyeongsangnam-do, Korea

* Correspondence: jungwoolee@pusan.ac.kr (J.W.L.); mhseo@pusan.ac.kr (M.-H.S.)

† These authors contributed equally to this work.

Abstract: The interest in biodegradable pressure sensors in the biomedical field is growing because of their temporary existence in wearable and implantable applications without any biocompatibility issues. In contrast to the limited sensing performance and biocompatibility of initially developed biodegradable pressure sensors, device performances and functionalities have drastically improved owing to the recent developments in micro-/nano-technologies including device structures and materials. Thus, there is greater possibility of their use in diagnosis and healthcare applications. This review article summarizes the recent advances in micro-/nano-structured biodegradable pressure sensor devices. In particular, we focus on the considerable improvement in performance and functionality at the device-level that has been achieved by adapting the geometrical design parameters in the micro- and nano-meter range. First, the material choices and sensing mechanisms available for fabricating micro-/nano-structured biodegradable pressure sensor devices are discussed. Then, this is followed by a historical development in the biodegradable pressure sensors. In particular, we highlight not only the fabrication methods and performances of the sensor device, but also their biocompatibility. Finally, we intoduce the recent examples of the micro/nano-structured biodegradable pressure sensor for biomedical applications.

Keywords: biodegradable electronics; pressure sensors; MEMS (microelectromechanical systems); NEMS (nanoelectromechanical systems); biomedical engineering

Citation: Shin, Y.-K.; Shin, Y.; Lee, J.W.; Seo, M.-H. Micro-/Nano-Structured Biodegradable Pressure Sensors for Biomedical Applications. *Biosensors* **2022**, *12*, 952. https://doi.org/10.3390/bios12110952

Received: 23 September 2022
Accepted: 27 October 2022
Published: 1 November 2022

1. Introduction

A pressure sensor is a device that can transduce mechanical pressure to an electrical or optical signal. It has gained considerable attention from various traditional industries such as automotive [1,2], aviation [3–5], and manufacturing industries [6–9] since it can help in controlling the system stability and process flow. Recently, owing to the advances in materials science, mechanical engineering, and fabrication technologies, the pressure sensor can be implemented with various materials and dimension, enabling its application in a wide range of emerging academia disciplines and industrial fields, such as robotics [10–17], artificial intelligence [18–21], and smart factory [22,23]. Among them, the use of pressure sensors in the biomedical engineering field is one of the representative promising applications of this technology because it can help to measure the pressure of the target tissue and organ that is essential for wide range of disease diagnosis and therapy (Figure 1a,b) [24–29].

Conventionally, bio-implantable pressure sensors have been made of materials being able to permanently exist in human-body with bio-compatibility. However, these types of devices require secondary surgical extraction after clinical use, and it intrinsically accompanies the risk, cost, distress, and pain for the patient [30]. In this regard, biodegradable (or, equivalently, bioabsorbable and bioresorbable) pressure sensor has recently received lots of attention from researchers because it can be degraded and disappear in human body spontaneously by bio-fluids after specific period [31–34].

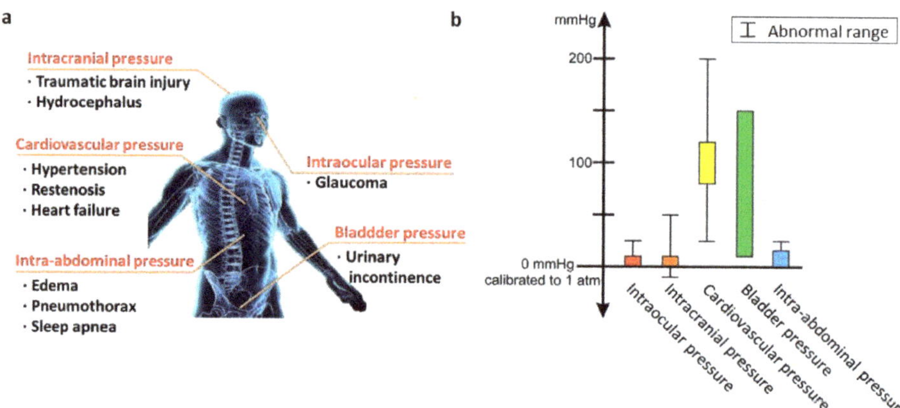

Figure 1. Pressure monitoring. (**a**) Pressure is linked to several diseases in various organs of the body [24]. (**b**) Relevant pressure ranges for in vivo pressure monitoring for diagnostic applications [29].

Earlier developed biodegradable pressure sensors were partially composed of biodegradable materials with a simple structure [35,36]. However, owing to the recent developments in materials science, mechanical and electrical engineering and fabrication technologies, biodegradable pressure sensors have recently become fully biodegradable with high scalability [37,38]. In particular, as the device is designed to exploit micro-/nano-structures, it can not only be manufactured in a miniature dimension making it easier to implant, but it also shows great improvement in performance when used for precise detection of bio-signals and diseases [30,39,40]. As a result, fully biodegradable and miniaturized pressure sensors with improved performance in terms of high sensitivity, fast response time, stability, and reliability have been realized, and this has made biodegradable pressure sensors usable in diagnosis and healthcare applications.

Even though there has been a sudden spurt in progress on these sensors, however, there has been less reports to summarize and review the recent progress and knowledge regarding the micro-/nano-structured biodegradable pressure sensors and their applications. Here, this review paper introduces the latest progress in the development of micro-/nano-structured biodegradable pressure sensors. We first introduce recent developments and libraries in biodegradable materials for the micro-/nano-structured biodegradable pressure sensors; in particular, materials are classified as wet and dry transient materials, depending on the degradation conditions. Next, we introduce details about the performance-enhancement in the biodegradable pressure sensor exploiting the micro-/nano-structure. Especially, we focus on the characteristics, performance, and fabrication methods of these sensors. Finally, the practical applications of the micro-/nano-structured biodegradable pressure sensor in biomedical engineering are introduced. We review the reports on the feasibility of the developed micro-/nano-structured biodegradable pressure sensors for healthcare and diagnosis of disease in wearable and implantable biomedical applications without any biocompatibility issue.

2. Transient Materials

To develop biodegradable electronics with a transient performance (Figure 2), appropriate selections of materials should be considered from the viewpoint of electronic components (e.g., conductors, semiconductors, insulators, encapsulations, and substrates).

These materials are required to satisfy designated dissolution rates, electrical/mechanical properties, and other demands depending on the desired application of the device. In the following sections, we summarize the various types of inorganic and organic biodegradable materials that are commonly utilized in recent research on transient electronic devices. The materials are classified into two parts namely wet and dry transient materials, according to their degrading mechanisms.

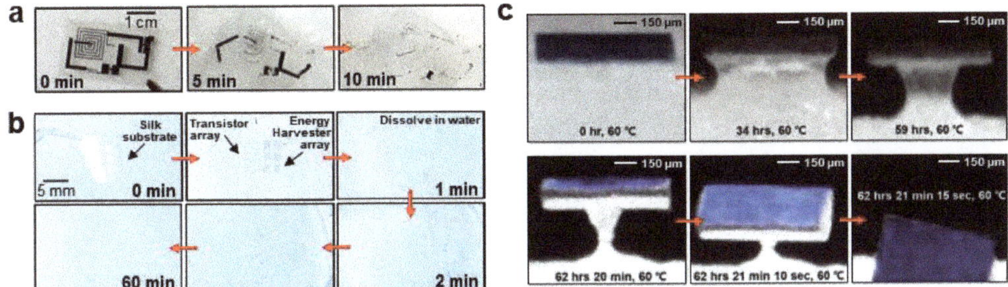

Figure 2. Examples of transient electronics. Red arrows indicate the elapse of time. (**a**) Optical images of Si diodes, Si/MgO/Mg transistors, Mg/MgO inductors and capacitors, and Mg resistors and interconnects on a silk substrate [32]. (**b**) Photographs of ZnO thin film transistor arrays and mechanical energy harvester arrays on silk substrate [41]. (**c**) Optical images of an array of solar cells on a dry transient CDD substrate [42].

2.1. Wet Transient Materials

2.1.1. Conductors

Alkaline earth metals including magnesium (Mg) and calcium (Ca) along with transition metals such as molybdenum (Mo), tungsten (W), iron (Fe), and Zinc (Zn) are considered as potential conductors for transient devices owing to their high hydrolysis/dissolution rates, easy fabrication process, and their biocompatibility along with their intrinsic electrical conductivity [32,43–47]. Mg and Fe are the most promising candidates that could be used as in vivo conductors in electronic devices because of their high biocompatibility [48,49]. These inorganic materials are essential to keep the human body healthy and functioning properly. According to the National Institute of Health (NIH), the recommended dietary allowance (RDA) values of Mg and Fe are 310–420 mg and 8–18 mg each for adults older than 19 [50,51]. In other words, when these metals are taken in or absorbed in the right amount they have no side effects on the human body. Thus, Mg, Mg alloy, and Fe are widely used as materials for bioresorbable implants such as vascular stents [44,48]. In general, the aforementioned biodegradable metals can dissolve in an aqueous solution and produce metal cations. Then, the metal cations may form metal hydroxides or metal oxides via hydrolysis as described in the following reaction procedures [43,52–56]:

$$Mg + 2H_2O \rightarrow Mg(OH)_2 + H_2$$

$$Fe + 2H_2O \rightarrow Fe(OH)_2 + H_2$$

$$Zn + 2H_2O \rightarrow Zn(OH)_2 + H_2$$

$$2W + 2H_2O + 3O_2 \rightarrow 2H_2WO_4$$

$$2Mo + 2H_2O + 3O_2 \rightarrow 2H_2MoO_4$$

Figure 3a illustrates the degrading behavior of the aforementioned metals. The dissolution rates in simulated body fluids of Mg, Fe, Zn, W are ≈0.05–0.5, ≈0.02, ≈0.005, ≈0.02–0.06 µm/h, respectively [43]. Each material shows different dissolution kinetics and distinctive advantages. In certain cases, specific electronics must disintegrate as soon as they start to react, whereas others are required to maintain their shapes and functions until the targeted time. Hence, appropriate materials should be selected for the transient devices depending on the aim of the reaction and the operating time. For instance, metals based on Mg have relatively high dissolution rate; hence, they would require encapsulation to reduce the rate of disintegration [57]. Moreover, methods to control the dissolution rate of Mg have been studied such as alloying with aluminum (Al) (e.g., AZ31B Mg alloy, Mg-Al-Zn alloy) or coating with biocompatible ceramics such as Ca-P or MgF_2 [58–60]. For example, the electrical dissolution rate of AZ31B Mg alloy is three times lower than Mg. By contrast, transition metals such as Mo have relatively low dissolution rate; hence, they can delay disintegration upto the targeted time and function normally in vivo without any encapsulation [43,52].

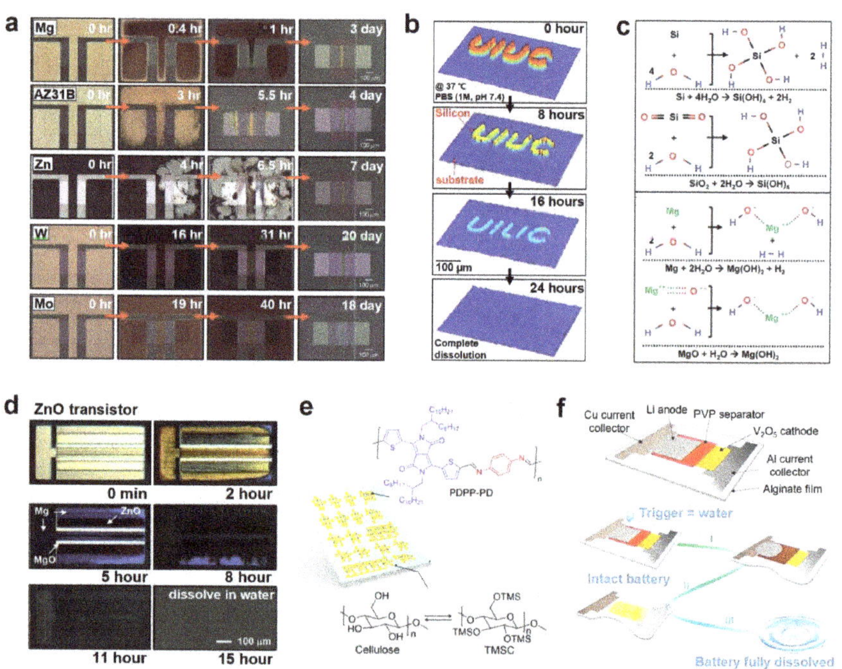

Figure 3. Examples of wet transient materials. (**a**) Optical images of metal dissolution behavior [43] (Mg, AZ31B Mg alloy, Zn, W, and Mo). Red arrows indicate the elapse of time. (**b**) Transmission-mode laser diffraction phase microscopic (DPM) image of Si NMs (∼100 nm-thick) at various times of immersion in phosphate-buffered solution (1 M, pH 7.4) at physiological temperature (37 °C) [61]. Black arrows indicate the elapse of time. (**c**) The chemical reaction for degrading mechanisms of Mg and Si in water [32]. (**d**) Optical images of a ZnO thin film transistor at various instants during the dissolution consisting of ZnO (active materials), Mg (electrodes, contacts, and interconnects), and MgO (gate and interlayer dielectrics) [41]. (**e**) Illustration and structures of biodegradable polymer (PDPP-PD) semiconductors on biodegradable cellulose substrate [49]. (**f**) Schematic process of water-triggered transient battery dissolution [62].

2.1.2. Semiconductors

Semiconducting materials are essential in most electronic devices, such as transistors, sensors, diodes, power harvesting devices [63–65]. However, traditional semiconducting materials when used for transient electronics are limited by their toxicity and rigidity [66]. For instance, gallium arsenide (GaAs) is a traditional compound semiconducting material that is widely utilized in microcircuits owing to its distinct advantage in increased electron mobility. However, several studies have demonstrated the acute and chronic toxicity of GaAs and the damage caused by GaAS to the lungs, reproductive organs, and kidneys of animals after inhalation exposure [67–69]. Therefore, transient semiconducting materials are still challenging and are many studies have been conducted to overcome these drawbacks.

Inorganic Materials

Among the various semiconducting materials, silicon (Si) is one of the most common. Bulk Si is usually considered to be non-degradable; traditional devices have been fabricated as thick wafers so that the integrated circuits do not decompose for several hundred years [53,70]. However, when its structure is scaled down to the nanoscale, it could be fully biodegraded in aqueous solutions or biofluids (Figure 3b). Furthermore, the compatibility of Si with widely used fabricating processes such as photolithography is an advantage in utilizing it for transient devices. Hwang et al. reported the dissolution property of a single-crystalline Si nanomembrane (Si NM), on a silicon-on-insulator wafer with a lateral dimension of 3 μm × 3 μm and a thickness of 70 nm, in phosphate-buffered saline (PBS; pH of 7.4). The dissolution rates at room temperature (25 °C) and body temperature (37 °C) were 2 nm/day and 4.5 nm/day, respectively [32]. These rates could vary in the range of 0.5–624 nm/day depending on the crystallinity, morphology and doping level of Si or on the surroundings such as the ambient temperature or composition of the solution [71]. In general, Si undergoes hydrolysis to produce orthosilicic acid ($Si(OH)_4$) as described in the following reaction (Figure 3c) [32,72].

$$Si + 4H_2O \rightarrow Si(OH)_4 + 2H_2$$

Besides Si NM, dissolution behavior and biocompatibility of other semiconducting materials such as amorphous silicon (a-Si), polycrystalline silicon (poly-Si), alloys of silicon and germanium (SiGe), and germanium (Ge) have also been investigated. By measuring the thickness of a patterned array of squares (3 μm × 3 μm × 100 nm) of each material, the dissolution rate of a-Si, poly-Si, SiGe, and Ge in buffer solutions (pH 7.4, 37 °C) are investigated as 4.1, 2.8, 0.1, and 3.1 nm/day, respectively. Compared to poly-Si, the dissolution rate of a-Si is accelerated owing to its low-density structure, which facilitates diffusion of aqueous solutions [52].

In addition, semiconducting oxides including zinc oxide (ZnO) are alternatives to Si owing to their high carrier mobility and outstanding transparency in the visible wavelength range. Furthermore, as ZnO shows degrading behavior in aqueous conditions, it is used as a semiconducting material in transient electronics. ZnO involves hydrolysis to form zinc hydroxide ($Zn(OH)_2$). For example, a 200 nm-thick ZnO completely disappeared in DI water at room temperature within 15 h (Figure 3d) [41]. Furthermore, ZnO can be dissolved in ammonia (pH of ~7.0 and ~9.0) and NaOH solutions (pH of 7.0 and ~9.0). Following are the chemical reactions of ZnO during dissolution in DI water, ammonia, and NaOH solution [73]:

$$ZnO + H_2O \leftrightarrow Zn(OH)_2$$

$$ZnO(s) + 2H^+ \rightarrow Zn^{2+} + H_2O$$

$$ZnO(s) + 4HN_3{\cdot}H_2O \rightarrow Zn(NH_3)_4^{2+} + 2OH^- + 3H_2O$$

$$ZnO(s) + 2OH^- \rightarrow ZnO_2^{2-} + H_2O$$

Zhou et al. reported that ZnO was visibly etched in about one hour in horse blood serum. This indicates that if a ZnO wire is trapped in a blood vessel or in the body, it would dissolve into Zn ions that could be absorbed by the surroundings and become a part of the nutrition of the body. Notably, Zn ions are needed every day for the human body to function properly [73].

Organic Materials

Polymer materials are also widely used as semiconducting material for transient devices because of their soft texture, flexible mechanical properties, low cost, and their suitability for large-scale production. These polymers could be categorized into synthetic and natural based polymers.

First, several synthetic biodegradable polymers such as poly(diketopyrrolopyrrole-*p*-phenylenediamine (PDPP-PD) and 5,5′-bis-(7-dodecyl-9*H*-fluoren-2-yl)-2,2′-bithiophene (DDFTTF) are currently utilized in the fabrication of soft electronics [74]. Diketopyrrolopyrrole(DPP) is a precursor of many synthetic polymers. Its monomer originated from a natural resource and hence, would also be degradable [49,75,76]. The PDPP-PD is a semiconducting polymer synthesized by a condensation reaction between diketopyrrolopyrrole-aldehyde (DPP-CHO) and *p*-phenylenediamine, under catalysis, using *p*-toluenesulfonic acid (PTSA). The chemical structure formula of PDPP-PD is illustrated in Figure 3e. PDPP-PD is fully disintegrable and biocompatible based on reversible imine chemistry. Under neutral-pH conditions, the imine bond (–C=N–) preserves a stable conjugated linker; however, it can be easily hydrolyzed by adding a trace amount of acid. A device consisting of this conjugated polymer degraded completely within 30 days in a pH 4.6 buffer solution [49]. DDFTTF is a robust small-molecule *p*-channel semiconducting material that indicates outstanding device performance. Bettinger et al. reported that a DDFTTF layer that was exposed in a citrate buffer delaminated in less than two days; thus, the device lost its functionality, irreversibly [34].

Natural polymers such as indigo and melanin are also utilized owing to their excellent biocompatibility. Indigo is an intrinsically ambipolar organic semiconducting material with a bandgap of 1.7 eV and high electron and hole mobilities of 1×10^{-2} cm^2/V·s [77]. Because of the natural existence of indigoids (derivatives of indigo), which have good semiconducting performance, low toxicity, and chemical stability, indigo is one of the most promising candidates for biocompatible semiconducting materials [78]. Eumelanins are a subclass of melanins. They exhibit excellent biocompatibility along with biodegradability via free radical degradation mechanisms. Furthermore, eumelanin can serve as a biocompatible electrode in high-density charge storage devices due to its unique properties. For instance, they show a hybrid ionic-electronic conduction behavior by the self-doping mechanism which is hydration-dependent [79,80].

2.1.3. Insulators, Encapsulations, and Substrates

Other important transient materials for electronic components are insulating materials that can isolate conductors and semiconductors [32,41]. In addition, substrate materials for most biodegradable electronics or devices, in thin geometry, should be suitable for handling and integrating all necessary components. The lifespan of transient electronic devices depends highly on the dissolution behavior of bioresorbable materials, which make up the device. Therefore, in some cases, encapsulation layers are necessary to enable stable and programmable operation of the devices by protecting the active layers from direct contact with the ambient aqueous environments, thus preserving the electrical functions of the devices [44,55,81]. Several inorganic/polymer materials have been widely adopted for this purpose. In general, oxides and nitrides of Si, Mg, and some polymers are used as insulators, encapsulations, or substrates for transient electronics owing to their bioresorbability and compatibility with the vacuum deposition processes and photolithography [39,53,82].

Yu et al. reported a fully bioresorbable, thin, flexible neural silicon electronic array, which could record electrophysiological signals in vivo. This embodiment uses SiO$_2$ for gate dielectrics, a tri-layer of SiO$_2$/Si$_3$N$_4$/SiO$_2$ for encapsulation, and a 30 μm-thick flexible sheet of poly(lactic-*co*-glycolic acid) (PLGA), a bioresorbable polymer, as the substrate. SiO$_2$ and Si$_3$N$_4$ dissolve at a rate of ~8.2 nm/day and ~5.1 nm/day, respectively, in biofluids at pH 7.4 and temperature 37 °C; PLGA completely dissolves within 4 or 5 weeks in biofluids at 37 °C [81]. SiO$_2$, SiN$_x$, and MgO are consumed by hydrolysis according to following reaction procedures [32,83]:

$$SiO_2 + 2H_2O \rightarrow Si(OH)_4$$

$$Si_3N_4 + 6H_2O \rightarrow 3SiO_2 + 4NH_3$$

$$MgO + H_2O \rightarrow Mg(OH)_2$$

MgO, specially, is widely used as an encapsulating oxide layer in implantable devices such as endovascular bioresorbable stents, to control the rate of degradation of the active agents, owing to its adequate biocompatibility [44]. The dissolution kinetics of MgO was investigated concerning the Mg^{2+} and H^+ concentration. For example, a solution where pH is larger than 7 at room temperature leads to a maximum dissolution rate [84]. Moreover, thermoplastic polyesters such as PLGA and water-soluble polymers including poly(vinyl alcohol) (PVA) are commonly used biodegradable polymers for medical implants and drug delivery systems [39,85,86]. As PGLA and PVA have the advantages of price, easy access for use, and easy fabrication, they have been used in a variety of devices for biomedical and biotechnology applications such as surface-mediated drug delivery and tissue engineering [87–89]. For instance, PLGA is used as a substrate for organic thin-film transistors, whereas PVA is used as the gate dielectric of the transistors [34]. In addition, natural silk fiber is also considered as a promising material for substrates or encapsulations owing to their outstanding mechanical robustness, biocompatibility, flexibility, ease of processing, and programmable biodegradability from minutes to years [49,66,90]. For example, Tao et al. reported a fully dissolvable wireless heating device operating in vivo to provide the necessary thermal therapy of trigger drug delivery. Here, an Mg resistor, an MgO dielectric layer, and an Mg coil are deposited onto a silk substrate. Further, all the layers were encapsulated in a silk overcoat, which could be used to adjust the lifespan of the device [91].

Besides the aforementioned biomedical applications, energy harvesting devices such as batteries that utilize the transient behavior of some materials have been studied (see Figure 3f) [62,92].The term "transient" here could be considered to mean that even though certain parts of the materials continue to exist, individual parts of the device could break apart and these devices do not function properly any longer. For example, Fu et al. worked on a transient battery that consisted of vanadium oxide (V_2O_5) as cathode, polyvinylpyrrolidone (PVP) as separator, sodium alginate (Na-AG) as battery encasement, and aluminum (Al), copper (Cu) as current collector [62].

2.2. Dry Transient Materials

In addition, materials that can spontaneously degrade in dry conditions or whose deterioration can be be triggered by a certain stimulus have been studied. They are called 'dry transient materials', whereas wet transient materials are materials that degrade when submerged in an aqueous solution or biofluid. Therefore, dry transient materials have the advantage of being suitable for use in a broad range of operating environments without the requirement of microfluidics. In general, they are utilized in encapsulations and substrates for integrating active electronics within transient devices. Beyond these characteristics, in the case of stimuli-triggered transient materials, they can provide further benefits, such as the control of the dissolution rate and the lifespan of the system, by triggering condition. Conversely, the dissolution rate of a wet transient device could be only determined by the materials selected and the fabrication procedure.

Cyclododecane (CDD) is a dry-degradable material, which disappears completely through sublimation when continuously exposed to air. Owing to its sublimating nature, it can be applied as a transient layer to protect fragile or sensitive surfaces. As CDD has a high vapor pressure of 0.1 hPa at 20 °C so it could sublimate adequately at room temperature, and has a low melting point (60.8 °C) as shown in Figure 4a. The sublimation rate of CDD could be affected by several conditions. For example, the rate is about 1.3 μm/h in the absence of ventilation, whereas it increases to 4.6 μm/h under ventilation (under air velocity at 1.9 m/s). In addition, it could be effectively tuned by adding 3–15% w/w of titanium dioxide nanoparticles (TiO_2 NPs, 10 nm of average diameter). This leads to an increase in sublimation enthalpy owing to the reduction of mesoscale extension of volatile molecular layers, resulting in reduced rate of mass loss. For example, the CDD samples doped with 8% TiO_2 NPs are approximately twice as thick as those of pure CDD after 500 h of sublimation [42].

In addition, poly(phthalaldehyde) (PPA) is an acid-sensitive metastable polymer, which has a low ceiling temperature (T_c = $-43\ °C$); therefore, it can be easily synthesized with various end-groups and it rapidly depolymerizes upon backbone bond cleavage [93,94]. Hernandez et al. demonstrated photo-triggerable or photo-degradable transient electronics fabricated on a cyclic PPA (cPPA) substrate with a photo-acid generator (PAG) as additive (Figure 4b,c). The electronics were disintegrated by stimulating the PAG/cPPA substrates with a UV source. The rates of transience were controlled by regulating the PAG concentration and the irradiance of UV light. To include UV light sensitivity, 2-(4-methoxystyryl)-4,6-bis(trichloromethyl)-1,3,5-triazine (MBTT) were utilized for PAG additives. When exposed to a UV light with maximum wavelength of 379 nm, the MBTT formed a highly reactive Cl· radical that extracts a hydrogen atom from ambient environment to form hydrochloric acid (HCl). Figure 4c illustrates the photoinduced disintegration mechanism of MBTT/cPPA substrate along with the electronics [93].

Park et al. reported the development of thermal-triggered transient electronics consisting of Mg electrodes and wax coatings, which contain triggerable-encapsulated acid microdroplets on acid-sensitive cPPA as substrate (Figure 4d,e). When exposed to sufficient heat (~55 °C), the device broke down rapidly as the protective wax coatings melted and released the encapsuled acid microdroplets. Rapid destruction due to the acidic depolymerization of cPPA can be induced using cPPA substrates. The overall mechanism of the Mg electrode degradation and cPPA substrate depolymerization are shown in Figure 4d. Furthermore, the destruction time can be controlled by tuning the thickness of the wax protection layer, acid concentration, and trigger temperature [94].

Depending on the materials used and the structure of devices, the range of the degrading temperature is controllable. Si NM electronics integrated with sufficiently thin and high-temperature degradable poly-α-methylstyrene (PAMS) was also suggested by Li et al. (Figure 4f–h). The PAMS layer releases volatile monomers (α-methylstyrene) when it is heated up to ~250–300 °C. The degradation mechanism of PAMS is illustrated in Figure 4h. Assuming that the PAMS layer would decompose completely to its monomers, the maximum pressure in a confined space could be estimated using the ideal gas law as follows:

$$P = \frac{nRT}{V} = \frac{\left(\frac{m}{M}\right)PT}{V} = \frac{\rho RT}{M}$$

where n, m, and M is the mole, mass, and molecular weight of the monomer, respectively. Here, density ρ = 1.075 g/cm^3, gas constant R = 8.31 $Pa·m^3/mol·K$, temperature T = 573 K, and M = 118.18 g/mol. In a confined space, the maximum gas pressure could be P = 43.34 MPa. The pressure from these gas products causes a high stress (~23 kPa) on the Si NM layer, which suffers a displacement of up to 11.9 μm (see Figure 4g). This distortion in the layers, results in a functional degradation of the device [95].

Owing to the advantages of each of the aforementioned materials, devices that integrate both dry and wet transient materials have been studied. For example, Camposeo et al. presented a dry-wet transient device in photonics, based on fully organic components. It consists of a water-soluble compound layer such as PVA on CDD as the sublimating substrate [96]. In conclusion, by utilizing the dry transient materials for some components of the device which operate outside of the body with bioresorbable materials for the in vivo part, a fully degradable sensor for healthcare monitoring can be developed.

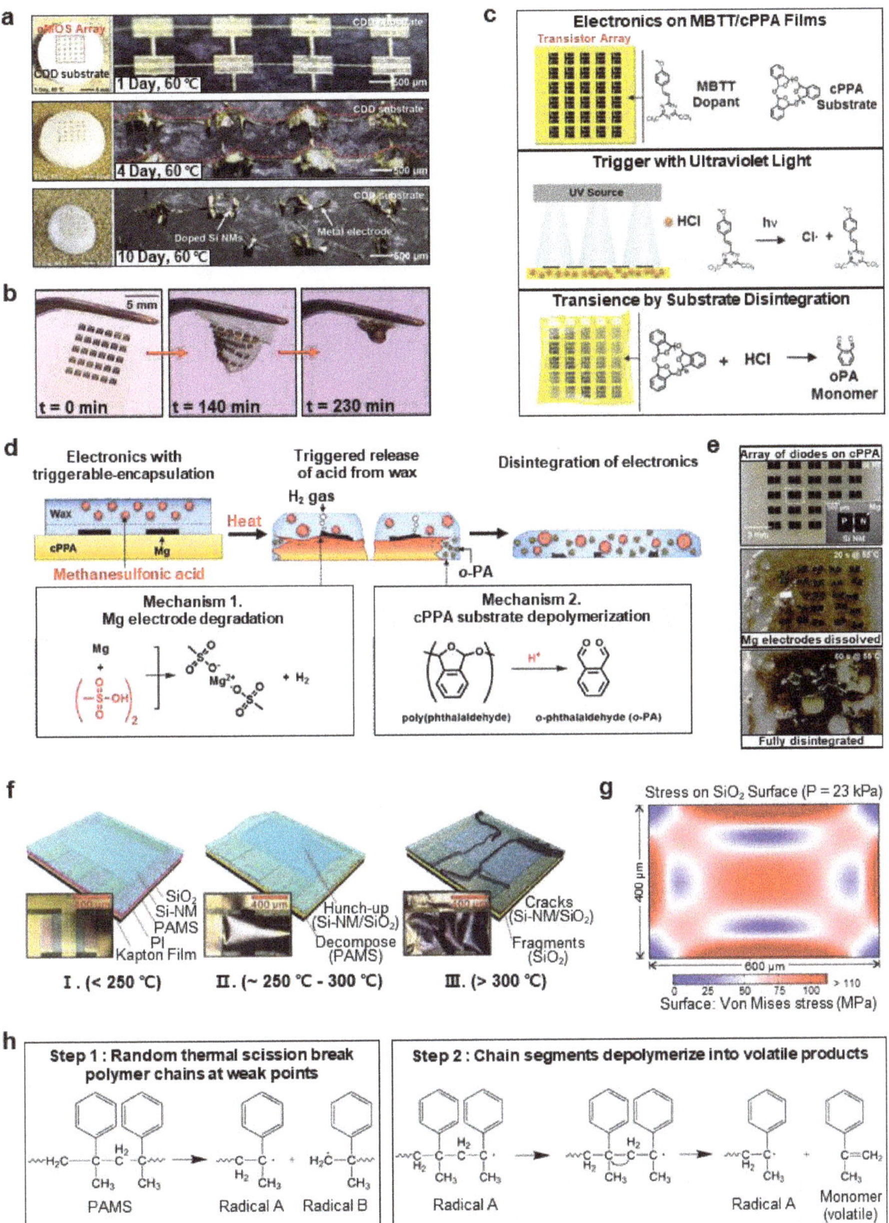

Figure 4. Examples of dry transient materials. (**a**) Microscopic images showing the time sequence of disintegration behavior of an array of transistors due to a sublimation of the supporting substrate of CDD [42]. Optical images (**b**) and mechanism with chemical structure (**c**) of UV light-triggered transient material [93]. (**d**) Illustrations and chemical structures of the disintegration process, and (**e**) optical images of the transient behavior of 55 °C heat-triggered transient device [94]. (**f**) Schematics of device structure failure and (**g**) FEA analysis image of stress distribution on SiO₂ layer surface of 330 °C heat-triggered transient device. (**h**) Degradation mechanism of PAMS [95].

3. Mechanism, Development, and Applications of Biodegradable Pressure Sensors

3.1. Mechanisms

Biodegradable pressure sensors can be classified, on the basis of their working principle, into piezoresistive, piezocapacitive, piezoelectric and piezo-optical sensors. A piezoresistive pressure sensor is a sensor that indicates a change in pressure applied to the resistor by a change in the resistance as an electrical signal (Figure 5a) [97–99]. The mechanism of the piezoresistive type is associated with two main factors. One mechanism is the resistance change by the deformation of a resistor. When an external physical force is applied to the resistor, the dimension of the resistor changes according to Poisson's ratio and is determined by the material of the resistor [100]. The other mechanism utilizes the changing resistivity of the resistor material, such as the conductive particle–polymer composite. When a force is applied to the conductive particle–polymer composite, the effective resistance value is lowered as the inter-particle spacing between the conductive particles is reduced and the contact area is increased to improve the electrical path (i.e., percolation effect) [101–103]. Typically, piezoresistive pressure sensors have the following advantages: low cost, simple structure and working principle, and a relatively easy fabrication process. They are one of the most widely used pressure sensors [104]. Based on the aforementioned advantages, piezoresistive pressure sensors have been widely used in many applications such as automobiles, medical and home appliances, industrial and aerospace instruments [105–108]. In spite of its many advantages, the piezoresistive type pressure sensor has the disadvantages of relatively slow response time, low durability, cross-talk between temperature and strain as the resistance changes, and higher power consumption than other types of sensors [109]. To develop a high-performance pressure sensor, piezoresistive pressure sensors are being actively investigated, through the use of micro-/nano-structures and material designs, to maximize the deformation and the porosity of the resistor.

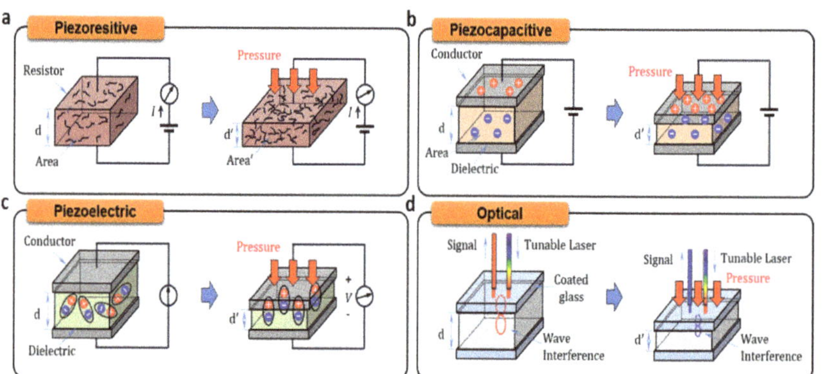

Figure 5. Schematic illustration of various mechanisms for biodegradable pressure sensors. (**a**) Piezoresistive type. (**b**) Piezocapacitive type. (**c**) Piezoelectric type. (**d**) Optical type.

The piezocapacitive pressure sensor measures the change in the capacitance (*C*) as electrical signal when pressure is applied to the device (Figure 5b) [110]. The capacitance is dependent on the dielectric constant (ε), the distance between the conductor plates (*d*), and the area of the conductor plates (*A*).

$$C = \varepsilon \frac{A}{d}$$

Generally, when a force is applied perpendicularly to the conducting plates of the capacitor, the distance between the plates decreases, the accumulated electric charge between the conductor plates increases and the capacitance value changes. The capacitive pressure sensors present a non-linear response as the sensitivity drops towards high pressure, owing to decreased compressibility at high pressure [111]. The piezocapacitive pressure sensors have the advantage of low hysteresis, good repeatability, fast response times, and insensitivity to temperature changes. In general, capacitive pressure sensors have been widely used in traditional industries such as automobiles and robots. Recent advances in microstructures, materials, and process design have made it possible to develop high performance piezocapacitive sensors that are small, light, and flexible. Therefore, recent capacitive pressure sensors can also be used in the healthcare and wearable devices [112–119].

Figure 5c is a schematic of the mechanism of working of the piezoelectric pressure sensor [110]. Piezoelectric effect is produced by a change in the arrangement of ions in the noncentrosymmetric structure of piezoelectric materials or the molecular structure of a polymer and its orientation (defined as "re-arragement of the material charges") when the dynamic pressure is applied. The electric charge can be measured as a voltage proportional to the applied pressure. Piezoelectric pressure sensors have the advantage of high frequency and fast response time which are very important in the automotive industry and aerospace fields [120]. However, it is difficult to measure static pressure because the output signal generated from the piezoelectric materials has a characteristic that the voltage gradually drops to zero in the case of constant pressure. Traditionally, piezoelectric pressure sensors have been developed using inorganic materials with high piezoelectric performance such as lead zirconate titanate (PZT). However, PZT, it has the property of toxicity and hence is not considered to be biocompatible. Although organic piezoelectric materials have recently attracted considerable attention use in biocompatible pressure sensors, this material often does not have a comparable piezoelectric output compared to inorganic piezoelectric materials due to intrinsically insufficient, normal or shear, piezoelectric effects. To improve the piezoelectric output of the organic pressure sensor, several research groups have attempted to design micro and nanostructures. For example, Curry et al. reported an organic nano-fiber based piezoelectric pressure sensor. The developed sensor exploits the aligned nanofiber of poly(L-lactic acid) (PLLA) that can maximize the piezoelectric performance with high flexibility based on the high alignment parallel to the input stress [121–123].

One of the other representative biodegradable pressure sensors for biomedical application is an optical type that measures the change of intensity or peak wavelength of the input light when pressure is applied (Figure 5d) [24]. Owing to its immunity to electromagnetic interference, the optical pressure sensor can be used to monitor pressure during an MRI scan or RF ablation procedure. The sensor also has other advantages such as the absence of potentially hazardous voltage, low weight, small size, high sensitivity, and large bandwidth. In addition, light intensity based sensors have the characteristic that they are very insensitive to temperature changes because the measurement and reference detectors are equally affected by the temperature and have very low hysteresis and repeatability errors. In particular, since the output signal of the optical type pressure sensor can be made sensitive to micro-/nano-scale changes, the device can be easily miniaturized and used in several applications that can adapt the micro-/nano-structures for the higher performances. In this regard, the miniaturized pressure sensor of the piezo-optical type is suitable for the medical field. The Fabry–Perot interferometer (FPI), using the interference of light reflected between two parallel glass plates, and the Fiber Bragg grating (FBG) using the principle that light propagating in an optical fiber reflects part of the light in a periodic grating, are widely used in medical applications to measure pressure in the blood vessels, lung pressure, bladder, brain, bone, and joint pressure [24,124–126].

3.2. Development of Various Biodegradable Pressure Sensor Devices

Based on the aforementioned pressure sensing mechanism, biodegradable pressure sensor technologies utilizing biodegradable (or bioresorbable, equivalently) materials have recently attracted considerable attention in the biomedical field, especially in bio-implantable applications. To adapt the biodegradable pressure sensor to biomedical applications, it is fundamentally important to have high sensitivity and fast response times to accurately measure pressure in the targeted bio-pressure range. In addition, a high flexibility for measuring the pressure on a curvilinear body, and power efficiency, stability, and reliability characteristics to reduce power consumption are also important. Although there are various studies to satisfy the above-mentioned requirements, studies using micro-/nano-structures are receiving a lot of attention especially because of their superiority in terms of their miniaturization and reproducibility. In this section, we will introduce the latest technologies for various high-performance biodegradable pressure sensors using micro-/nano-structures technology.

3.2.1. Biodegradable Piezoresistive Pressure Sensor

The first biodegradable piezoresistive pressure sensor was a piezoresistive porous foam using a multi-walled carbon nanotube (MWCNT)-Poly (glycerol sebacate) (PGS) (Figure 6a) [127]. The developed sensor was basically a composite of PGS, a biodegradable material, and MWCNT, which is conductive, as pressure sensing material. In this sensor, the MWCNT-PGS composite deforms according to the pressure, and inter-particle spacing decreases resulting in a change in resistance (percolation effect). Since the developed composite material has a microporous structure inside, it can easily cause mechanical deformation with respect to the applied pressure, thus high sensitivity can be expected (Figure 6b) [127]. One advantage of this method is that the sensitivity can be a design factor. The micro-porous density can be adjusted by the hydrolytic degradation time of the MWCNT-PGS composite, and the sensitivity can be adjusted from 0.12 ± 0.03 kPa^{-1} to 8.00 ± 0.20 kPa^{-1} (Figure 6c) [127].

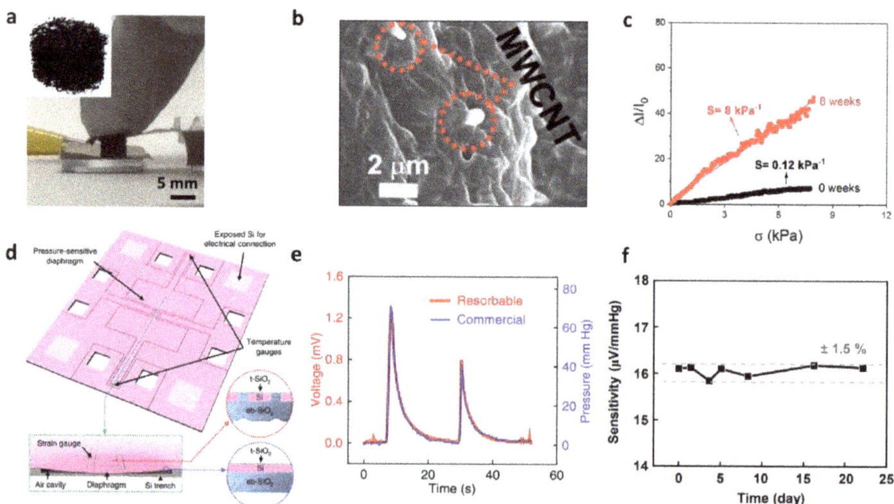

Figure 6. Biodegradable resistive pressure sensors. (**a**) Optical image of the piezoresistive foam sensor (inset: multi-walled carbon nanotube (MWCNT)-Poly (glycerol sebacate) (PGS). (**b**) Scanning electron microscope (SEM) image of the MWCNT-PGS foam. (**c**) Pressure-current response curve of the MWCNT-PGS foam device black), and its transition in the PBS solution (red) [127]. (**d**) Schematic illustration of bioresorbable pressure sensors protected with a thermally grown silicon dioxide (t-SiO₂) nano-layer. Inset: cross-section across the diaphragm revealing the air cavity and silicon (Si) trench located underneath. The cross-sections across the strain gauge (red inset) and non-strain gauge (blue) regions showing the tri-layer composition. (**e**) Responses of the fabricated device (red) and commercial sensor (blue) to time-varying pressures in in vitro evaluations. (**f**) Comprehensive results from continuous in vitro operation over a period of 22 days (variations in pressure sensitivity within ±1.5%) [30].

The material-based piezoresistive pressure sensor has advantages such as a simple fabrication process and structure. However, in this method, the deformation of the material containing the conductive fillers easily reaches the saturation state in the range of higher pressures; hence, the resistance change according to the pressure is very non-linear, and there are limitations in performance such as operation range. To improve the performance of the sensor, researchers have recently conducted a study on designing biodegradable materials in the shape of micro-/nano-structures [30,39,40].

A piezoresistive pressure sensor using a typical micro-/nano-structure was fabricated [39]. The developed device has a structure comprising a silicon cavity, a thin PLGA membrane, and a strain gauge. When pressure is applied, the PLGA membrane is bent into the Si substrate cavity. At this time, strain is applied to the Si nanomembrane strain gauge in the PLGA membrane, and the resistance changes. In order to maximize the sensitivity, the Si nanomembrane (NM) was placed on the clamp of the PLGA. To implement the proposed idea, a device was fabricated using microfabrication and transfer methods [39]. To characterize the piezoresistive performance of the device, the device was placed in a syringe filled with artificial cerebrospinal fluid (CSF) at physiological temperature (37 °C). The pressure was manually controlled by moving the plunger part of the syringe with the hands. The sensitivity of the device was 83 Ω/mmHg (622.55 Ω/kPa^{-1}) [39]. The biodegradability of the device was also tested by placing the device in a poly(dimethylsiloxane) (PDMS) chamber filled with an aqueous buffer solution of pH 12 at room temperature. In the test, the Si NM and SiO_2 components dissolved within 15 h, and the nanoporous silicon disappeared within 30 h [39].

Although the research on the piezoresistive pressure sensor using micro-/nano-structures greatly improved the performance, it is also necessary to improve the lifetime of the device in order to be able to use these devices in bioimplantable applications. When actual biofluids come into contact with active materials, water permeates through the material of the device or the interface between materials which leads to the deterioration in the performance of the device. Therefore, these functional lifetime issues must be resolved to meet clinical needs. Initially, passive polymer layers were added to delay the time during which the biofluids and the active area are in contact. However, biodegradable polymers such as PLGA have the limitations of poor stability because the hydrophilic nature of the material causes swelling and water permeation, which causes premature fracture, buckling and/or dissolution in materials in the active area. In a more recent study, a method that uses a thermally grown silicon dioxide (t-SiO_2) layer was reported. A t-SiO_2 layer has a low decomposition rate (rates of several hundredths of a nanometre per day) and can serve as a defect-free biofluid barrier in a wide area, resulting in a longer functional lifetime. A piezoresistive silicon pressure sensor with t-SiO_2 as encapsulation was facbricated using a silicon on insulator (SOI) wafer (Figure 6d) [30]. Since the developed device also consisted of an Si NM resistor and a Si cavity similar to the previous device, the pressure could be measured through the change in the resistance of the Si NM. In vitro evaluations that mimic the thermodynamic conditions inside (artificial CSF; pH 7.4 at 37 °C (physiological temperature)) the device show that comparison of the voltage responses of the sensor with the measured pressures of a commercial sensor reveal a linear correspondence (Figure 6e) [30]. In the reliability test, the functional lifetime of the device was maintained up to 22 days, and the sensitivity showed a reliability of $\pm1.5\%$ (Figure 6f) [30].

Although the t-SiO_2 can increase the functional lifetime by encapsulating the device, the rate of degradation of the device is very slow (10^{-3}–10^{-1} nm/day) [40]. It has a timespan in years which is much longer than clinically relevant operational requirements required to achieve complete dissolution of the t-SiO_2 layer. Therefore, the researchers conducted a study to reduce the physical lifetime of the device, while continuing to have the required functional lifetime, and an encapsulation method combining the thin, lightly doped micromembrane of monocrystalline Si (Si MM) and natural wax was developed. The lightly doped Si MM is impermeable to water and is very thin; thus, the functional lifetime is improved, and the physical lifetime is also relatively shortened. Therefore, a piezoresistive pressure sensor, taking the different lifetimes into consideration, was fabricated. This device is mechanically stable and its lifetime is appropriately controlled using an Si MM encapsulation on the pressure-sensing part and covering the other part with natural wax [40]. The fabrication method for the device begins with fabricating the lightly doped Si MM that acts as a biofluid barrier. Then, using the conventional semiconductor processes, such as photolithography and etching, the Si MM, Si NM, and Mg substrate are patterned, and the components are assembled with the PLGA. Finally, the surface of device excepted the sensing area is covered with wax for the device-encapsulation. For the encapsulation, a natural wax material, with excellent lifetime, using Beeswax (CB01) and Candelilla wax (CB10) was also developed. Various ratios of the wax mixtures were tested to identify the optimal natural wax for encapsulation, and CB32 (ratio of Candelilla wax to Beeswax: 3:2) was identified as the best as it prevented water from penetrating into the device for more than 22 days [40]. Another advantage of the developed sensor is that there is little change in characteristics due to device biodegradation. Earlier pressure sensors based on a similar principle showed a change in the membrane thickness, due to degradation of the encapsulation layer with time, which results in a performance change such as a change in sensitivity. However, the fabricated device was designed considering a mechanical neutral line, minimizing the change in performance resulting from degradation of the encapsulation layer [40].

3.2.2. Biodegradable Piezo-Capacitive Pressure Sensor

The Bao group focused on a capacitive biodegradable pressure sensor that exploits the microstructured dielectric material. The material used as the dielectric was PGS. In general, elastomers having a smaller mechanical modulus exhibit greater viscoelasticity; thus, their mechanical response is relatively slow. However, PGS has a relatively high mechanical modulus of ≈ 1 MPa, and the viscoelasticity can also be controlled during the manufacturing processes, such as through reaction, curing temperature and time. The microstructure which is a square pyramidal structure of the PGS helps to achieve a high performance as a pressure sensor. This square pyramidal structure of the PGS has a large mechanical deformation and the fast response time because of the structure having an air layer inside [47,118,119,128–130]. In practice, the fabricated device shows a high sensitivity and fast response time. In the pressure ranges of p < 2 kPa and 2 kPa < p < 10 kPa, the sensitivity was 0.76 ± 0.14 kPa^{-1} and 0.11 ± 0.07 kPa^{-1}, respectively. Moreover, the response time was approximately 100–200 ms [47]. In vitro, the biodegradation studies of the device were also performed for 7 weeks in an incubation filled with PBS solution (pH 7.4) at 37 °C. The device consisted of PGS (dielectric layer), polyhydroxybutyrate/polyhydroxyvalerate (PHB/PHV, substrate layer), PVA (Adhesive layer), and iron-magnesium (Fe-Mg, electrode). After 7 weeks, the weight of the device remains at about 85% of its initial value. Fe-Mg of the device was dissolved for the first time, but both PGS films and PHB/PHV was not fully dissolved. These materials require a duration of at least a few months to fully dissolve [47].

A two-axis pressure sensor integrating pressure and strain and a simple pressure sensor was developed using biodegradable material. The sensor was designed to measure pressure in the vertical direction and the strain in the horizontal direction to distinguish between the two physical stimuli, independently. The developed two-axis pressure sensor also employs a capacitive mechanism. The pressure sensing and strain sensing was performed by measuring the capacitance between two electrodes while varying the distance between two different electrodes and by varying the overlapped area, respectively (Figure 7a) [131]. To fabricate the proposed two-axis sensor, a pressure sensor and a strain sensor are stacked together and then combined using a UV-curable, biodegradable polymer (poly (octamethylene maleate (anhydride) citrate); POMaC) [132]. The pressure sensor consists of a square pyramidal microstructure of PGS and a pair of Mg electrodes to detect pressure with high sensitivity. The varying overlapped electrode areas for the strain sensor were fabricated by designing separate thin-film comb electrodes with soft polymer. Figure 7b shows the strain response of the fabricated device for five consecutive cycles.

The applied strain is 0–15%, and the capacitance of the device changes between 3.5 and 7 pF [131]. The device shows negligible hysteresis. Figure 7c shows the pressure response characteristics for six consecutive linear loading–unloading cycles. In the pressure range from 0 to 100 kPa, the capacitance of the device changes between 3.5 and 11.5 pF, and the device shows the negligible hysteresis [131].

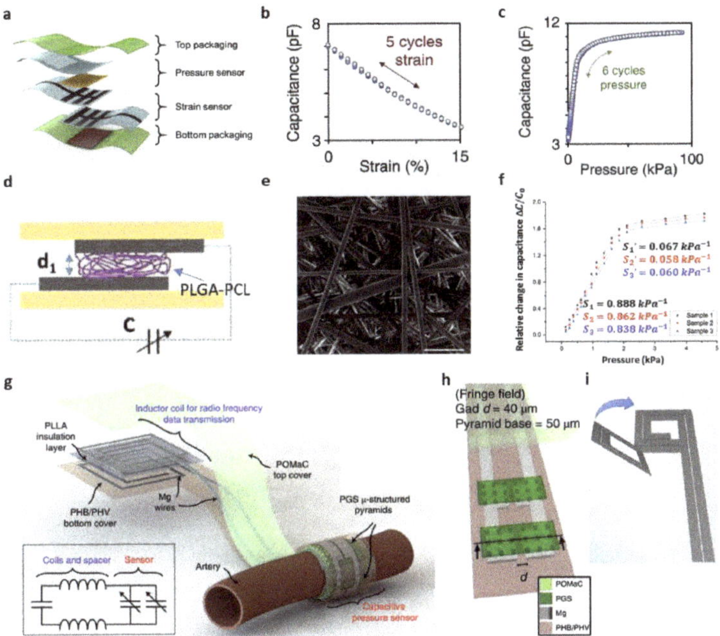

Figure 7. Biodegradable capacitive pressure sensors. (**a**) Schematic illustration of an integrated strain and pressure sensor. (**b**) Strain and capacitance-response in loading-unloading cycle (5-cycle). (**c**) Cyclic test of pressure and capacitance-response (6-cycle) [131]. (**d**) Schematic illustration of a biodegradable pressure sensor based on nano-fibrous dielectric. (**e**) SEM image of the polylactic-co-glycolic acid (PLGA)-polycaprolactone (PCL) layer (Scale bar, 10 µm). (**f**) Pressure and capacitance-response curve of the device [133]. (**g**) Schematic illustration of a biodegradable, flexible and passive arterial-pulse sensor. Inset: equivalent circuit of the device (left). (**h**) Flexible design of the pressure-sensitive region of the device. (**i**) Optimal design of the inductor coil (asymmetrical; the top coil turns clockwise, whereas the bottom coil turns anticlockwise) [134].

In addition to the microstructured dielectric approach, there is also a method to construct the nanostructured dielectric layer using composite nanofiber membranes (CNMs). A typical method for fabricating the CNMs is the electrospinning process. The electrospinning process is to continuously draw out fibers from a viscous polymeric solution through rapid solvent evaporation using an electric field. The advantage of this approach is that it can produce fibers with very thin diameters of a few micrometers or nanometers. In addition, it has the advantages of excellent flexibility, porosity, lightweight and compatibility with the printing process. This method has been actively used to fabricate the active layer of the strain sensor and the dielectric layer of the pressure sensor. CNMs using biodegradable composite polymer (poly(lactic-co-glycolic acid)-poly(caprolactone); PLGA-PCL) as a dielectric material for the biodegradable pressure sensor was reported (Figure 7d) [133]. CNMs can have very low mechanical modulus because they have a considerable number of air pores owing to the fiber network inside (Figure 7e) [133]. The fabricated sensor has a sensitivity of 0.863 ± 0.025 kPa^{-1} in a specific pressure range ($0 < p \leq 1.86$ kPa) and high detectivity (1.24 Pa at 10 mgf) (Figure 7f) [133]. In addition, this sensor has a low pressure-detection value of 1.24 Pa, and good average response and recovery times of 251 ms and 170 ms, respectively. In vitro the biodegradation studies of the device were also performed for 18 days in an incubation filled with PBS solution at 37 °C. After 14 days, the weight of the device remains about 40% of its initial weight. The device continued to degrade until 18 days [133].

Thanks to the emergence of micro-/nanofabrication, various biodegradable materials have come to be fabricated with high designability and scalability, and it enables the development of wireless biodegradable pressure sensors exploiting various passive biodegradable components. The biodegradable wireless pressure sensor can be fabricated by combining a variable capacitive sensor (C) and an inductor (L). This LC combination will have a resonance frequency, and the change in resonance frequency caused by an external stimulus can be measured by an external secondary coil connected to the impedance analyzer wirelessly. A wireless pressure sensor was fabricated by building a simple LC circuit using microstructured biodegradable materials, metal alloys (iron-zinc alloy; Fe-Zn) and polymers (polylactic acid; PLA, polycaprolactone; PCL) [64]. The Fe-Zn alloy was designed to promote the corrosion of Zn. This is because pure Zn decomposes slowly in a biological environment. The microfabricated coil acts as an inductor, and the two microplates and air cavity act as a capacitor. The principle of wireless measurement of the sensor is that when pressure is applied to the sensor, the distance between the plates decreases, so the capacitance value changes. The phase and magnitude of the RF signal with respect to the resonant frequency can be measured [64]. The performance of the fabricated sensor was evaluated in a sealed chamber. The resonant frequency signal of the sensor, which depends on the applied pressure, was measured in wireless manner in air and in a saline (0.9% NaCl in deionized water) environment. When the applied pressure was increased, the resonant frequency significantly decreased. In the pressure range of 0–20 kPa, the resonant frequency shift of the device is -39 MHz/kPa and -35 kHz/kPa in air and saline, respectively [64]. In vitro the biodegradation studies of the Fe-Zn were also performed for 300 h in 0.9% saline (NaCl, pH 6.8) at 37 ± 0.5 °C. The degradation rate was divided 3 stages. The degradation rate of stage 1 (between 0 and 72 h), 2 (between 84 and 180 h), and 3 (after 200 h) was 0.15 mg/h, 0.05–0.1 mg/h, and under 0.04 mg/h, respectively [64].

Recently, the functional lifetime of a biodegradable wireless pressure sensor has been extensively studied. Lu et al. attempted to increase the functional lifetime by encapsulating the device with silicon nitride (Si_3N_4) micromembranes and natural wax in the wireless sensor [135]. The thin micromembranes (2 μm), Si_3N_4, help to use the mechanical properties of the flexible top electrode. In practical, the bioresorbable water barrier structures, Si_3N_4 micromembranes and natural wax, prevent water permeation and ensure stable operation electrode [135]. The rate of degradation of the Si_3N_4 and wax are 4.5–30 nm per month and 10 μm per month, respectively. In addition, there were attempts to reduce the sensor's parasitic capacitance (C_p) and increase the sensitivity as well as the functional lifetime of the sensor. The sensor was designed by removing a part of the bottom electrode from the existing parallel plate capacitor structure, creating an additional trench around the cavity, and adding wax to the edge between the two electrodes. This design reduced the C_p by relaxing the overlap between the top and bottom electrodes, which is one of the main sources of C_p, and increased the sensitivity by increasing the deformation. A sensor designed in this manner can reduce noise compared to the capacitance sensor built with an intuitive approach. As a result of measuring the applied pressure and the resonant frequency over time, using the external coil in vitro environment, the performance of the sensor was found to be similar to the performance of a clinical standard device used for measuring intracranial pressure (ICP) monitoring electrode [135]. In vitro the biodegradation studies of the device were also performed for 44 days in PBS solution at 37 °C. The device consisted of Mg, PLGA, Si_3N_4, Zn, and wax. Over a period of 44 days, the device was gradually dissolved. After 44 days, leaving only the Si_3N_4 and the wax. The degradation rate of Si_3N_4 and wax are 4.5–30 nm/month and 10 μm/month, respectively. The full degradation of the device materials will require a duration of at least a few months [135].

Recently, a cuff-type biodegradable wireless piezocapacitive pressure sensor having high flexibility has been studied (Figure 7g) [134]. The sensor consists of a capacitive pressure sensor, which detects the applied pressure by observing changes in the electrical fringe field. Owing to the microstructure of pyramidal dielectric, the device deformation causing the fringe field, and the change in capacitance are easily generated, when an object comes into contact and a pressure is applied. The sensor was also made with soft encapsulation material, a biodegradable polymer (POMaC) and comb-designed metal electrode, so that the device could have a flexible configuration (Figure 7h). In particular, the developed sensor was designed to have a micro-bilayer structure as the coil (Figure 7i). The micro-bilayer coil structure is favorable to obtain a lower resonant frequency owing to the high mutual inductance. This causes the signal to undergo a lower attenuation within the body, resulting in a longer operating distance of the wireless sensor. Moreover, this design can provide a larger frequency shift for a given applied pressure, which means higher sensitivity. The device was made entirely of biodegradable materials, including PGS, POMaC, PHB/PHV, PLLA, and Mg [134].

3.2.3. Biodegradable Piezoelectric Pressure Sensor

With the recent development of piezoelectric biodegradable materials, it has become possible to realize biodegradable piezoelectric pressure sensors. A piezoelectric pressure sensor based on β-glycine-chitosan (β-Gly/CS) microfilm, a biogradale organic material, with a noncentrosymmetric polar structure was fabricated (Figure 8a) [136]. This β-Gly/CS microfilm was fabricated by embedding grown bio-organic glycine crystals (β-glycine) inside a chitosan polymer used as a matrix material (i.e., self-assembly). The fabrication method using the self-assembly makes the unstable β-glycine with a piezoelectric coefficient (d_{16} = 174 pmV^{-1}) uniform and thermodynamically stable, making it possible to act as a piezoelectric film (Figure 8b) [136]. To fabricate the piezoelectric pressure sensor, a simple solvent-casting method is employed. The sensor was fabricated by depositing Au electrode using a hard mask on both sides of the β-Gly/CS microfilm fabricated earlier. The fabricated piezoelectric pressure sensor was characterized using a vibration system (S50018) to measure the performance of the output voltage of the device. Under dynamic vibration, the device stably generated the open circuit output voltage with a fast response time (<100 ms). In particular, the measured sensitivity of the sensor was 2.82 ± 0.2 mV/kPa in the pressure range of 5–60 kPa (Figure 8c).

To increase the sensitivity in a specific pressure range of the biodegradable piezoelectric pressure sensors, Curry et al. fabricated a biodegradable piezoelectric pressure sensor based on stretched poly(L-lactic acid) (PLLA) microfilm (Figure 8d) [122]. Stretched PLLA microfilm was fabricated using PLLA, a biodegradable polymer approved by the Food and Drug Administration (FDA) [137,138]. In particular, they developed the high performance PLLA piezoelectric film through thermal annealing, mechanical stretching, and cutting processes. Thermal annealing and mechanical stretching of the PLLA microfilm helped form the piezoelectric property in the PLLA film because these methods could create a 45° alignment of the carbon-oxygen double bonds (C=O) that intrinsically have the piezoelectric property [139,140]. Then, by cutting the film with a 45° angle along the stretching direction, the optimized PLLA piezoelectric film is fabricated. To complete the fabrication of the piezoelectric pressure sensor, Fe and Mg alloy electrode is deposited on both sides of the PLLA microfilm using an electron beam evaporator. Then, Mo wires are placed on each surface of the electrode and it is encapsulated within the PLA layers. Finally, it was enclosed using biodegradable PLLA glue and a thermal bag sealer. Figure 8e shows the optical image of the fabricated device. Figure 8f shows that the output voltage increased as the applied pressure was increased from 0 kPa to 18 kPa. The evaluated sensitivity was about 0.12 V/kPa in the range of 0 kPa < p < 2 kPa, and a sensitivity is about 0.013 V/kPa in the range of 2 kPa < p < 18 kPa [122]. In vitro the biodegradation studies of the device were also performed in the buffered solution at an accelerated-degradation temperature of 74 °C. The device consisted of biodegradable materials (PLA [141], Mo [142], and PLLA), and completely degraded after 56 days [122].

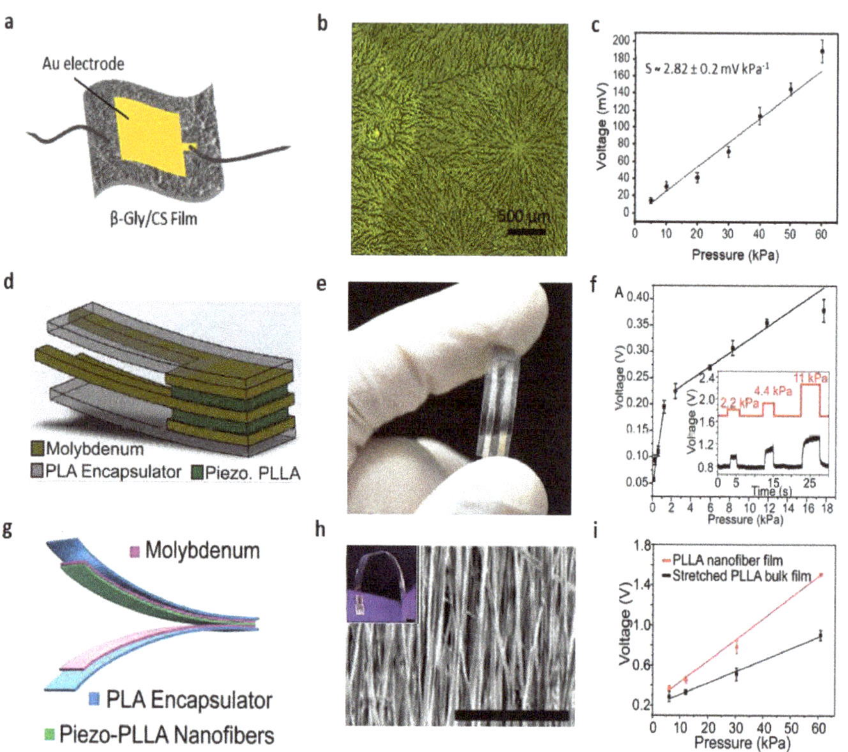

Figure 8. Biodegradable piezoelectric pressure sensors. (**a**) Design of β-glycine-chitosan (Gly/CS)-based flexible biodegradable piezoelectric pressure sensor. (**b**) Optical microscope image of the β-Gly/CS film. (**c**) Piezoelectric sensitivity of the sensor as a function of applied pressure [136]. (**d**) Schematic illustration of a biodegradable piezoelectric PLLA pressure sensor. (**e**) Optical image of the fabricated device (5 mm × 5 mm and 200 µm thick). (**f**) Pressure-response curve generated by a poly (L-lactic acid) (PLLA) sensor. Inset shows output voltage signals from different input pressure [122]. (**g**) Schematic illustration of a biodegradable nanofiber-based piezoelectric pressure sensor. (**h**) SEM image of piezo-PLLA nanofibers (Scale bar, 40 µm) (inset: the flexible fabricated device). (**i**) Comparison of calibration curves for a biodegradable sensor using stretched, bulk piezo-PLLA film (black) and a PLLA nanofiber film (red) [123].

To improve the piezoelectric response of PLLA, the same research group attempted to fabricate a PLLA nanofiber film (Figure 8g) [123]. The PLLA nanofiber film was made of nanofiber PLLA material using a rotating drum. The rotating drum aligned the crystal domains of the PLLA nanofibers and improved the piezoelectric properties because of the uniform and scalable unidirectional polarized structure that was created. Figure 8h shows the orientation of the crystal domains inside 4000 rpm electrospun PLLA nanofibers, and the inset shows the flexibility of the sensor. Moreover, Figure 8i shows a 1.8 times higher piezoelectric response of the PLLA nanofibers film compared to the stretched PLLA microfilm.

3.2.4. Biodegradable Optical Pressure Sensor

A measurement method using resonant peak positions in the reflection spectra is one of the representative methods of biodegradable optical pressure sensors. Shin et al. fabricated a biodegradable optical micro-pressure sensor using the principle of Fabry–Perot interferometers (FBI) (Figure 9a) [143–146]. To measure the pressure using the sensor, a tunable laser source illuminates the sensor through a PLGA optical fiber. A photodetector measures the resonator signals corresponding to the thickness of Si slab microcavity, the volume of which changes depending on the applied pressure (Figure 9b) [143]. The amorphous silica layer (thickness: ~200 nm), the Si slab microcavity (thickness: 10 μm), and t-SiO$_2$ layer (thickness: ~250 nm) serve as glass, which can reflect the light, a resonator, and a passivation film, respectively. For in vitro evaluations, the FPI based biodegradable pressure sensor is immersed in PBS (pH 7.4). In the PBS solution, the device shows a sensitivity of −3.8 nm/mmHg with an accuracy of ±0.40 mmHg in the pressure range of 0–15 mmHg (Figure 9c) [143]. The lifetime of the device is also evaluated. For the evaluation of the functional lifetime, the sensor was immersed in PBS at 37 °C for 8 days. The device did not show any significant degradation in performance.

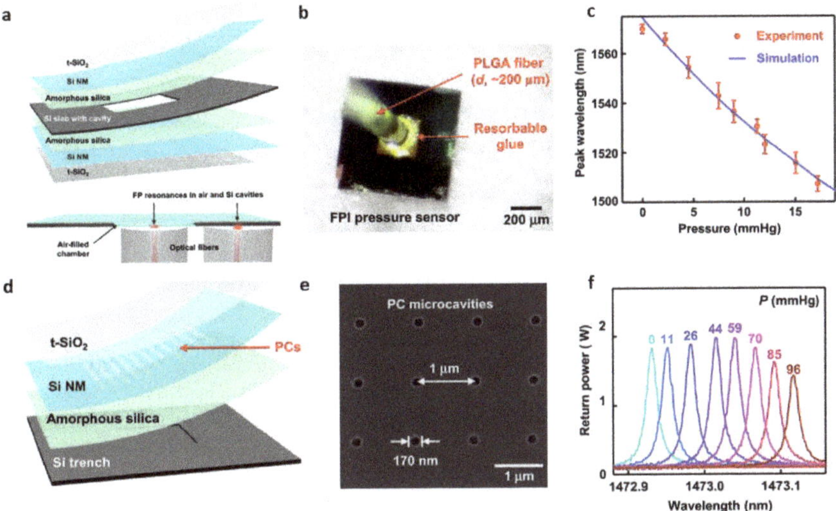

Figure 9. Biodegradable optical pressure sensors. (**a**) Schematic illustration of a biodegradable Fabry–Perot interferometers (FBI) pressure sensor (top). A cross-sectional view of the sensor integrated with two optical fibers (bottom). (**b**) Optical image of a PLGA fiber and the device. (**c**) Pressure-peak wavelength-response curve for the device (red) compared with simulation results (blue). (**d**) Schematic illustration of a biodegradable photonic crystal structure (PCs) pressure sensor. (**e**) SEM image of microcavities of PCs. (**f**) Pressure–wavelength response of the device (red) compared with simulation data (blue) [143].

The same group further fabricated a sensor that measures pressure using microcavities through photonic crystal structures (PCs) [125,147–149] that is a periodic structure in the nanometer range (Figure 9d) [143]. The sensor with nanostructures of PCs on a flexible diaphragm yields sharp resonance peaks with a high Q factor. When pressure is applied to the sensor, the size of the nanostructure changes, and then the resonance peak shifts. The nanostructures of PCs also make wireless measurements possible, it can reduce the possibility of problems such as infection [143]. The principle of this pressure sensor is that when the pressure increases, the grating size of the PC microcavities in SI NM increases; the grating size of PC microcavities in SI NM increases with the increased pressure, resulting in the increase in the resonant wavelength. The fabrication process of the sensor consists of fabricating SOI-A with PC microcavities and SOI-B with silicon trench via e-beam lithography and deep reactive-ion etching (DRIE), respectively, and combining with PDMS between SOI-A and SOI-B (Figure 9e). In particular, PDMS is transformed into silica by calcination [143]. In in vitro evaluation, the environment has a tunable laser source, tilt stage (chamber), circulator, and photodetector. The sensor has a sensitivity of 1.9 pm/mmHg and an accuracy of ± 0.64 mmHg in the wide range of pressure (0 to 100 mmHg) (Figure 9f) [143].

3.3. Biomedical Applications

Attempts have been made to apply the developed pressure sensors to practical biomedical applications. This section introduces the biomedical applications using biodegradable pressure sensor such as ICP, blood pressure, tendon, and intra-abdominal pressure.

3.3.1. Application for Intracranial Pressure (ICP) Monitoring

In the body, ICP monitoring is necessary to diagnose diseases such as traumatic brain injury, aneurysms, brain tumors, hydrocephalus, stroke, and meningitis. Tests on animals (rat model), using developed microstructured Si piezoresistive biodegradable pressure sensor, were conducted in Section 3.2.1 [39]. In the rat model, the initial piezoresistive-type biodegradable pressure sensor was connected to a wireless transmitter with a potentiostat through a degradable wire [39]. The wireless transmitter was used to measure pressure after surgery; it was protected using a head protector, allowing the rat to freely move around [39]. In in vivo evaluation, to measure ICP, the sensor was mounted on the top of the skull. A comparative test of the clinical intracranial pressure sensor (Integra Life Sciences, Princeton, HJ, USA) transplanted in parallel with a biodegradable pressure sensor encapsulated with polyanhydride showed stable results for three days [39]. Further, in vivo observation of the changes in ICP as a function of time in the Trendelenburg and reverse Trendelenburg positions were made. Trendelenburg position and reverse Trendelenburg position are associated with accumulation and depletion of blood in the brain, respectively. ICP increases in Trendelenburg position ($30°$ head-down) as compared with the supine position, and decreases in reverse Trendelenburg ($30°$ head-up) [39].

Biodegradable resistive pressure sensors focused on the encapsulation (t-SiO$_2$ nanolayer for device-protection) in Section 3.2.1 also showed highly stable results in the in vivo test, using a male Lewis rat model [30]. For the ICP monitoring, a flank squeeze (contract, release), Trendelenburg position (flat, $30°$ head-up, down), and a mannitol infusion (dose: 2 g per kg weight) were performed. In the flank squeeze test, the ICP increased up to 20 mmHg in the contract, and decreases below 10 mmHg in the release. In the Trendelenburg position test, the ICP decreased below 7 mmHg in the head-up, and increased above 9 mmHg in the head-down. For mannitol infusion in the saphenous vein, the ICP decreased from 8 to 1 mmHg after intravenous drug infusion. In addition, the fabricated sensor device also collected the varying ICP continuously for 18 days. The device shows the stable functional lifetime as a signal at 18 days after surgery (red: bioresorbable sensor; blue: commercial sensor). Over a period of 25 days, the response of the sensor showed an absolute accuracy within ± 2 mmHg, baseline drift within ± 1 mmHg, and negative drift of ~3 mmHg [30].

The biodegradable wax-encapsulated piezoresistive pressure sensor was also tested for ICP measurement [40]. The procedure for implantation consisted of exposing the skull area, opening a craniotomy defect, attaching a sensor using dental cement over the defect and commercial bioresorbable glue (COSEAL surgical sealant) on the edge, and sealing the cavity. At this time, the opening of the skull ensures contact with the cerebrospinal fluid (CSF) to enable ICP monitoring. The in vivo sensor was tested for Trendelenburg position ($30°$ head-down) and reverse Trendelenburg position ($30°$ head-up), compressing and releasing the rat's flank for a short period (<3 min). Baseline and sensitivity drift of the sensor through the flank contracting and releasing was measured on 0, 7, 14, and 21 days postsurgery. Baseline and sensitivity showed a variation of ± 1.0 mmHg and 2.1% of 15.0 Ω mmHg^{-1} over 3 weeks, respectively [40].

In addition to the resistive type, there was also an in vivo evaluation of the rat model for the optical FPI type biodegradable pressure sensor (Figures 9a and 10a) [143]. The implantation procedure, which was approved by the Institutional Animal Care and Use Committee (IACUC) of Northwestern University, consisted of drilling a small defect hole inside the skull, and implanting the sensor inside, putting a film of PLGA with a hole in the center (dimensions, 5 mm × 5 mm × 10 μm; hole diameter, ~400 μm) on the top, and bonding the PLGA film and the skull by applying a layer of bioresorbable glue to form an airtight seal on the intracranial cavity. The hardened glue helps to prevent changes in the sensor–fiber alignment due to shear or slanting effects during and after implantation (Figure 10b) [143]. In in vivo testing, the sensor measured the ICP by squeezing and holding the rat's flank. Figure 10c shows the optical spectra and pressure calibration curves obtained from the FPI pressure sensor. For contracting the flank, the sensor's ICP was measured in the range 3–13 mm Hg, and the sensitivity was 3.1 nm/mmHg [143].

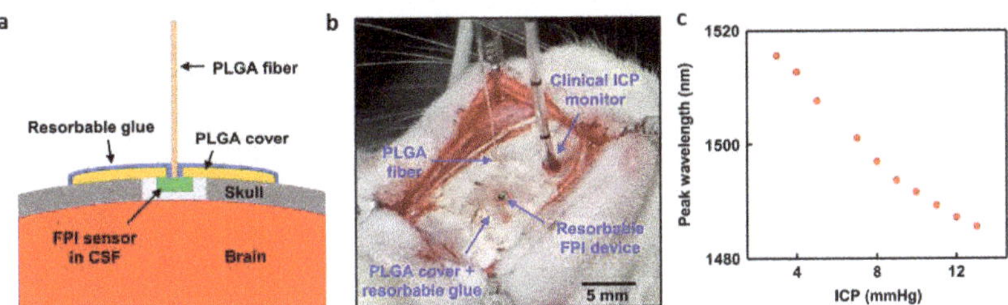

Figure 10. Application of biodegradable pressure sensor for the intracranial pressure (ICP) monitoring. (**a**) Cross-sectional schematic illustration of the device setup for animal-test. (**b**) Photograph of a bioresorbable FPI sensor implanted in the intracranial space of a rat for monitoring ICP. (**c**) Pressure calibration curve obtained from a bioresorbable FPI pressure sensor [143].

A study on the biodegradable wireless sensor for ICP monitoring was also conducted. The biodegradable wireless piezocapacitive pressure sensor with wax encapsulation, including an LC circuit (primary coil), was tested in vivo [135]. The implantation procedure includes sterilization of the head, creation of craniectomy over the head, and implantation of a bioresorbable sensor over the skull. Applying dental cement (Fusio Liquid Dentin) and curing under ultraviolet light fixed the implant on the skull and made the sensor airtight [135]. In the animal test, the ICP was increased by 5–10 mmHg for 15 s by gripping the rat's flank by hand, under anesthesia, and then released. When measuring the ICP, the body temperature of the rat measured using a rectal probe was 35.5 °C. A result of the ICP monitoring in rats for 4 days after implantation shows a stable functional lifetime. The readout system consists of a single turn coil (external secondary coil) connected to an Agilent E5062A or Agilent portable N9923A vector network analyzer (VNA), and the real and imaginary parts of the S-matrix element S_{11} were measured by setting the VNA to reflective mode. At this time, the variable resonance frequency of the LC circuit was measured between 309 and 312 MHz [135].

3.3.2. Application for Pressure of Blood Vessels, Tendon, and Intra-Abdominal

In addition to the ICP, the biodegradable pressure sensor can be utilized to measure the pressure of other tissues, for various in vivo applications, such as the pressure of blood vessels, tendon, and intra-abdomina. Cardiovascular diseases are the leading cause of premature death worldwide today, and hypertension has been defined as the main risk factor and leading cause of cardiovascular diseases. It is important to detect hypertension and continue monitoring before too much damage occurs. Measurement of the carotid-femoral pulse wave velocity (cf-PWV) is currently the "gold standard" for measuring aortic arterial stiffness which is directly related to hypertension, and is a reference in the international cardiology/hypertension treatment guidelines [27]. A biodegradable piezocapacitive pressure sensor with a pyramid microstructured dielectric is suitable for cardiovascular monitoring such as arterial tonometry and cf-PWV measurement due to its high sensitivity and fast response time [47]. To measure the cf-PWV, the arterial pulse wave and an electrocardiogram (ECG) were recorded simultaneously for time reference. At this time, the sensor was first attached to the adult's neck (carotid artery) and then to the groin (femoral artery). The cf-PWV (v) was measured by dividing the time delay (t) between the feet of the carotid and the femoral pulse signals by the distance (d) between the neck and the groin. The measured cf-PWV is 7.5, a result typically obtained in healthy subjects [47].

After injury to the tendons, ligaments, and joints, body tissues such as the hard tissues (bones) and soft tissues (tendon, skin, muscle) undergo changes in their native biomechanical properties to repair themselves. The purpose of surgery and rehabilitation is to restore tissues to their pre-injury function. Monitoring the biomechanical characteristics of the repair site in real time can be a diagnostic tool that can predict the healing process for personalized rehabilitation. The level of strain and strain rate of tissues during the rehabilitation are the most important parameters that can characterize the biomechanical properties and healing stage of soft tissues. An implantable sensor should be able to measure the typical tendon strains (<10%) without interfering with the natural movement of the tendon. In addition, it should be possible to measure the pressure applied to the repaired area that directly affects the healing profile [131]. The previously developed sensor (Figure 7a) that can measure strain and pressure independently without interfering with each other is suitable for this biomedical engineering application [131]. For in vivo testing, three Sprague Dawley rats were treated in accordance with the regulation of the animal care and use committee of Veteran Affair palo Alto Health Care System Research Administration. The sensor was implanted into the subcutaneous paravertebral pocket under isoflurane inhalation anaesthesia (Figure 11a). Figure 11b shows the signals of pressure (top) and strain (bottom) successfully recorded 3.5 weeks after sensor insertion for pressure and strain detection. The strain and pressure were measured using a E4980A Agilent Precision LCR meter to measure the sensor capacitance in the vertical and horizontal directions, respectively. The sensor could distinguish between strains as small as 0.4% and the pressure exerted by the salt particles (12 Pa) without interfering with each other [131].

Failure after reconstructive surgery, including microsurgical anastomosis, may occur due to the formation of haematoma or thrombosis in the artery or vein. To ensure successful recovery after the surgery, it is important to measure blood vessel movement. Therefore, it is important to detect failure such as anastomosis early by monitoring the surgical site. Using the developed flexible cuff-type biodegradable piezocapacitive pressure sensor (Figure 7g), an in vivo testing was performed to wirelessly observe the movement of blood vessels [134]. This sensor measured the pulse rate of the femoral artery in Sprague Dawley rats (300–350 g, male; ENVIGO) in compliance with the regulation of the animal care and use committee of Veteran Affair Palo Alto Health Care System Research Administration.

The implantation procedure was performed under isoflurane inhalation anaesthesia. The sensor was wrapped around the femoral vessel and mounted on the abductor muscles with sutures. As we mentioned in the above section, this biodegradable sensor device can detect the vessel movement wirelessly. For the wireless sensor testing, the coil structure of the device was placed on a groin fat pad. Wireless measurement is performed by inductively coupling the implanted sensor coil structure with an external reader coil connected to a vector network analyzer. The reader coil measures the Δf_0 value of the LCR resonator through the scattering parameter S11. A pulse rate of 3.47 beat per second (b.p.s) was recorded (Figure 11c). Using this sensor, an occlusion test (femoral artery was blocked for 1 min and then released) was further performed; this test mimicked early clot formation in the vessel by placing two nylon sutures in the femoral artery on either side of the sensor and applying tension (which slows the blood flow through the artery and reduces the extension of the artery diameter). From the experimental results, it was clearly confirmed that the blood vessels have different movements during tension and release, and a pulse rate of 4.29 b.p.s was recorded (Figure 11d) [134].

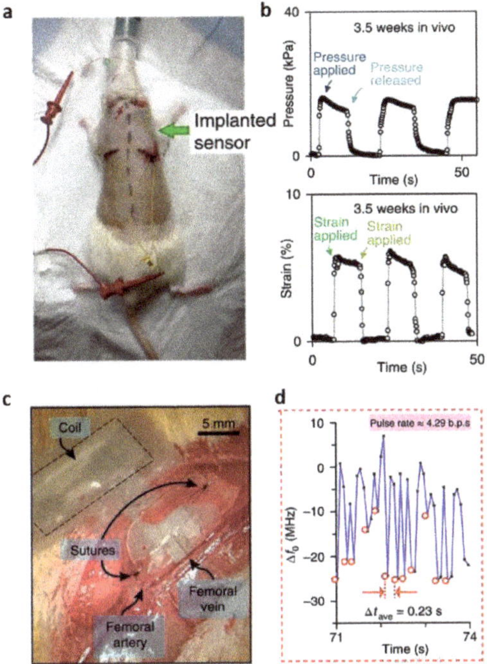

Figure 11. Other biomedical applications of the biodegradable pressure sensor. (**a**) Optical image of animal-test (rat model) implanted with the biodegradable strain and pressure sensor. (**b**) Measured pressure (top) and strain (bottom) signal recorded 3.5 weeks after sensor implantation [131]. (**c**) Optical image of an animal-test (rat model) implanted with a biodegradable and flexible arterial-pulse sensor. (**d**) Δf_0 measured on the femoral artery displaying an average pulse rate of 4.29 b.p.s. [134].

It was confirmed that the piezoelectric pressure sensor with the stretched PLLA could detect the breathing pattern by measuring intra-abdominal pressure [122]. The in vivo test was performed under the approval of the Connecticut Health Center's IACUC. A sensor coated with a very thin medical glue is inserted into a small incision (8 mm) made under the diaphragm of the rat in the abdomen. Small Mo/PLA wires from the sensing patch were connected to an external charge amplifier circuit connected to an oscilloscope, through a sutured wound, to measure the voltage. The respiration of rats under normal anesthesia after resting them for 15 min after surgery was monitored. Signals generated in the mice were completely suppressed after euthanasia due to an overdose of anesthetics. When the signal was alive, a frequency of ~2 Hz and ~0.1 N/cm^2 (~1 kPa) were measured. The sensor also detected abnormal respiration with a lower frequency and greater pressure until the animal died after an anesthetic overdose [122].

3.3.3. Biocompatibility

Biocompatibility can be defined as the compatibility between a material and biological system [150]. Therefore, biocompatibility of the devices through all stages of the life cycle is essential. In previous studies of biodegradable pressure sensors, immunohistochemistry (IHC), complete blood count (CBC) and blood chemistry, and histopathology studies were conducted. Immunohistochemistry is an immunostaining method that uses the reaction between antigen and antibody to identify substances present in cells or tissues. After implantation of the microstructured Si piezoresistive type sensor, comprehensive studies on the IHC of brain tissues several times (2, 4 and 8 weeks) prove that the byproducts of the sensor and dissolved device in the space within the skull are biocompatible through the absence of an inflammatory response in confocal fluorescence microscopy (CFM) images [39]. CFM images were double-immunostained for glial fibrillary acidic protein (GFAP) (red) to detect astrocytes and ionized calcium-binding adapter molecule 1 (Iba1) (green) to identify microglia/macrophages (dashed line is the implant site). This indicates that there is no concentrated aggregation of glial cells at the implantation site for all time ranges and no apparent response of brain glial cells. The astrocytosis and microglial activity on the cortical surface are within the normal range [39].

The CBC is a blood test used to evaluate overall health and detect a wide range of disorders. A blood chemistry test is a blood test that measures specific chemical content in a sample of blood. This can help make sure that certain organs are working well and locate disorders if any. Figure 12a shows the results of CBC and blood chemistry for the biodegradable resistive pressure sensor protected with t-SiO$_2$ nanolayer (Figure 6d) [30]. The CBC (left in Figure 12a) shows no significant difference between implanted and control mice over five weeks in the average counts of white blood cells (WBC), red blood cells (RBC), platelets (PLT), and in the levels of haemoglobin (HGB) and hematocrit (HCT). Blood chemistry (right in Figure 12a) shows that the blood levels of enzymes and electrolytes fall within the confidence intervals of control values. Normal levels of alanine aminotransferase (ALT), cholesterol (CHOL), triglyceride (TRIG), phosphorus (PHOS), blood urea nitrogen (BUN), glucose (GLU), calcium (CAL), albumin (ALB), and total proteins (TP) indicates the absence of disorders in the liver, heart, kidney, bone and nerve and good overall health. Control data were provided from a mouse supplier (grey) or collected from 22–24 mice (cyan) [30].

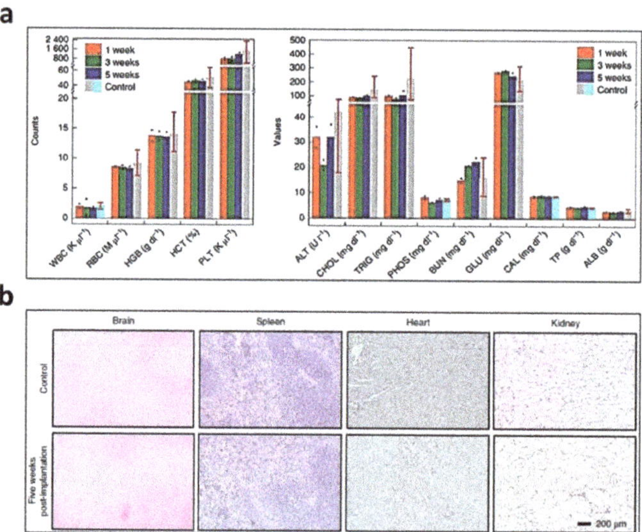

Figure 12. Studies of biocompatibility of biodegradable pressure sensors. (**a**) Results of the complete blood count and blood chemistry tests for the mice implanted with the biodegradable resistive pressure sensor. (**b**) Histopathologic evaluation of the biodegradable resistive pressure sensor in various tissue [30].

Histopathology is the diagnosis and study of diseases of the tissues, and involves examining tissues and/or cells under a microscope. Histopathologic evaluation of tissues obtained from a control mouse and a mouse implanted with the device protected with the t-SiO$_2$ nanolayer (Figure 6d) for five weeks showed absence of an inflammatory response, ischaemia/tissue necrosis, and architectural/histologic abnormalities indicating no significant changes within major organs such as the brain, spleen, heart and kidney (Figure 12b) [30,151].

4. Summary and Outlook

This paper highlights the recent advances in micro-/nano-structured biodegradable pressure sensor devices through discussions on material options, sensing mechanisms, device design, biocompatibility aspects, and biomedical applications. Various classes of materials including conductors, semiconductors, and insulators are summarized. The materials are categorized into wet and dry transient materials based on certain aspects in the biodegradation mechanisms; the materials are, generally, degraded based on hydrolysis in wet conditions, but a few materials can be spontaneously degraded in dry conditions or when triggered by a certain stimulus, resulting in wider material selection and utilization at the device level. Pressure sensing mechanisms, including piezoresistive, piezocapacitive, piezoelectric, and optical, were also introduced. The general theory of each mechanism is not only reviewed, but the structural strategy to improve device performances, such as sensitivity, response time, stability, and reliability, is also discussed in detail. An in depth review of micro-/nano-structured biodegradable pressure sensors that demonstrates their excellent pressure sensing performances, based on physical phenomena induced by the associated miniaturized structure is presented (Table 1).

Table 1. Summary of characteristics of biodegradable pressure sensors.

Ref.	Key Features	Mechanism	Key Structure	Target	Sensitivity (Pressure Range)	Functional Lifetime
[127]	Ultra sensitivity (8 kPa^{-1})	Resistive	MWCNT-PGS, Nanocomposite, Morphology and Porosity	E-skin	8 kPa^{-1} ($0 < p < 8.5$ kPa) (In a PBS solution)	N/A
[39]	Multi Si sensors	Resistive	Si-Nanomembrane	Intracranial pressure; Intra-abdominal cavity; Leg cavity	622.55 Ω/kPa ($0 < p < 10.67$ kPa)	3 days (In vivo)
[30]	Long-Functional lifetime (25 days)	Resistive	Si-Nanomembrane, t-SiO$_2$ layer	Intracranial pressure	0.98 Ω/kPa ($0 < p < 5.33$ kPa) (In vivo)	Fine on day 18; 4 mmHg drifts on day 25 (In vivo)
[40]	Fast dissolution of Si MM barrier; Wax edge barrier; Warning indicators	Resistive	Si-Nanomembrane	Intracranial pressure	118.51 Ω/kPa ($0 < p < 13.87$ kPa) (In vitro)	3 weeks (In vivo)
[64]	Wireless; RF	Capacitive	Zn/Fe bilayer conductor	Implantation application	39 kHz/kPa ($0 < p < 20$ kPa) (In saline)	86 h (In saline)
[47]	Microstructured pyramid; Sensing array	Capacitive	PGS pyramid structure of dielectric	Cardiovascular pressure	0.76 ± 0.14 kPa^{-1} ($0 < p < 2$ kPa) 0.11 ± 0.07 kPa^{-1} ($2 < p < 10$ kPa)	N/A
[134]	Microstructured pyramid; Contact/Non-contact mode; Wireless	Capacitive	PGS pyramid structure of dielectric	Arterial pulse	N/A	N/A
[133]	Nanofibers	Capacitive	PLGA-PCL nanofibers	E-skin	0.863 ± 0.025 kPa^{-1} ($0 < p < 1.86$ kPa) 0.062 ± 0.005 kPa^{-1} ($1.86 < p < 4.6$ kPa)	N/A
[37]	Wireless	Capacitive	Mg Coil & electrode	Orthopedic application	-45.00 ± 3.75 kHz/kPa ($0 < p < 26.66$ kPa) (In vitro)	8 h (In vitro)
[135]	long-Functional lifetime of Wireless sensor (4 days)	Capacitive	Mg Coil	Intracranial pressure	1500.12 kHz/kPa ($0 < p < 4$ kPa) (In vitro)	4 days (In vivo)

Table 1. *Cont.*

Ref.	Key Features	Mechanism	Key Structure	Target	Sensitivity (Pressure Range)	Functional Lifetime
[131]	Microstructured pyramid; Strain and pressure sensor	Capacitive	PGS pyramid structure of dielectric	Orthopedic application	0.7 ± 0.4 kPa^{-1} ($0 < p < 1$ kPa) / 0.13 ± 0.03 kPa^{-1} ($5 < p < 10$ kPa)	3.5 weeks (In vivo)
[122]	Maximal piezoelectric output of PLLA; bilayer	Piezoelectric	treated PLLA	Intra-abdominal cavity	~0.12 V/kPa ($0 < p < 2$ kPa) ~0.013 V/kPa ($2 < p < 18$ kPa)	4 days (In PBS)
[123]	Bilayer	Piezoelectric	PLLA nanofiber film	Intra-abdominal cavity	N/A	N/A
[136]	Organic piezoelectric material	Piezoelectric	β-Gly/CS film	E-skin	2.82 ± 0.2 mV/kPa ($5 < p < 60$ kPa)	N/A
[143]	Compatibility with MRI Wireless	Optical	Si-Nanomembrane; PC-microcavities; t-SiO$_2$ layer	Intracranial pressure	23.25 mm/kPa ($0.40 < p < 1.73$ kPa) (In vitro)	7 days (In vitro)

We discussed the recently developed micro-/nano-structured biodegradable pressure sensors including their concept, fabrication, and performances. Finally, we also introduced real cases showing the feasibility of the developed micro-/nano-structured biodegradable pressure sensors for diagnosis of disease and healthcare in wearable and implantable biomedical applications without any biocompatibility issues. Despite of the recent development in biodegradable pressure sensors, there are still some issues, such as sensitivity, operating range, and controllability of biodegradability, essentially to be improved for accurate pressure monitoring of various tissues and organs. In other words, regarding there are huge recent demands for healthcare and diagnosis of diseases in early stage through the pressure sensing of diverse tissues and organs, such as intracranial pressure [152–157], intraocular pressure [158–163], heartbeat [164–166], and bladder pressure [167–170], we expect that the high-performance biodegradable pressure sensors will be much more important for diverse biomedical applications in the future.

Author Contributions: Conceptualization, Y.-K.S., Y.S., J.W.L. and M.-H.S.; investigation, Y.-K.S., Y.S., J.W.L. and M.-H.S.; writing, Y.-K.S., Y.S., J.W.L. and M.-H.S., supervision, J.W.L. and M.-H.S. All authors have read and agreed to the published version of the manuscript.

Funding: This work was supported by a 2-Year Research Grant of Pusan National University.

Institutional Review Board Statement: Not applicable.

Informed Consent Statement: Not applicable.

Data Availability Statement: Not applicable.

Conflicts of Interest: The authors declare no conflict of interest.

References

1. Matsuzaki, R.; Todoroki, A. Wireless Monitoring of Automobile Tires for Intelligent Tires. *Sensors* **2008**, *8*, 8123–8138. [CrossRef] [PubMed]
2. Hu, Y.; Xu, C.; Zhang, Y.; Lin, L.; Snyder, R.L.; Wang, Z.L. A nanogenerator for energy harvesting from a rotating tire and its application as a self-powered pressure/speed sensor. *Adv. Mater.* **2011**, *23*, 4068–4071. [CrossRef] [PubMed]
3. Anderson, A.; Menguc, Y.; Wood, R.J.; Newman, D. Development of the Polipo Pressure Sensing System for Dynamic Space-Suited Motion. *IEEE Sens. J.* **2015**, *15*, 6229–6237. [CrossRef]
4. Culshaw, B.; Zhang, X.; Wang, A.; Jiang, J.; Liu, T.; Liu, K.; Wang, S.; Yin, J.; Zhao, B.; Zhang, J.; et al. Development of optical fiber sensing instrument for aviation and aerospace application. In Proceedings of the 2013 International Conference on Optical Instruments and Technology: Optical Sensors and Applications, Beijing, China, 17–19 November 2013. [CrossRef]
5. Sparks, D. MEMS pressure and flow sensors for automotive engine management and aerospace applications. In *MEMS for Automotive and Aerospace Applications*; Woodhead Publishing Limited: Sawston, Cambridge, UK, 2013; pp. 78–105. [CrossRef]
6. Bai, Y.; Qi, Y.; Dong, Y.; Jian, S. Highly Sensitive Temperature and Pressure Sensor Based on Fabry–Perot Interference. *IEEE Photonics Technol. Lett.* **2016**, *28*, 2471–2474. [CrossRef]

7. Du, H.H.; Pickrell, G.; Udd, E.; Baldwin, C.S.; Benterou, J.J.; Wang, A.; Baldwin, C.S. Brief history of fiber optic sensing in the oil field industry. In Proceedings of the Fiber Optic Sensors and Applications XI, Baltimore, MD, USA, 5–9 May 2014. [CrossRef]
8. Han, J.-H.; Kim, K.B.; Kim, J.H.; Min, N.K. High withstand voltage pressure sensors for climate control systems. *Sens. Actuators A Phys.* **2021**, *321*, 112410. [CrossRef]
9. Kanoun, O.; Bouhamed, A.; Ramalingame, R.; Bautista-Quijano, J.R.; Rajendran, D.; Al-Hamry, A. Review on Conductive Polymer/CNTs Nanocomposites Based Flexible and Stretchable Strain and Pressure Sensors. *Sensors* **2021**, *21*, 341. [CrossRef]
10. Ahmed, A.; Zhang, S.L.; Hassan, I.; Saadatnia, Z.; Zi, Y.; Zu, J.; Wang, Z.L. A washable, stretchable, and self-powered human-machine interfacing Triboelectric nanogenerator for wireless communications and soft robotics pressure sensor arrays. *Extrem. Mech. Lett.* **2017**, *13*, 25–35. [CrossRef]
11. Almassri, A.M.; Wan Hasan, W.Z.; Ahmad, S.A.; Ishak, A.J.; Ghazali, A.M.; Talib, D.N.; Wada, C. Pressure Sensor: State of the Art, Design, and Application for Robotic Hand. *J. Sens.* **2015**, *2015*, 846487. [CrossRef]
12. Chang, T.H.; Tian, Y.; Li, C.; Gu, X.; Li, K.; Yang, H.; Sanghani, P.; Lim, C.M.; Ren, H.; Chen, P.Y. Stretchable Graphene Pressure Sensors with Shar-Pei-like Hierarchical Wrinkles for Collision-Aware Surgical Robotics. *ACS Appl. Mater. Interfaces* **2019**, *11*, 10226–10236. [CrossRef]
13. Gao, Y.; Ota, H.; Schaler, E.W.; Chen, K.; Zhao, A.; Gao, W.; Fahad, H.M.; Leng, Y.; Zheng, A.; Xiong, F.; et al. Wearable Microfluidic Diaphragm Pressure Sensor for Health and Tactile Touch Monitoring. *Adv. Mater.* **2017**, *29*, 1701985. [CrossRef]
14. Larimi, S.R.; Rezaei Nejad, H.; Oyatsi, M.; O'Brien, A.; Hoorfar, M.; Najjaran, H. Low-cost ultra-stretchable strain sensors for monitoring human motion and bio-signals. *Sens. Actuators A Phys.* **2018**, *271*, 182–191. [CrossRef]
15. Oh, D.; Seo, J.; Kim, H.G.; Ryu, C.; Bang, S.-W.; Park, S.; Kim, H.J. Multi-height micropyramids based pressure sensor with tunable sensing properties for robotics and step tracking applications. *Micro Nano Syst. Lett.* **2022**, *10*, 7. [CrossRef]
16. Sekine, T.; Abe, M.; Muraki, K.; Tachibana, S.; Wang, Y.-F.; Hong, J.; Takeda, Y.; Kumaki, D.; Tokito, S. Microporous Induced Fully Printed Pressure Sensor for Wearable Soft Robotics Machine Interfaces. *Adv. Intell. Syst.* **2020**, *2*, 2000179. [CrossRef]
17. Zhan, Z.; Lin, R.; Tran, V.T.; An, J.; Wei, Y.; Du, H.; Tran, T.; Lu, W. Paper/Carbon Nanotube-Based Wearable Pressure Sensor for Physiological Signal Acquisition and Soft Robotic Skin. *ACS Appl. Mater. Interfaces* **2017**, *9*, 37921–37928. [CrossRef]
18. Luo, Z.; Hu, X.; Tian, X.; Luo, C.; Xu, H.; Li, Q.; Li, Q.; Zhang, J.; Qiao, F.; Wu, X.; et al. Structure-Property Relationships in Graphene-Based Strain and Pressure Sensors for Potential Artificial Intelligence Applications. *Sensors* **2019**, *19*, 1250. [CrossRef] [PubMed]
19. Shi, Z.; Meng, L.; Shi, X.; Li, H.; Zhang, J.; Sun, Q.; Liu, X.; Chen, J.; Liu, S. Morphological Engineering of Sensing Materials for Flexible Pressure Sensors and Artificial Intelligence Applications. *Nanomicro Lett.* **2022**, *14*, 141. [CrossRef]
20. Zang, Y.; Zhang, F.; Di, C.-a.; Zhu, D. Advances of flexible pressure sensors toward artificial intelligence and health care applications. *Mater. Horiz.* **2015**, *2*, 140–156. [CrossRef]
21. Zhang, Z.; Shi, Q.; He, T.; Guo, X.; Dong, B.; Lee, J.; Lee, C. Artificial intelligence of toilet (AI-Toilet) for an integrated health monitoring system (IHMS) using smart triboelectric pressure sensors and image sensor. *Nano Energy* **2021**, *90*, 106517. [CrossRef]
22. Kalsoom, T.; Ramzan, N.; Ahmed, S.; Ur-Rehman, M. Advances in Sensor Technologies in the Era of Smart Factory and Industry 4.0. *Sensors* **2020**, *20*, 6783. [CrossRef]
23. Pech, M.; Vrchota, J.; Bednar, J. Predictive Maintenance and Intelligent Sensors in Smart Factory: Review. *Sensors* **2021**, *21*, 1470. [CrossRef]
24. Poeggel, S.; Tosi, D.; Duraibabu, D.; Leen, G.; McGrath, D.; Lewis, E. Optical Fibre Pressure Sensors in Medical Applications. *Sensors* **2015**, *15*, 17115–17148. [CrossRef] [PubMed]
25. Jeong, Y.; Park, J.; Lee, J.; Kim, K.; Park, I. Ultrathin, Biocompatible, and Flexible Pressure Sensor with a Wide Pressure Range and Its Biomedical Application. *ACS Sens.* **2020**, *5*, 481–489. [CrossRef] [PubMed]
26. Park, J.; Kim, J.K.; Patil, S.J.; Park, J.K.; Park, S.; Lee, D.W. A Wireless Pressure Sensor Integrated with a Biodegradable Polymer Stent for Biomedical Applications. *Sensors* **2016**, *16*, 809. [CrossRef] [PubMed]
27. Mancia, G.; Fagard, R.; Narkiewicz, K.; Redon, J.; Zanchetti, A.; Bohm, M.; Christiaens, T.; Cifkova, R.; De Backer, G.; Dominiczak, A.; et al. 2013 ESH/ESC Guidelines for the management of arterial hypertension: The Task Force for the management of arterial hypertension of the European Society of Hypertension (ESH) and of the European Society of Cardiology (ESC). *J. Hypertens.* **2013**, *31*, 1281–1357. [CrossRef] [PubMed]
28. Rangel-Castillo, L.; Robertson, C.S. Management of intracranial hypertension. *Crit. Care Clin.* **2006**, *22*, 713–732, abstract ix. [CrossRef] [PubMed]
29. Yu, L.; Kim, B.J.; Meng, E. Chronically implanted pressure sensors: Challenges and state of the field. *Sensors* **2014**, *14*, 20620–20644. [CrossRef]
30. Shin, J.; Yan, Y.; Bai, W.; Xue, Y.; Gamble, P.; Tian, L.; Kandela, I.; Haney, C.R.; Spees, W.; Lee, Y.; et al. Bioresorbable pressure sensors protected with thermally grown silicon dioxide for the monitoring of chronic diseases and healing processes. *Nat. Biomed. Eng.* **2019**, *3*, 37–46. [CrossRef]
31. Boutry, C.M.; Chandrahalim, H.; Streit, P.; Schinhammer, M.; Hanzi, A.C.; Hierold, C. Towards biodegradable wireless implants. *Philos. Trans. A Math. Phys. Eng. Sci.* **2012**, *370*, 2418–2432. [CrossRef]
32. Hwang, S.W.; Tao, H.; Kim, D.H.; Cheng, H.; Song, J.K.; Rill, E.; Brenckle, M.A.; Panilaitis, B.; Won, S.M.; Kim, Y.S.; et al. A physically transient form of silicon electronics. *Science* **2012**, *337*, 1640–1644. [CrossRef]

33. Irimia-Vladu, M. "Green" electronics: Biodegradable and biocompatible materials and devices for sustainable future. *Chem. Soc. Rev.* **2014**, *43*, 588–610. [CrossRef]

34. Bettinger, C.J.; Bao, Z. Organic thin-film transistors fabricated on resorbable biomaterial substrates. *Adv. Mater.* **2010**, *22*, 651–655. [CrossRef] [PubMed]

35. Elsayes, A.; Sharma, V.; Yiannacou, K.; Koivikko, A.; Rasheed, A.; Sariola, V. Plant-Based Biodegradable Capacitive Tactile Pressure Sensor Using Flexible and Transparent Leaf Skeletons as Electrodes and Flower Petal as Dielectric Layer. *Adv. Sustain. Syst.* **2020**, *4*, 2000056. [CrossRef]

36. Hou, C.; Xu, Z.; Qiu, W.; Wu, R.; Wang, Y.; Xu, Q.; Liu, X.Y.; Guo, W. A Biodegradable and Stretchable Protein-Based Sensor as Artificial Electronic Skin for Human Motion Detection. *Small* **2019**, *15*, e1805084. [CrossRef] [PubMed]

37. Palmroth, A.; Salpavaara, T.; Lekkala, J.; Kellomäki, M. Fabrication and Characterization of a Wireless Bioresorbable Pressure Sensor. *Adv. Mater. Technol.* **2019**, *4*, 1900428. [CrossRef]

38. Yan, Y.; Potts, M.; Jiang, Z.; Sencadas, V. Synthesis of highly-stretchable graphene—Poly(glycerol sebacate) elastomeric nanocomposites piezoresistive sensors for human motion detection applications. *Compos. Sci. Technol.* **2018**, *162*, 14–22. [CrossRef]

39. Kang, S.K.; Murphy, R.K.; Hwang, S.W.; Lee, S.M.; Harburg, D.V.; Krueger, N.A.; Shin, J.; Gamble, P.; Cheng, H.; Yu, S.; et al. Bioresorbable silicon electronic sensors for the brain. *Nature* **2016**, *530*, 71–76. [CrossRef]

40. Yang, Q.; Lee, S.; Xue, Y.; Yan, Y.; Liu, T.L.; Kang, S.K.; Lee, Y.J.; Lee, S.H.; Seo, M.H.; Lu, D.; et al. Materials, Mechanics Designs, and Bioresorbable Multisensor Platforms for Pressure Monitoring in the Intracranial Space. *Adv. Funct. Mater.* **2020**, *30*, 1910718. [CrossRef]

41. Dagdeviren, C.; Hwang, S.-W.; Su, Y.; Kim, S.; Cheng, H.; Gur, O.; Haney, R.; Omenetto, F.G.; Huang, Y.; Rogers, J.A. Transient, Biocompatible Electronics and Energy Harvesters Based on ZnO. *Small* **2013**, *9*, 3398–3404. [CrossRef]

42. Kim, B.H.; Kim, J.H.; Persano, L.; Hwang, S.W.; Lee, S.; Lee, J.; Yu, Y.; Kang, Y.; Won, S.M.; Koo, J.; et al. Dry Transient Electronic Systems by Use of Materials that Sublime. *Adv. Funct. Mater.* **2017**, *27*, 1606008. [CrossRef]

43. Yin, L.; Cheng, H.; Mao, S.; Haasch, R.; Liu, Y.; Xie, X.; Hwang, S.-W.; Jain, H.; Kang, S.-K.; Su, Y.; et al. Dissolvable Metals for Transient Electronics. *Adv. Funct. Mater.* **2014**, *24*, 645–658. [CrossRef]

44. Son, D.; Lee, J.; Lee, D.J.; Ghaffari, R.; Yun, S.; Kim, S.J.; Lee, J.E.; Cho, H.R.; Yoon, S.; Yang, S.; et al. Bioresorbable Electronic Stent Integrated with Therapeutic Nanoparticles for Endovascular Diseases. *ACS Nano* **2015**, *9*, 5937–5946. [CrossRef] [PubMed]

45. Mueller, P.P.; Arnold, S.; Badar, M.; Bormann, D.; Bach, F.-W.; Drynda, A.; Meyer-Lindenberg, A.; Hauser, H.; Peuster, M. Histological and molecular evaluation of iron as degradable medical implant material in a murine animal model. *J. Biomed. Mater. Res. Part A* **2012**, *100A*, 2881–2889. [CrossRef] [PubMed]

46. Dorozhkin, S.V. Calcium orthophosphate coatings on magnesium and its biodegradable alloys. *Acta Biomater.* **2014**, *10*, 2919–2934. [CrossRef] [PubMed]

47. Boutry, C.M.; Nguyen, A.; Lawal, Q.O.; Chortos, A.; Rondeau-Gagné, S.; Bao, Z. A Sensitive and Biodegradable Pressure Sensor Array for Cardiovascular Monitoring. *Adv. Mater.* **2015**, *27*, 6954–6961. [CrossRef]

48. Pierson, D.; Edick, J.; Tauscher, A.; Pokorney, E.; Bowen, P.; Gelbaugh, J.; Stinson, J.; Getty, H.; Lee, C.H.; Drelich, J.; et al. A simplified in vivo approach for evaluating the bioabsorbable behavior of candidate stent materials. *J. Biomed. Mater. Res. Part B Appl. Biomater.* **2012**, *100B*, 58–67. [CrossRef]

49. Lei, T.; Guan, M.; Liu, J.; Lin, H.-C.; Pfattner, R.; Shaw, L.; McGuire, A.F.; Huang, T.-C.; Shao, L.; Cheng, K.-T.; et al. Biocompatible and totally disintegrable semiconducting polymer for ultrathin and ultralightweight transient electronics. *Proc. Natl. Acad. Sci. USA* **2017**, *114*, 5107–5112. [CrossRef]

50. National Institude of Health. 2021. Available online: https://ods.od.nih.gov/factsheets/Magnesium-HealthProfessional/#en1 (accessed on 29 March 2021).

51. Iron—Health Professional. National Institutes of Health. Available online: https://ods.od.nih.gov/factsheets/Iron-HealthProfessional/ (accessed on 29 March 2021).

52. Kang, S.-K.; Park, G.; Kim, K.; Hwang, S.-W.; Cheng, H.; Shin, J.; Chung, S.; Kim, M.; Yin, L.; Lee, J.C.; et al. Dissolution Chemistry and Biocompatibility of Silicon- and Germanium-Based Semiconductors for Transient Electronics. *ACS Appl. Mater. Interfaces* **2015**, *7*, 9297–9305. [CrossRef]

53. Cha, G.D.; Kang, D.; Lee, J.; Kim, D.H. Bioresorbable Electronic Implants: History, Materials, Fabrication, Devices, and Clinical Applications. *Adv. Healthc. Mater.* **2019**, *8*, 1801660. [CrossRef]

54. Katarivas Levy, G.; Goldman, J.; Aghion, E. The Prospects of Zinc as a Structural Material for Biodegradable Implants—A Review Paper. *Metals* **2017**, *7*, 402. [CrossRef]

55. Yu, X.; Shou, W.; Mahajan, B.K.; Huang, X.; Pan, H. Materials, Processes, and Facile Manufacturing for Bioresorbable Electronics: A Review. *Adv. Mater.* **2018**, *30*, 1707624. [CrossRef]

56. Ng, W.F.; Chiu, K.Y.; Cheng, F.T. Effect of pH on the in vitro corrosion rate of magnesium degradable implant material. *Mater. Sci. Eng. C* **2010**, *30*, 898–903. [CrossRef]

57. Hwang, S.-W.; Park, G.; Edwards, C.; Corbin, E.A.; Kang, S.-K.; Cheng, H.; Song, J.-K.; Kim, J.-H.; Yu, S.; Ng, J.; et al. Dissolution Chemistry and Biocompatibility of Single-Crystalline Silicon Nanomembranes and Associated Materials for Transient Electronics. *ACS Nano* **2014**, *8*, 5843–5851. [CrossRef] [PubMed]

58. Fu, K.; Liu, Z.; Yao, Y.; Wang, Z.; Zhao, B.; Luo, W.; Dai, J.; Lacey, S.D.; Zhou, L.; Shen, F.; et al. Transient Rechargeable Batteries Triggered by Cascade Reactions. *Nano Lett.* **2015**, *15*, 4664–4671. [CrossRef] [PubMed]

59. Kirkland, N.T.; Birbilis, N.; Staiger, M.P. Assessing the corrosion of biodegradable magnesium implants: A critical review of current methodologies and their limitations. *Acta Biomater.* **2012**, *8*, 925–936. [CrossRef] [PubMed]

60. Hornberger, H.; Virtanen, S.; Boccaccini, A.R. Biomedical coatings on magnesium alloys—A review. *Acta Biomater.* **2012**, *8*, 2442–2455. [CrossRef]

61. Kaabi Falahieh Asl, S.; Nemeth, S.; Tan, M.J. Novel biodegradable calcium phosphate/polymer composite coating with adjustable mechanical properties formed by hydrothermal process for corrosion protection of magnesium substrate. *J. Biomed. Mater. Res. Part B Appl. Biomater.* **2016**, *104*, 1643–1657. [CrossRef]

62. Walker, J.; Shadanbaz, S.; Kirkland, N.T.; Stace, E.; Woodfield, T.; Staiger, M.P.; Dias, G.J. Magnesium alloys: Predicting in vivo corrosion with in vitro immersion testing. *J. Biomed. Mater. Res. Part B Appl. Biomater.* **2012**, *100B*, 1134–1141. [CrossRef]

63. Irimia-Vladu, M.; Troshin, P.A.; Reisinger, M.; Shmygleva, L.; Kanbur, Y.; Schwabegger, G.; Bodea, M.; Schwödiauer, R.; Mumyatov, A.; Fergus, J.W.; et al. Biocompatible and Biodegradable Materials for Organic Field-Effect Transistors. *Adv. Funct. Mater.* **2010**, *20*, 4069–4076. [CrossRef]

64. Luo, M.; Martinez, A.W.; Song, C.; Herrault, F.; Allen, M.G. A Microfabricated Wireless RF Pressure Sensor Made Completely of Biodegradable Materials. *J. Microelectromech. Syst.* **2014**, *23*, 4–13. [CrossRef]

65. Hwang, S.-W.; Huang, X.; Seo, J.-H.; Song, J.-K.; Kim, S.; Hage-Ali, S.; Chung, H.-J.; Tao, H.; Omenetto, F.G.; Ma, Z.; et al. Materials for Bioresorbable Radio Frequency Electronics. *Adv. Mater.* **2013**, *25*, 3526–3531. [CrossRef]

66. Cao, Y.; Uhrich, K.E. Biodegradable and biocompatible polymers for electronic applications: A review. *J. Bioact. Compat. Polym.* **2018**, *34*, 3–15. [CrossRef]

67. Tanaka, A. Toxicity of indium arsenide, gallium arsenide, and aluminium gallium arsenide. *Toxicol. Appl. Pharmacol.* **2004**, *198*, 405–411. [CrossRef] [PubMed]

68. Flora, S.J.S. Possible Health Hazards Associated with the Use of Toxic Metals in Semiconductor Industries. *J. Occup. Health* **2000**, *42*, 105–110. [CrossRef]

69. Bogner, E.; Dominizi, K.; Hagl, P.; Bertagnolli, E.; Wirth, M.; Gabor, F.; Brezna, W.; Wanzenboeck, H.D. Bridging the gap— Biocompatibility of microelectronic materials. *Acta Biomater.* **2006**, *2*, 229–237. [CrossRef]

70. Li, W.; Liu, Q.; Zhang, Y.; Li, C.a.; He, Z.; Choy, W.C.H.; Low, P.J.; Sonar, P.; Kyaw, A.K.K. Biodegradable Materials and Green Processing for Green Electronics. *Adv. Mater.* **2020**, *32*, 2001591. [CrossRef]

71. Hwang, S.-W.; Park, G.; Cheng, H.; Song, J.-K.; Kang, S.-K.; Yin, L.; Kim, J.-H.; Omenetto, F.G.; Huang, Y.; Lee, K.-M.; et al. 25th Anniversary Article: Materials for High-Performance Biodegradable Semiconductor Devices. *Adv. Mater.* **2014**, *26*, 1992–2000. [CrossRef]

72. Morita, M.; Ohmi, T.; Hasegawa, E.; Kawakami, M.; Ohwada, M. Growth of native oxide on a silicon surface. *J. Appl. Phys.* **1990**, *68*, 1272–1281. [CrossRef]

73. Zhou, J.; Xu, N.S.; Wang, Z.L. Dissolving Behavior and Stability of ZnO Wires in Biofluids: A Study on Biodegradability and Biocompatibility of ZnO Nanostructures. *Adv. Mater.* **2006**, *18*, 2432–2435. [CrossRef]

74. Tran, H.; Feig, V.R.; Liu, K.; Wu, H.-C.; Chen, R.; Xu, J.; Deisseroth, K.; Bao, Z. Stretchable and Fully Degradable Semiconductors for Transient Electronics. *ACS Cent. Sci.* **2019**, *5*, 1884–1891. [CrossRef]

75. Grzybowski, M.; Gryko, D.T. Diketopyrrolopyrroles: Synthesis, Reactivity, and Optical Properties. *Adv. Opt. Mater.* **2015**, *3*, 280–320. [CrossRef]

76. Qu, S.; Tian, H. Diketopyrrolopyrrole (DPP)-based materials for organic photovoltaics. *Chem. Commun.* **2012**, *48*, 3039–3051. [CrossRef] [PubMed]

77. Irimia-Vladu, M.; Głowacki, E.D.; Troshin, P.A.; Schwabegger, G.; Leonat, L.; Susarova, D.K.; Krystal, O.; Ullah, M.; Kanbur, Y.; Bodea, M.A.; et al. Indigo—A Natural Pigment for High Performance Ambipolar Organic Field Effect Transistors and Circuits. *Adv. Mater.* **2012**, *24*, 375–380. [CrossRef] [PubMed]

78. Klimovich, I.V.; Leshanskaya, L.I.; Troyanov, S.I.; Anokhin, D.V.; Novikov, D.V.; Piryazev, A.A.; Ivanov, D.A.; Dremova, N.N.; Troshin, P.A. Design of indigo derivatives as environment-friendly organic semiconductors for sustainable organic electronics. *J. Mater. Chem. C* **2014**, *2*, 7621–7631. [CrossRef]

79. Kim, Y.J.; Wu, W.; Chun, S.-E.; Whitacre, J.F.; Bettinger, C.J. Biologically derived melanin electrodes in aqueous sodium-ion energy storage devices. *Proc. Natl. Acad. Sci. USA* **2013**, *110*, 20912–20917. [CrossRef]

80. Mostert, A.B.; Powell, B.J.; Pratt, F.L.; Hanson, G.R.; Sarna, T.; Gentle, I.R.; Meredith, P. Role of semiconductivity and ion transport in the electrical conduction of melanin. *Proc. Natl. Acad. Sci. USA* **2012**, *109*, 8943–8947. [CrossRef] [PubMed]

81. Yu, K.J.; Kuzum, D.; Hwang, S.-W.; Kim, B.H.; Juul, H.; Kim, N.H.; Won, S.M.; Chiang, K.; Trumpis, M.; Richardson, A.G.; et al. Bioresorbable silicon electronics for transient spatiotemporal mapping of electrical activity from the cerebral cortex. *Nat. Mater.* **2016**, *15*, 782–791. [CrossRef]

82. Koo, J.; MacEwan, M.R.; Kang, S.K.; Won, S.M.; Stephen, M.; Gamble, P.; Xie, Z.; Yan, Y.; Chen, Y.Y.; Shin, J.; et al. Wireless bioresorbable electronic system enables sustained nonpharmacological neuroregenerative therapy. *Nat. Med.* **2018**, *24*, 1830–1836. [CrossRef]

83. Chang, J.-K.; Fang, H.; Bower, C.A.; Song, E.; Yu, X.; Rogers, J.A. Materials and processing approaches for foundry-compatible transient electronics. *Proc. Natl. Acad. Sci. USA* **2017**, *114*, E5522–E5529. [CrossRef]

84. Fruhwirth, O.; Herzog, G.W.; Hollerer, I.; Rachetti, A. Dissolution and hydration kinetics of MgO. *Surf. Technol.* **1985**, *24*, 301–317. [CrossRef]

85. Hines, D.J.; Kaplan, D.L. Poly(lactic-co-glycolic) acid-controlled-release systems: Experimental and modeling insights. *Crit. Rev. Drug Carr. Syst.* **2013**, *30*, 257–276. [CrossRef]

86. Jensen, B.E.; Dávila, I.; Zelikin, A.N. Poly(vinyl alcohol) Physical Hydrogels: Matrix-Mediated Drug Delivery Using Spontaneously Eroding Substrate. *J. Phys. Chem. B* **2016**, *120*, 5916–5926. [CrossRef] [PubMed]

87. Hosta-Rigau, L.; Jensen, B.E.; Fjeldsø, K.S.; Postma, A.; Li, G.; Goldie, K.N.; Albericio, F.; Zelikin, A.N.; Städler, B. Surface-adhered composite poly(vinyl alcohol) physical hydrogels: Polymersome-aided delivery of therapeutic small molecules. *Adv. Healthc. Mater.* **2012**, *1*, 791–795. [CrossRef] [PubMed]

88. Fejerskov, B.; Smith, A.A.A.; Jensen, B.E.B.; Hussmann, T.; Zelikin, A.N. Bioresorbable Surface-Adhered Enzymatic Microreactors Based on Physical Hydrogels of Poly(vinyl alcohol). *Langmuir* **2013**, *29*, 344–354. [CrossRef] [PubMed]

89. Makadia, H.K.; Siegel, S.J. Poly Lactic-co-Glycolic Acid (PLGA) as Biodegradable Controlled Drug Delivery Carrier. *Polymers* **2011**, *3*, 1377–1397. [CrossRef] [PubMed]

90. Zhu, B.; Wang, H.; Leow, W.R.; Cai, Y.; Loh, X.J.; Han, M.-Y.; Chen, X. Silk Fibroin for Flexible Electronic Devices. *Adv. Mater.* **2016**, *28*, 4250–4265. [CrossRef] [PubMed]

91. Tao, H.; Hwang, S.-W.; Marelli, B.; An, B.; Moreau, J.E.; Yang, M.; Brenckle, M.A.; Kim, S.; Kaplan, D.L.; Rogers, J.A.; et al. Silk-based resorbable electronic devices for remotely controlled therapy and in vivo infection abatement. *Proc. Natl. Acad. Sci. USA* **2014**, *111*, 17385–17389. [CrossRef] [PubMed]

92. Fu, K.; Wang, Z.; Yan, C.; Liu, Z.; Yao, Y.; Dai, J.; Hitz, E.; Wang, Y.; Luo, W.; Chen, Y.; et al. All-Component Transient Lithium-Ion Batteries. *Adv. Energy Mater.* **2016**, *6*, 1502496. [CrossRef]

93. Hernandez, H.L.; Kang, S.-K.; Lee, O.P.; Hwang, S.-W.; Kaitz, J.A.; Inci, B.; Park, C.W.; Chung, S.; Sottos, N.R.; Moore, J.S.; et al. Triggered Transience of Metastable Poly(phthalaldehyde) for Transient Electronics. *Adv. Mater.* **2014**, *26*, 7637–7642. [CrossRef]

94. Park, C.W.; Kang, S.-K.; Hernandez, H.L.; Kaitz, J.A.; Wie, D.S.; Shin, J.; Lee, O.P.; Sottos, N.R.; Moore, J.S.; Rogers, J.A.; et al. Thermally Triggered Degradation of Transient Electronic Devices. *Adv. Mater.* **2015**, *27*, 3783–3788. [CrossRef]

95. Li, G.; Song, E.; Huang, G.; Guo, Q.; Ma, F.; Zhou, B.; Mei, Y. High-Temperature-Triggered Thermally Degradable Electronics Based on Flexible Silicon Nanomembranes. *Adv. Funct. Mater.* **2018**, *28*, 1801448. [CrossRef]

96. Camposeo, A.; D'Elia, F.; Portone, A.; Matino, F.; Archimi, M.; Conti, S.; Fiori, G.; Pisignano, D.; Persano, L. Naturally Degradable Photonic Devices with Transient Function by Heterostructured Waxy-Sublimating and Water-Soluble Materials. *Adv. Sci.* **2020**, *7*, 2001594. [CrossRef]

97. Gao, L.; Zhu, C.; Li, L.; Zhang, C.; Liu, J.; Yu, H.D.; Huang, W. All Paper-Based Flexible and Wearable Piezoresistive Pressure Sensor. *ACS Appl. Mater. Interfaces* **2019**, *11*, 25034–25042. [CrossRef]

98. Li, X.; Li, X.; Ting, L.; Lu, Y.; Shang, C.; Ding, X.; Zhang, J.; Feng, Y.; Xu, F.J. Wearable, Washable, and Highly Sensitive Piezoresistive Pressure Sensor Based on a 3D Sponge Network for Real-Time Monitoring Human Body Activities. *ACS Appl. Mater. Interfaces* **2021**, *13*, 46848–46857. [CrossRef] [PubMed]

99. Lv, B.; Chen, X.; Liu, C. A Highly Sensitive Piezoresistive Pressure Sensor Based on Graphene Oxide/Polypyrrole@Polyurethane Sponge. *Sensors* **2020**, *20*, 1219. [CrossRef] [PubMed]

100. Wang, Z.; Ye, X. An investigation on piezoresistive behavior of carbon nanotube/polymer composites: II. Positive piezoresistive effect. *Nanotechnology* **2014**, *25*, 285502. [CrossRef] [PubMed]

101. Kim, S.; Choi, S.; Oh, E.; Byun, J.; Kim, H.; Lee, B.; Lee, S.; Hong, Y. Revisit to three-dimensional percolation theory: Accurate analysis for highly stretchable conductive composite materials. *Sci. Rep.* **2016**, *6*, 34632. [CrossRef] [PubMed]

102. Fiorillo, A.S.; Critello, C.D.; Pullano, S.A. Theory, technology and applications of piezoresistive sensors: A review. *Sens. Actuators A Phys.* **2018**, *281*, 156–175. [CrossRef]

103. Yoo, J.-Y.; Seo, M.-H.; Lee, J.-S.; Choi, K.-W.; Jo, M.-S.; Yoon, J.-B. Industrial Grade, Bending-Insensitive, Transparent Nanoforce Touch Sensor via Enhanced Percolation Effect in a Hierarchical Nanocomposite Film. *Adv. Funct. Mater.* **2018**, *28*, 1804721. [CrossRef]

104. He, J.; Zhang, Y.; Zhou, R.; Meng, L.; Chen, T.; Mai, W.; Pan, C. Recent advances of wearable and flexible piezoresistivity pressure sensor devices and its future prospects. *J. Mater.* **2020**, *6*, 86–101. [CrossRef]

105. Song, P.; Si, C.; Zhang, M.; Zhao, Y.; He, Y.; Liu, W.; Wang, X. A Novel Piezoresistive MEMS Pressure Sensors Based on Temporary Bonding Technology. *Sensors* **2020**, *20*, 337. [CrossRef]

106. Feng, R.; Mu, Y.; Zeng, X.; Jia, W.; Liu, Y.; Jiang, X.; Gong, Q.; Hu, Y. A Flexible Integrated Bending Strain and Pressure Sensor System for Motion Monitoring. *Sensors* **2021**, *21*, 3969. [CrossRef] [PubMed]

107. Stehlin, E.F.; McCormick, D.; Malpas, S.C.; Pontre, B.P.; Heppner, P.A.; Budgett, D.M. MRI interactions of a fully implantable pressure monitoring device. *J. Magn. Reson. Imaging* **2015**, *42*, 1441–1449. [CrossRef] [PubMed]

108. Tian, H.; Shu, Y.; Wang, X.F.; Mohammad, M.A.; Bie, Z.; Xie, Q.Y.; Li, C.; Mi, W.T.; Yang, Y.; Ren, T.L. A graphene-based resistive pressure sensor with record-high sensitivity in a wide pressure range. *Sci. Rep.* **2015**, *5*, 8603. [CrossRef] [PubMed]

109. Aryafar, M.; Hamedi, M.; Ganjeh, M.M. A novel temperature compensated piezoresistive pressure sensor. *Measurement* **2015**, *63*, 25–29. [CrossRef]

110. Li, Y.; Chen, W.; Lu, L. Wearable and Biodegradable Sensors for Human Health Monitoring. *ACS Appl. Biol. Mater.* **2020**, *4*, 122–139. [CrossRef]

111. Wu, J.; Yao, Y.; Zhang, Y.; Shao, T.; Wu, H.; Liu, S.; Li, Z.; Wu, L. Rational design of flexible capacitive sensors with highly linear response over a broad pressure sensing range. *Nanoscale* **2020**, *12*, 21198–21206. [CrossRef]

112. Wang, H.; Li, Z.; Liu, Z.; Fu, J.; Shan, T.; Yang, X.; Lei, Q.; Yang, Y.; Li, D. Flexible capacitive pressure sensors for wearable electronics. *J. Mater. Chem. C* **2022**, *10*, 1594–1605. [CrossRef]
113. Park, S.W.; Das, P.S.; Chhetry, A.; Park, J.Y. A Flexible Capacitive Pressure Sensor for Wearable Respiration Monitoring System. *IEEE Sens. J.* **2017**, *17*, 6558–6564. [CrossRef]
114. Kumar, A. Flexible and wearable capacitive pressure sensor for blood pressure monitoring. *Sens. Bio-Sens. Res.* **2021**, *33*, 100434. [CrossRef]
115. Zhao, S.; Ran, W.; Wang, D.; Yin, R.; Yan, Y.; Jiang, K.; Lou, Z.; Shen, G. 3D Dielectric Layer Enabled Highly Sensitive Capacitive Pressure Sensors for Wearable Electronics. *ACS Appl. Mater. Interfaces* **2020**, *12*, 32023–32030. [CrossRef]
116. Masihi, S.; Panahi, M.; Maddipatla, D.; Hanson, A.J.; Bose, A.K.; Hajian, S.; Palaniappan, V.; Narakathu, B.B.; Bazuin, B.J.; Atashbar, M.Z. Highly Sensitive Porous PDMS-Based Capacitive Pressure Sensors Fabricated on Fabric Platform for Wearable Applications. *ACS Sens.* **2021**, *6*, 938–949. [CrossRef] [PubMed]
117. Keum, K.; Eom, J.; Lee, J.H.; Heo, J.S.; Park, S.K.; Kim, Y.-H. Fully-integrated wearable pressure sensor array enabled by highly sensitive textile-based capacitive ionotronic devices. *Nano Energy* **2021**, *79*, 105479. [CrossRef]
118. Mannsfeld, S.C.; Tee, B.C.; Stoltenberg, R.M.; Chen, C.V.; Barman, S.; Muir, B.V.; Sokolov, A.N.; Reese, C.; Bao, Z. Highly sensitive flexible pressure sensors with microstructured rubber dielectric layers. *Nat. Mater.* **2010**, *9*, 859–864. [CrossRef]
119. Ruth, S.R.A.; Bao, Z. Designing Tunable Capacitive Pressure Sensors Based on Material Properties and Microstructure Geometry. *ACS Appl. Mater. Interfaces* **2020**, *12*, 58301–58316. [CrossRef] [PubMed]
120. Svete, A.; Kutin, J. Identifying the high-frequency response of a piezoelectric pressure measurement system using a shock tube primary method. *Mech. Syst. Signal Process.* **2022**, *162*, 108014. [CrossRef]
121. Chorsi, M.T.; Curry, E.J.; Chorsi, H.T.; Das, R.; Baroody, J.; Purohit, P.K.; Ilies, H.; Nguyen, T.D. Piezoelectric Biomaterials for Sensors and Actuators. *Adv. Mater.* **2019**, *31*, e1802084. [CrossRef]
122. Curry, E.J.; Ke, K.; Chorsi, M.T.; Wrobel, K.S.; Miller, A.N., 3rd; Patel, A.; Kim, I.; Feng, J.; Yue, L.; Wu, Q.; et al. Biodegradable Piezoelectric Force Sensor. *Proc. Natl. Acad. Sci. USA* **2018**, *115*, 909–914. [CrossRef]
123. Curry, E.J.; Le, T.T.; Das, R.; Ke, K.; Santorella, E.M.; Paul, D.; Chorsi, M.T.; Tran, K.T.M.; Baroody, J.; Borges, E.R.; et al. Biodegradable nanofiber-based piezoelectric transducer. *Proc. Natl. Acad. Sci. USA* **2020**, *117*, 214–220. [CrossRef]
124. Okatani, T.; Sekiguchi, S.; Hane, K.; Kanamori, Y. Surface-plasmon-coupled optical force sensors based on metal-insulator-metal metamaterials with movable air gap. *Sci. Rep.* **2020**, *10*, 14807. [CrossRef]
125. Karrock, T.; Gerken, M. Pressure sensor based on flexible photonic crystal membrane. *Biomed. Opt. Express* **2015**, *6*, 4901–4911. [CrossRef]
126. Ghildiyal, S.; Ranjan, P.; Mishra, S.; Balasubramaniam, R.; John, J. Fabry–Perot Interferometer-Based Absolute Pressure Sensor With Stainless Steel Diaphragm. *IEEE Sens. J.* **2019**, *19*, 6093–6101. [CrossRef]
127. Sencadas, V.; Tawk, C.; Alici, G. Environmentally Friendly and Biodegradable Ultrasensitive Piezoresistive Sensors for Wearable Electronics Applications. *ACS Appl. Mater. Interfaces* **2020**, *12*, 8761–8772. [CrossRef] [PubMed]
128. Tee, B.C.K.; Chortos, A.; Dunn, R.R.; Schwartz, G.; Eason, E.; Bao, Z. Tunable Flexible Pressure Sensors using Microstructured Elastomer Geometries for Intuitive Electronics. *Adv. Funct. Mater.* **2014**, *24*, 5427–5434. [CrossRef]
129. Ruth, S.R.A.; Feig, V.R.; Tran, H.; Bao, Z. Microengineering Pressure Sensor Active Layers for Improved Performance. *Adv. Funct. Mater.* **2020**, *30*, 2003491. [CrossRef]
130. Ruth, S.R.A.; Beker, L.; Tran, H.; Feig, V.R.; Matsuhisa, N.; Bao, Z. Rational Design of Capacitive Pressure Sensors Based on Pyramidal Microstructures for Specialized Monitoring of Biosignals. *Adv. Funct. Mater.* **2019**, *30*, 1903100. [CrossRef]
131. Boutry, C.M.; Kaizawa, Y.; Schroeder, B.C.; Chortos, A.; Legrand, A.; Wang, Z.; Chang, J.; Fox, P.; Bao, Z. A stretchable and biodegradable strain and pressure sensor for orthopaedic application. *Nat. Electron.* **2018**, *1*, 314–321. [CrossRef]
132. Tran, R.T.; Thevenot, P.; Gyawali, D.; Chiao, J.C.; Tang, L.; Yang, J. Synthesis and characterization of a biodegradable elastomer featuring a dual crosslinking mechanism. *Soft Matter* **2010**, *6*, 2449–2461. [CrossRef]
133. Khalid, M.A.U.; Ali, M.; Soomro, A.M.; Kim, S.W.; Kim, H.B.; Lee, B.-G.; Choi, K.H. A highly sensitive biodegradable pressure sensor based on nanofibrous dielectric. *Sens. Actuators A Phys.* **2019**, *294*, 140–147. [CrossRef]
134. Boutry, C.M.; Beker, L.; Kaizawa, Y.; Vassos, C.; Tran, H.; Hinckley, A.C.; Pfattner, R.; Niu, S.; Li, J.; Claverie, J.; et al. Biodegradable and flexible arterial-pulse sensor for the wireless monitoring of blood flow. *Nat. Biomed. Eng.* **2019**, *3*, 47–57. [CrossRef]
135. Lu, D.; Yan, Y.; Deng, Y.; Yang, Q.; Zhao, J.; Seo, M.H.; Bai, W.; MacEwan, M.R.; Huang, Y.; Ray, W.Z.; et al. Bioresorbable Wireless Sensors as Temporary Implants for In Vivo Measurements of Pressure. *Adv. Funct. Mater.* **2020**, *30*, 2003754. [CrossRef]
136. Hosseini, E.S.; Manjakkal, L.; Shakthivel, D.; Dahiya, R. Glycine-Chitosan-Based Flexible Biodegradable Piezoelectric Pressure Sensor. *ACS Appl. Mater. Interfaces* **2020**, *12*, 9008–9016. [CrossRef] [PubMed]
137. Capuana, E.; Lopresti, F.; Ceraulo, M.; La Carrubba, V. Poly-l-Lactic Acid (PLLA)-Based Biomaterials for Regenerative Medicine: A Review on Processing and Applications. *Polymers* **2022**, *14*, 1153. [CrossRef] [PubMed]
138. Casalini, T.; Rossi, F.; Castrovinci, A.; Perale, G. A Perspective on Polylactic Acid-Based Polymers Use for Nanoparticles Synthesis and Applications. *Front. Bioeng. Biotechnol.* **2019**, *7*, 259. [CrossRef] [PubMed]
139. Minary-Jolandan, M.; Yu, M.F. Nanoscale characterization of isolated individual type I collagen fibrils: Polarization and piezoelectricity. *Nanotechnology* **2009**, *20*, 085706. [CrossRef]
140. Yoshida, M.; Onogi, T.; Onishi, K.; Inagaki, T.; Tajitsu, Y. High piezoelectric performance of poly(lactic acid) film manufactured by solid-state extrusion. *Jpn. J. Appl. Phys.* **2014**, *53*, 09PC02. [CrossRef]

141. Singhvi, M.; Gokhale, D. Biomass to biodegradable polymer (PLA). *RSC Adv.* **2013**, *3*, 13558–13568. [CrossRef]
142. Fernandes, C.; Taurino, I. Biodegradable Molybdenum (Mo) and Tungsten (W) Devices: One Step Closer towards Fully-Transient Biomedical Implants. *Sensors* **2022**, *22*, 3062. [CrossRef]
143. Shin, J.; Liu, Z.; Bai, W.; Liu, Y.; Yan, Y.; Xue, Y.; Kandela, I.; Pezhouh, M.; MacEwan, M.R.; Huang, Y.; et al. Bioresorbable optical sensor systems for monitoring of intracranial pressure and temperature. *Sci. Adv.* **2019**, *5*, eaaw1899. [CrossRef]
144. Islam, M.R.; Ali, M.M.; Lai, M.H.; Lim, K.S.; Ahmad, H. Chronology of Fabry-Perot interferometer fiber-optic sensors and their applications: A review. *Sensors* **2014**, *14*, 7451–7488. [CrossRef]
145. Yu, Q.; Zhou, X. Pressure sensor based on the fiber-optic extrinsic Fabry-Perot interferometer. *Photonic Sens.* **2010**, *1*, 72–83. [CrossRef]
146. Domingues, M.F.; Rodriguez, C.A.; Martins, J.; Tavares, C.; Marques, C.; Alberto, N.; André, P.; Antunes, P. Cost-effective optical fiber pressure sensor based on intrinsic Fabry-Perot interferometric micro-cavities. *Opt. Fiber Technol.* **2018**, *42*, 56–62. [CrossRef]
147. Zhou, W.; Zhao, D.; Shuai, Y.-C.; Yang, H.; Chuwongin, S.; Chadha, A.; Seo, J.-H.; Wang, K.X.; Liu, V.; Ma, Z.; et al. Progress in 2D photonic crystal Fano resonance photonics. *Prog. Quantum Electron.* **2014**, *38*, 1–74. [CrossRef]
148. Vijaya Shanthi, K.; Robinson, S. Two-dimensional photonic crystal based sensor for pressure sensing. *Photonic Sens.* **2014**, *4*, 248–253. [CrossRef]
149. De, M.; Gangopadhyay, T.K.; Singh, V.K. Prospects of Photonic Crystal Fiber as Physical Sensor: An Overview. *Sensors* **2019**, *19*, 464. [CrossRef] [PubMed]
150. Kirsten Peters, R.E.U.; James Kirkpatrick, C. *Biomedical Materials*; Roger, N., Ed.; Springer Nature: Cham, Switzerland, 2021; Volume 2, Chapter 13; pp. 423–453.
151. *ISO-10993-11:2017(E)*; Biological Evaluation of Medical Devices-Part 11: Tests for systemic toxicity International Organization for Standardization. ISO: Geneva, Switzerland, 2017.
152. Andrade, R.d.A.P.; Oshiro, H.E.; Miyazaki, C.K.; Hayashi, C.Y.; de Morais, M.A.; Brunelli, R.; Carmo, J.P. A Nanometer Resolution Wearable Wireless Medical Device for Non Invasive Intracranial Pressure Monitoring. *IEEE Sens. J.* **2021**, *21*, 22270–22284. [CrossRef]
153. Jiang, H.; Guo, Y.; Wu, Z.; Zhang, C.; Jia, W.; Wang, Z. Implantable Wireless Intracranial Pressure Monitoring Based on Air Pressure Sensing. *IEEE Trans. Biomed. Circuits Syst.* **2018**, *12*, 1076–1087. [CrossRef]
154. Khan, M.W.A.; Bjorninen, T.; Sydanheimo, L.; Ukkonen, L. Remotely Powered Piezoresistive Pressure Sensor: Toward Wireless Monitoring of Intracranial Pressure. *IEEE Microw. Wirel. Compon. Lett.* **2016**, *26*, 549–551. [CrossRef]
155. Harary, M.; Dolmans, R.G.F.; Gormley, W.B. Intracranial Pressure Monitoring-Review and Avenues for Development. *Sensors* **2018**, *18*, 465. [CrossRef]
156. Zhang, X.; Medow, J.E.; Iskandar, B.J.; Wang, F.; Shokoueinejad, M.; Koueik, J.; Webster, J.G. Invasive and noninvasive means of measuring intracranial pressure: A review. *Physiol. Meas.* **2017**, *38*, R143–R182. [CrossRef]
157. Moraes, F.M.; Silva, G.S. Noninvasive intracranial pressure monitoring methods: A critical review. *Arq. Neuro-Psiquiatr.* **2021**, *79*, 437–446. [CrossRef]
158. Campigotto, A.; Leahy, S.; Zhao, G.; Campbell, R.J.; Lai, Y. Non-invasive Intraocular pressure monitoring with contact lens. *Br. J. Ophthalmol.* **2020**, *104*, 1324–1328. [CrossRef] [PubMed]
159. Chen, P.-J.; Saati, S.; Varma, R.; Humayun, M.S.; Tai, Y.-C. Wireless Intraocular Pressure Sensing Using Microfabricated Minimally Invasive Flexible-Coiled LC Sensor Implant. *J. Microelectromech. Syst.* **2010**, *19*, 721–734. [CrossRef]
160. Ha, D.; de Vries, W.N.; John, S.W.; Irazoqui, P.P.; Chappell, W.J. Polymer-based miniature flexible capacitive pressure sensor for intraocular pressure (IOP) monitoring inside a mouse eye. *Biomed. Microdevices* **2012**, *14*, 207–215. [CrossRef] [PubMed]
161. Ittoop, S.M.; SooHoo, J.R.; Seibold, L.K.; Mansouri, K.; Kahook, M.Y. Systematic Review of Current Devices for 24-h Intraocular Pressure Monitoring. *Adv. Ther.* **2016**, *33*, 1679–1690. [CrossRef]
162. Katuri, K.C.; Asrani, S.; Ramasubramanian, M.K. Intraocular Pressure Monitoring Sensors. *IEEE Sens. J.* **2008**, *8*, 9–16. [CrossRef]
163. Saxby, E.; Mansouri, K.; Tatham, A.J. Intraocular pressure monitoring using an intraocular sensor before and after glaucoma surgery. *J. Glaucoma* **2021**, *30*, 941–946. [CrossRef]
164. Chen, S.; Wu, N.; Ma, L.; Lin, S.; Yuan, F.; Xu, Z.; Li, W.; Wang, B.; Zhou, J. Noncontact Heartbeat and Respiration Monitoring Based on a Hollow Microstructured Self-Powered Pressure Sensor. *ACS Appl. Mater. Interfaces* **2018**, *10*, 3660–3667. [CrossRef]
165. Jiang, Y.; Deng, S.; Sun, H.; Qi, Y. Unconstrained Monitoring Method for Heartbeat Signals Measurement using Pressure Sensors Array. *Sensors* **2019**, *19*, 368. [CrossRef]
166. Koyama, Y.; Nishiyama, M.; Watanabe, K. Smart Textile Using Hetero-Core Optical Fiber for Heartbeat and Respiration Monitoring. *IEEE Sens. J.* **2018**, *18*, 6175–6180. [CrossRef]
167. Senthil Kumar, K.; Xu, Z.; Sivaperuman Kalairaj, M.; Ponraj, G.; Huang, H.; Ng, C.F.; Wu, Q.H.; Ren, H. Stretchable Capacitive Pressure Sensing Sleeve Deployable onto Catheter Balloons towards Continuous Intra-Abdominal Pressure Monitoring. *Biosensors* **2021**, *11*, 156. [CrossRef]
168. Zhong, Y.; Qian, B.; Zhu, Y.; Ren, Z.; Deng, J.; Liu, J.; Bai, Q.; Zhang, X. Development of an Implantable Wireless and Batteryless Bladder Pressure Monitor System for Lower Urinary Tract Dysfunction. *IEEE J. Transl. Eng. Health Med.* **2020**, *8*, 2500107. [CrossRef] [PubMed]

169. Soebadi, M.A.; Weydts, T.; Brancato, L.; Hakim, L.; Puers, R.; De Ridder, D. Novel implantable pressure and acceleration sensor for bladder monitoring. *Int. J. Urol.* **2020**, *27*, 543–550. [CrossRef] [PubMed]
170. Liao, C.H.; Cheng, C.T.; Chen, C.C.; Wang, Y.H.; Chiu, H.T.; Peng, C.C.; Jow, U.M.; Lai, Y.L.; Chen, Y.C.; Ho, D.R. Systematic Review of Diagnostic Sensors for Intra-Abdominal Pressure Monitoring. *Sensors* **2021**, *21*, 4824. [CrossRef] [PubMed]

 *biosensors*

Review

Recent Advances in Multicellular Tumor Spheroid Generation for Drug Screening

Kwang-Ho Lee and Tae-Hyung Kim *

School of Integrative Engineering, Chung-Ang University, 84 Heukseuk-ro, Dongjak-gu, Seoul 06974, Korea; dlrhkdgh17@naver.com
* Correspondence: thkim0512@cau.ac.kr

Abstract: Multicellular tumor spheroids (MCTs) have been employed in biomedical fields owing to their advantage in designing a three-dimensional (3D) solid tumor model. For controlling multicellular cancer spheroids, mimicking the tumor extracellular matrix (ECM) microenvironment is important to understand cell–cell and cell–matrix interactions. In drug cytotoxicity assessments, MCTs provide better mimicry of conventional solid tumors that can precisely represent anticancer drug candidates' effects. To generate incubate multicellular spheroids, researchers have developed several 3D multicellular spheroid culture technologies to establish a research background and a platform using tumor modelingvia advanced materials science, and biosensing techniques for drug-screening. In application, drug screening was performed in both invasive and non-invasive manners, according to their impact on the spheroids. Here, we review the trend of 3D spheroid culture technology and culture platforms, and their combination with various biosensing techniques for drug screening in the biomedical field.

Keywords: MCTs; ECM; drug screening; anticancer drug; 3D spheroid culture technology; culture platform; biosensing techniques

Citation: Lee, K.-H.; Kim, T.-H. Recent Advances in Multicellular Tumor Spheroid Generation for Drug Screening. *Biosensors* **2021**, *11*, 445. https://doi.org/10.3390/bios11110445

Received: 30 September 2021
Accepted: 10 November 2021
Published: 11 November 2021

Publisher's Note: MDPI stays neutral with regard to jurisdictional claims in published maps and institutional affiliations.

1. Introduction

Cancer is currently widespread and is the cause of many deaths, regardless of gender and age. Unlike other diseases, cancer is believed to have a variety of causes, such as genetics, environmental factors, and acquired factors. Remarkably, factors that induce the formation of cancer are closely linked to intracellular interactions, creating an in vivo microenvironment. A tumor microenvironment (TME) consists of tumor cells; tumor stromal cells including stromal fibroblasts, endothelial cells, and microglia; immune cells such as macrophages and lymphocytes; and non-cellular components of the extracellular matrix such as collagen, fibronectin, hyaluronic acid, and laminin [1–6]. At the heart of the TME are tumor cells, which control the functions of both cellular and non-cellular components through complex signaling networks, using non-malignant signals to control them. As a result of this crosstalk, tumorigenesis and responses to multidrug resistance are linked [7,8]. Non-malignant cells in the TME are known to promote tumorigenesis at all stages of cancer development and metastasis [9–11]. In addition to the growth of the cancer, secondary tumors that develop in areas of the body that are far from the primary cancer are called "metastasis" [12–14]. The development of metastasis is a series of processes in which cancer cells leave the primary site, circulate through the bloodstream, withstand the pressure of blood vessels, and escape from combat with immune cells, i.e., as enemies, to the new cellular environment at the secondary site. Although metastasis is the major cause of cancer treatment failure and death, it is still poorly understood.

Researchers are currently studying cancer mechanisms and drug resistance using three-dimensional (3D) cell culture [15,16]. The existing two-dimensional (2D) culture system has enabled better understanding of complex cell physiology, and it can be used as the basis for biotechnology. However, 2D culture has a disadvantage that it cannot reflect cell-to-cell or cell-to-matrix interactions, and cell-matrix components that are important for differentiation, phagocytosis, and cell function in vivo [17–19]. Cell population produced through 3D culture systems exhibit characteristics that are more compatible with complex in vivo conditions [20]. Cultivation through 3D culture models not only leads to behaviors closer to natural conditions, but also exhibits novel and unexpected results on the tumorigenesis mechanisms [21]. The current trend is to utilize multi-cell culture rather than single-cell culture to simulate the microcellular environment and ensure cell interactions. Three-dimensional multicellular spheroids (MSCs) are gaining considerable attention in the biomedical field as they can simulate the interaction between cells and the environment of the extracellular matrix by emulating the structure and function of cellular tissues. In addition, multicellular tumor spheroids are used as 3D tumor models for anticancer drug screening due to their similar metabolic and proliferation gradient distribution to tumor tissue in vivo [22,23].

Therefore, multi-cultured cells in a 3D environment can serve as a cost-effective multi-drug screening platform for drug development and testing with in vivo mimicking models. This reduces the use of the existing animal clinical models and creates opportunities to evaluate the effects of drugs directly on humans. In terms of engineering, the configuration of a platform capable of realizing 3D cancer spheroids, appropriate cytotoxicity testing, and drug-screening methods should be considered in clinical cancer research [24]. In implementing 3D spheroids, factors such as appropriate material, cell adhesion strength, nutrient uptake, size, and extracellular matrix (ECM) components should be reflected, considering the TME. However, in terms of not harming 3D cells, difficulties persist in the process of cytotoxicity evaluation and drug screening using optical and electrochemical viability analysis kits and assays. In this review, a platform using a 3D multicellular spheroid was considered (Figure 1). First, the 3D cell culture technology was investigated to understand the technology being applied, and the series of processes in manufacturing the 3D co-culture spheroids using various methods was confirmed. In addition, the current trend of biosensing techniques was identified by discussing drug screening and cytotoxicity evaluation methods.

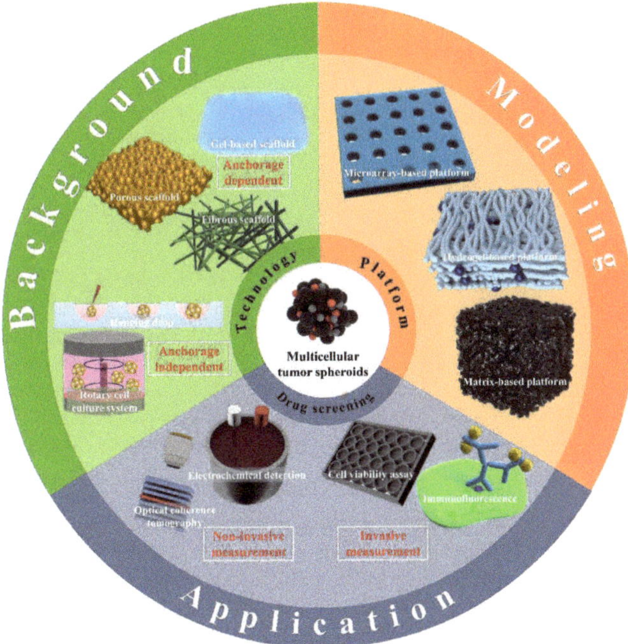

Figure 1. Multicellular tumor spheroid formation and drug screening through different approaches.

2. 3D Cell Culture Technology

Unlike the 2D cultures, which grow by attaching to the bottom as a monolayer, 3D cell culture refers to cells aggregated and expressed as a single tissue or form. Moreover, the 3D-cultured cells are attached to an artificially created ECM environment to interact with or grow with the surrounding environment. Therefore, unlike 2D cell cultures, cell growth in a 3D environment allows cells to grow in multiple directions rather than in a single direction in vitro, which is similar to in vivo conditions [25–27]. Upon comparison, the 3D cell culture exhibits several advantages: (1) A similar biomimetic model, which is more physiologically relevant. (2) A 3D culture exhibits a high level of structural complexity and maintains homeostasis for longer. (3) 3D models can indicate how different types of cells interact. (4) 3D cultures can reduce the use of animal models. (5) They are a good simulator for the treatment of disease groups including cancer tumors. The cell lines used in incubating multicellular spheroids are listed as below (Table 1).

Table 1. Spheroid formation technologies and platforms for cancer cells.

Cell Line	Culture Method	Substrate Type	Ref.
MCF-7, MDA-MB231	Hanging drop	-	[28]
BT474	Hanging drop	-	[29]
MCF-7	Hanging drop	Poly(N-isopropylacrylamide) (p(NIPA))	[30]
MDA-MB-231	Rotary cell culture system	-	[31]
HGC-27	Rotary cell culture system	-	[32]
HeLa, MCF-7, HUVECs	Porous scaffold	PDMS/CMC/PEDOT/Pt composites	[33]
PC3	Porous scaffold	Cellulose nanocrystals/poly(oligoethylene glycol methacrylate)	[34]
DU 145, A549, MCF-7, MDA-MB-231	Fibrous scaffold	Core–shell silk fibroin/rice paper	[35]
LNCaP	Gel-based scaffold	bQ13 (Ac-QQKFQFQFEQEQQ-Am) peptide	[36]
MCF-7, A549, A2780, P19, Panc02, UN-KC-6141	Hydrogel-based platform	Poly dimethyl siloxane (PDMS) derivatives (acoustic devices)	[37]
MCF-7	Hydrogel-based platform	ClO^-/SCN^-/carboxymethyl cellulose	[38]
NIH3T3, HepG2, HUVECs	Hydrogel-based platform	PEG-SH/Gela-SH/Gly-Tyr/D-PBS/HRP	[39]
A549, T24, Huh-7	Microwell-based platform	Polystyrene slides/PDMS	[40]
A549, MG-63, HLFs	Microwell-based hydrogel platform	N-isopropylacrylamide (NIPAM)	[41]
HepG2, HEK 293T	Microarray-based platform	Droplet microarray slides (DMA)	[42]
HUVEC, MCF-7	Matrix-based platform	Poly dimethyl siloxane (PDMS)/collagen	[43]
PANC-1, PS-1, HMEC	Matrix-based platform	Polyurethane/fibronectin (FN)/collagen I (COL)	[44]

2.1. Anchorage-Independent Approaches

Anchorage-independent 3D cell culture is a non-adherent cell culture system that is collectively referred to as a liquid-based system [45]. This system maintains cancer cells in a suspension to generate a self-assembly of tumor cells into a compact 3D aggregate known as a tumor spheroid or a cancer spheroid. Its main feature is that it enables the exchange of culture media to a certain point, although sophisticated handling is necessary due to the absence of an anchorage system to seize tumor spheroids. There are two pivotal liquid-based systems: (1) hanging-drop culture technology, (2) rotary-based culture system using spinner flasks and rotation culture system (RCCS) [46–49].

2.1.1. Hanging-Drop Method

MCTs can be easily formed using non-adhesive cell culture methods such as the hanging-drop method or centrifugation of cells in a suspension culture method [46,50], in which the cells in the suspension culture medium are located at the bottom due to the presence of a meniscus in the middle layer [51]. Conventional approaches to generate cancer cell aggregation are hindered by variations in the cell number and spheroid size, high-shear force, and their labor-intensive nature [52]. Recently, different microfabrication methods, including microfluidics and microwell, have exhibited the potential to form a large number of well-organized spheroids. Zhao et al., (2019) first introduced a 3D-printed hanging-drop dripper (3D-phd) device that enables long-term production of uniform cancer spheroids (Figure 2) [28]. In addition to culturing MCF-7 and MDA-MB-231 human breast cancer cells, this device performs a series of biomedical assessments, including imaging analysis, gene expression mapping, and anticancer drug assays. They proved that this device has some merits such as ease of design to perform 3D tumor migration analysis and can extend to organ-on-chip engineering, as well as assessing the anticancer drug efficiency on the co-culture spheroids. Some use the microfluidic technology, which is useful owing to its simple operation and enables point-of-care testing. Through this method, Park et al., (2020) devised a finger-actuated microfluidic device for spheroid cultivation and analysis that facilitates programmed media exchange and media injection for further analysis [29]. Different sizes of BT474 spheroids were generated after seven days of growing and further analyzed in a LIVE/DEAD assay, indicating its spheroid growth in a manipulated ECM-mimicking environment. The proposed microfluidic-based device can be widely applied in biomedical laboratories by combining it with automated machinery. For the formation of the MCTs, the manipulation of cell size uniformity and long-term cultivation is difficult to control. To overcome this limitation, thermoresponsive copolymers with a poly(N-isopropylacrylamide) (p(NIPA)) backbone were manufactured [30]. In this study, small-size spheroids of human breast adenocarcinoma cells were generated up to 2000 cells per drop. This demonstrated the superior performance of an in vivo 3D cell culture system. Through an immunofluorescence assay with the LIVE/DEAD analysis, these spheroids exhibited suitable drug penetration, making it a proper model for future drug-screening platforms. Hanging-drop is a conventional technology and has some limitations: (1) drops may fall off by mistake and (2) large amount of cells cannot be contained in one drop [53].

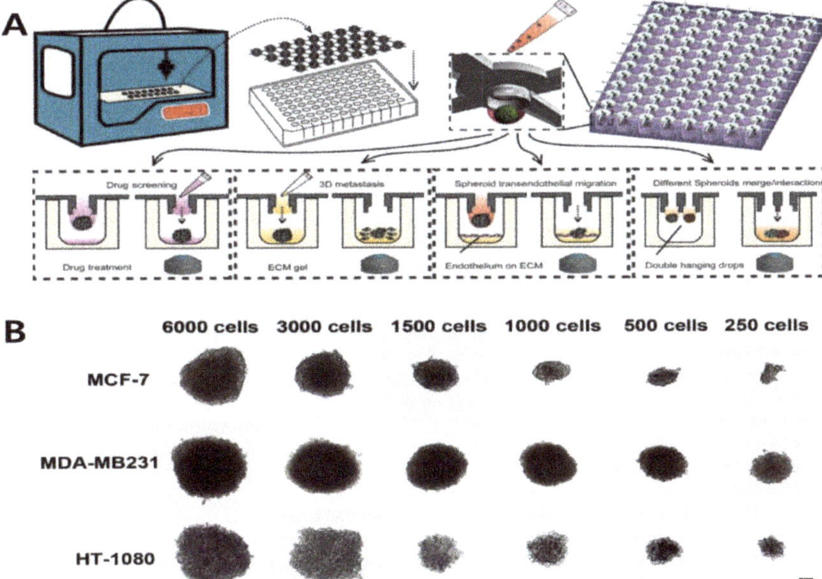

Figure 2. (**A**) Designing 3D-printed hanging-drop derivatives to investigate multicellular tumor spheroids. The device was printed for cancer spheroid formation on a 96/384 culture plate. Different types of assays are performed: drug screening, 3D metastasis, spheroid transendothelial migration, and spheroids merge/interactions. (**B**) Characterization of different cell lines cultured over two days with varying ratios. Reprinted with permission from [28]. Copyright 2019, Springer Nature.

2.1.2. Rotary Cell Culture System

The rotary cell culture system (RCCS) is a typical 3D cell culture system that involves both suspension and anchorage-independent cells. It is designed as a bioreactor system to simultaneously incorporate the ability to culture multiple types of cells with low turbulence and high mass transfer of nutrients. Jiang et al., (2019) investigated human MDA-MB-231 breast cancer cells in the RCCS to examine microgravity on the ultrastructure [31]. They aimed to determine the changes in the apoptosis, ultrastructure, and cycle progression for seven days. Through this trial, they pioneered new mechanisms and methods for preventing cancer cell metastasis and provided a deeper understanding of the treatment of malignant tumors. Recently, the RCCS bioreactor has been widely accepted as a microgravity simulation device. In one study, Chen et al., (2020) manipulated human HGC-27 gastric cancer cells cultured in an RCCS bioreactor system by simulating weightlessness [32]. Under this system, the effects of simulated microgravity (SMG) on the RCCS bioreactor were examined using liquid chromatography-mass spectrometry. Through this trial, the RCCS proved the importance of SMG, which has a major impact on lipid metabolism in cancer proliferation. This might be a novel target for treating gastric cancer disease. The RCCS enables uniform size and morphology of the spheroids. However, if the rotational speed is too high, the shear force becomes strong, which can affect the physiological response of the cells [54].

2.2. Anchorage-Independent Approaches

Mimicking the tumor microenvironment in culturing 3D cancer spheroids is the foremost consideration for guiding a successful 3D cell culture technology. Some studies utilize biomaterials to confine and attach cells three-dimensionally, such as encapsulating cells in hydrogels or growing cells in scaffolds [55–58]. Biomaterials are widely designed to facilitate cell adhesion, differentiation, and proliferation. They may comprise natural polymers, such as gelation, alginate, hyaluronic acid (HA), chitosan, and collagen or synthetic polymers, such as polycaprolactone (PCL), poly-L-lactic acid, poly (ethylene glycol) (PEG), polydimethylsiloxane (PMDS), and poly (lactic acid-co-caprolactone). A liquid mixture of the ECM can also be added directly to the culture media or numerous other ECM coating protocols can be employed to aid cell adhesion and stabilize cell aggregation to from 3D spheroids [59].

2.2.1. Porous Scaffold

A porous scaffold exhibits the following characteristics: porosity, pore size, morphology, which influences nutrient uptake for cell proliferation [60–63]. It has been widely accepted as a biodegradable polymer-based scaffold in tissue engineering; therefore, it needs to exhibit mechanical strength and flexibility [64]. Zhang et al., (2019) proposed a scaffold-based 3D cell culture system exploiting conductive polymer [33]. The importance of incorporating electroconductive material lies in its capability to sense electrochemical signals in a 3D cancer spheroid in a promising biocompatible polymer-based scaffold. A 3D porous PDMS scaffold was utilized to provide a favorable environment for a 3D cell culture. Then, a 3D PCP/Pt scaffold and PDMS scaffold were fabricated that exhibited stability and effectiveness in the 3D cell culture and tissue engineering. These scaffolds performed reactive oxygen species (ROS) monitoring of cancer cells, indicating great promise for future biomedical research. Different applications of the porous scaffold have been reported, which involved incorporating aerogel films and through pattering technology. Or et al., (2019) exhibited aerogel films with covalently cross-linked cellulose nanocrystals (CNCs) of designated dimensions and internal porous structures (Figure 3A) [34]. The usage of aerogel films has some merits in accelerating diffusion and reaction kinetics as a highly efficient catalyst matrix. Optimization of aerogel thickness and micropatterning were performed, and confocal imaging of human prostate cancer epithelial (PC3) cells showed its biocompatibility. In this work, they demonstrated an aerogel-based porous scaffold for a successful 3D cell culture technology.

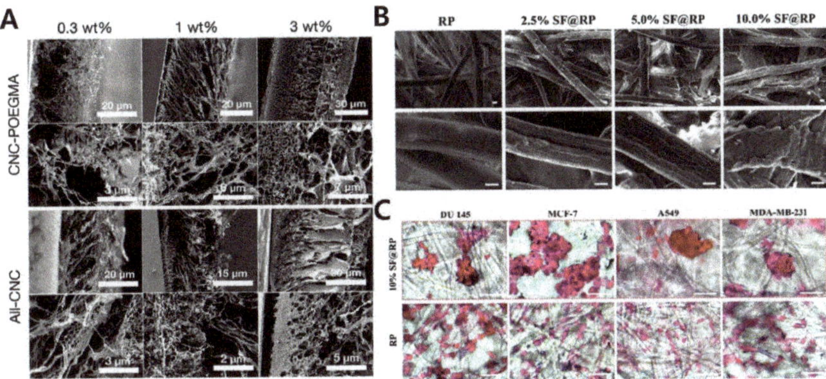

Figure 3. (**A**) Cross-sectional SEM images of the fabricated aerogel composite films with its porous 3D scaffolds. (**B**) FE-SEM microscopy of silk fibroin (SF)/rice paper (RP) composites with different concentrations indicating fibrous scaffolds. (**C**) Bright-field images of DU 145, MCF-7, A549, and MDA-MB-231 stained with H&E (red) in SF scaffolds. Reprinted with permission from [34]. Copyright 2019, American Chemical Society; Reprinted with permission from [35]. Copyright 2020, American Chemical Society.

2.2.2. Fibrous Scaffold

Fibrous type scaffolds are an attractive biomaterial in tissue engineering owing to their ECM-mimicking structures, such as fibrous proteins, that exist in a native ECM [65,66]. Depending on the cell type, the fibrous scaffold can be manipulated to fabricate a suitable TME through physical attachments. In its universal application, fiber-based scaffolds can be fabricated into composite/hybrid scaffolds, microfluidic-based fibrous scaffolds, nanofibrous structures, or electrospun fibrous scaffolds [67,68]. Fu et al., (2020) developed silk fibroin (SF) scaffolds derived from silkworms in a fibrous protein in a cocoon. In this study, SF-coated rice paper (RP) was fabricated, and it was prepared through a one-step dip-coating protocol (Figure 3B,C) [35]. To prove its biocompatibility, human breast, lung, and liver cancer cells were cultured on this platform, successfully demonstrating spheroid formation. To confirm its cell viability assessment, MTT assay and cell staining were performed, each demonstrating its potential for large-scale clinical application. Drug sensitivity was also investigated.

2.2.3. Gel-Based Scaffolds

Gel-based scaffolds can be modeled directly into an unstructured molded tissue. However, their stability is weak; therefore, they can be used with tissues under load such as bones. Biomaterials for gel-type 3D cell culture are derived from natural and artificially modified materials [69]. Since gels allow independent control of matrix and ECM functionalization, Ashworth et al., (2020) demonstrated a self-assembling peptide gel for designing the 3D cell culture [70]. By controlling stiffness through peptide concentration and pH condition, the peptide gel was fully utilized with further experiments on culturing breast cancer. An analysis of the results obtained through immunofluorescence staining and quantitative reverse transcription polymerase chain reaction (qRT-PCR) indicated that the peptide gel can be used to model the progression of breast cancer, independent of the matrix microenvironment regulation. Some researchers reported a self-assembling peptide, bQ13, that can be useful for obtaining the 3D culture of prostate cancer cells [36]. These self-assembled peptides have been known to help stabilize 3D culture and provide a user-defined matrix that can be tailored with different experimental conditions [71]. Investigation of the rheological properties of the peptides proved its maintenance in an ungelled state at a basic pH. Additional examinations of cell encapsulation and survival through immunostaining showed a well-organized, non-polarized morphology within the prostate spheroids, indicating an attractive scaffold for further modification. Hence, hydrogel scaffolds allow minimal damage to the spheroids and facilitate transport of nutrients and water as compared to other porous and fibrous scaffolds.

3. Multicellular Cancer Spheroid Formation Platform

Through 3D cell culture technologies, 3D cell culture platforms have been developed to recapitulate an in vivo microenvironment of solid tumors. Mimicking physiological characteristics of the TMEs, such as tumor metastasis, angiogenesis, and tumor-stromal interaction, enables researchers to achieve a cellular behavior closer to natural conditions [72–74]. Further, 3D cancer spheroid models have received wide attention for their potential applications in drug-screening assessments. Notably, 3D multicellular tumor spheroids can be used in the investigation of the TME regulation of tumor physiology and therapeutic obstacles associated with the proliferative and metabolic gradients in a 3D spheroid context [75]. In molecular biology approaches, the genomic stability of multicellular spheroids is superior to that observed in a 2D monolayer culture in terms of gene expression, DNA, and RNA level [76]. Since the MCTs can exhibit sophisticated in vivo solid tumor behaviors, researchers utilized a clinical drug-screening tool based on the 3D MCTs formation platform.

3.1. Hydrogel-Based Platforms

Hydrogels are widely studied for bioengineering applications, such as regenerative tissue engineering, 3D cell culture, and drug delivery, due to their ECM-mimicking structures [66,77–81]. The hydrogels exhibit 3D hydrophilic networks that show high water content and are advantageous in transporting nutrients, oxygen, and other water-soluble metabolites. In the formation of MCTs, an aqueous droplet or gel solution encompassing spheroids are guided to reduce the number of preparation steps such as culturing cells, maintenance, controlling cell aggregation, and delivery of reagents [82–84]. Recently, numerous acoustofluidic devices have been devised to load a single type of cell into aqueous droplets or to generate MTS assemblies of microchannels, an aqueous two-phase system (ATPS) [85,86]. ATPSs allow simplified culture preparation, maintenance of cancer cells, and aggregation of MCTs. In this way, Chen, Bin et al., (2019) developed a microfluidic platform for synthesizing dextran/alginate (DEX/ALG) hydrogel spheres that enable templated fabrication of multicellular spheroids in an ATPS [37]. The principle of fabricated acoustofluidic device lies in the following: (1) PEG- and DEX-enriched phases including the gel-forming agent ALG are pumped into the device. (2) Amplified acoustic flow stream is applied to the inner fluid with frequency control. (3) Droplets are cross-linked in a calcium bath to fabricate a hydrogel and transferred to the suspension culture system where the MTSs are formed. In order to culture mouse mammary carcinoma (EMT6) multicellular spheroids for a lengthy duration, this platform ensured long-term cultivation, uniform multicellular spheroids, and suspension culture conditions with growth factors to enable further analysis of organoid development. However, using enzymes or light ligands on the hydrogels is reported to be harmful to spheroid integrity [87,88]. To overcome these issues, researchers implemented anions that are reversibly responsive luminescent nanocellulose hydrogels for efficient formation and release of multicellular spheroids (Figure 4A–E) [38]. By mixing Eu(III) complex laden carboxymethyl cellulose (Eu(III) complex-CMC) and 2,6-pyridinedicarboxylic acid functionalized CMC (K-DPY-CMC), efficient regulation of hydrogel formation and release of entrapped MCF-7 breast cancer spheroids was accessible. The proposed hydrogels have a class of ClO^-/SCN^- reversibly responsive anions with fluorescence activation/deactivation. First, addition of a ClO^--induced destruction of the nanocellulose hydrogel network, accompanying fluorescent quenching. Upon addition of SCN^-, fluorescence hydrogel was recovered by cross-linking of ClO^-/SCN^- with precise regulation. Further, the 2,5-diphenyl-2H-tetrazolium bromide (MTT)x assay was performed to test in vitro cytotoxicity. Additionally, Fourier-transform infrared spectroscopy (FT-IR), scanning electron microscopy (SEM), and a rheometer were implemented to optimize the concentration of the hydrogel. Upon formation of MCF-7 multicellular spheroids (human breast cancer cells), it could be easily released through ClO^-/SCN^- regulation and monitored in a time-dependent manner through fluorescence imaging. However, there still exist some problems associated with the incomplete adherence of the spheroid on the substrate and difficulties in regulating the cellular movement and cell density on the outer membrane of the encapsulated cells [89]. Ramadhan et al., (2020) suggested a redox-responsive hydrogel that enables self-wrapping co-culture [39]. Its mechanism involves decomposition of hydrogels under mild reductive microenvironments, and the peeling-off of a monolayer of cells cultured on a redox-responsive hydrogel surface that self-folds to wrap other cell lines. The decomposition of a redox-degradable PEG-based hydrogel can be controlled via the concentration of cysteine (CYs), indicating that the detachment of cell membrane can be regulated. Optimization of the self-folding process of a fibroblast cell line (NIH3T3) to warp liver hepatocellular carcinoma (HepG2) and human umbilical vein endothelial cells (HUVECs) into higher-order microstructures. Since the ECM component is crucial for 3D tumor formation, a collagen bead was added and the result was remarkably favorable in that the size of the tumor increased gradually and the necrotic area was diminished dramatically, indicating its enhanced spheroid cell viability.

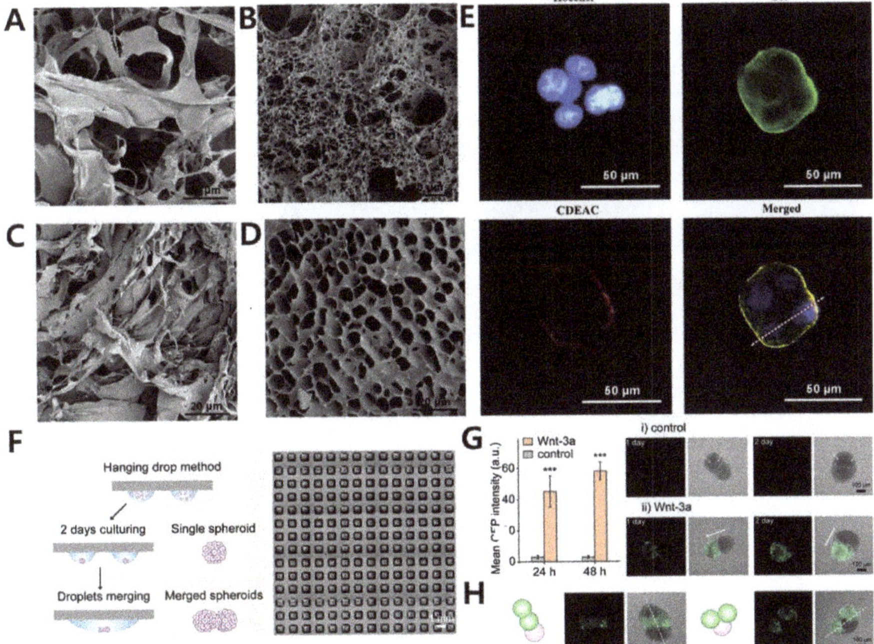

Figure 4. Morphology of the hydrogel platform after freeze-drying, investigated using scanning electron microscopy (SEM), with the SEM images of Eu (III) complex and carboxymethyl cellulose (CMC) backbone (**A**), the formed colloidally stable suspensions (CDEAC) (**B**), ClO^--added aqueous solution (**C**), and recovered state of the hydrogel upon addition of SCN^- (**D**). (**E**) Multicellular cancer spheroids of MCF-7 from the CDEAC hydrogel stained with Hoechst (Blue), Dio (green), and CDEAC hydrogel (red). (**F**) Culture mechanism of the programmable assembly of spheroids with hydrophobic borders. (**G**) Calculated activation intensity of the multicellular complex of Wnt-3a and GFP-labeled HEK spheroids. (Intensity of GFP was estimated from at least 10 spheroids. ***, $p < 0.001$, one-way ANOVA). (**H**) Fluorescence microscopy of triple spheroids with Wnt producer spheroid and Wnt reporter spheroids. Reprinted with permission from [38]. Copyright 2019, Elsevier; reprinted with permission from [42]. Copyright 2020, Wiley Online Library.

There still exist some issues with controlling cell aggregation and growth of multicellular spheroid. However, there are advantages that offset those challenges in terms of its diverse usage, biocompatibility, user-friendly suspension culture platform.

3.2. Microarray-Based Platform

Cells cultured in a micro-patterned array format are attracting tremendous attention in screening drug candidates for toxicity and efficacy in clinical trials [90]. Recent studies have proved that an in vitro microarray culture platform is an effective drug-screening tool with reduced cost and time that dramatically reduces the need to perform animal tests [91,92]. Moreover, 3D microarray culture platforms enable spheroid analysis in terms of drug treatment response, cell–extracellular matrix, and cell–cell interactions in a high-throughput manner. Thus, 3D microarrays provide an excellent alternative to conventional 2D plate-based assays [93–95]. Several methods have been developed to fabricate 3D microarray platforms including surface patterning, soft lithography, cell printing, and microfluidic-based application. The spatial position and morphology of the microarray is important for successful cell aggregation, and for multicellular spheroids to form stable co-cultures of multiple types of cells. To overcome these technical obstacles, different controlled arrangements of the microarray-based 3D spheroid culturing method have been reported.

The application of CO_2 laser ablation in the fabrication of the microarray has provided a considerably rapid and economical technique for generating multicellular spheroids utilizing size-controlled microwells [96–98]. Wu et al., (2021) published a reproducible U-shaped microwell array that facilitates high-throughput 3D tumor spheroid culture [40]. The size of the microwells is considered for precise manipulation of horizontal spacing (d_x) and vertical spacing (d_y) of an array for preventing cell loss during cell seeding. A549, Huh-7, and T24 multicellular spheroids were cultured in different sizes of the microwell, indicating the importance of optimization of the microwell size and seeding cell amount from different cell lines. Following this method, a size-controlled MCTs microarray culture platform was designed as an in vitro tumor-mimicking model to probe the drug-screening target. Microwells arrays on hydrogel assays form an interesting research topic as the 3D architecture of the array can be controlled. Dhamecha et al., (2021) suggested thermoresponsive hydrogel microwell array platforms that facilitate stress-free generation and isolation of the multicellular cancer spheroids [41]. In this assay, the poly N-isopropylacrylamide-based hydrogel microwell array (PHMA) was used, enabling the growth and aggregation of spheroids at 37 °C and convenient isolation of spheroids at room temperature (25 °C). A549, HeLa, and MG-63 cancer cell lines with human lung fibroblasts (HLF) were incubated in PHMA, forming multicellular spheroids with a spherical morphology with hypoxic cores. The swelling and de-swelling behavior of the PHMA allowed detachment of spheroids at room temperature, indicating its potential as a disease modeling platform and for drug-screening assessments. Cui et al., (2021) used the droplet-fusion technique to construct various multicellular structures in a miniaturized high-density assay format (Figure 4F–H) [42]. The droplet microarray (DMA) platform enables production of nanoliter droplet microarrays where the size, shape, and density of droplets depend on the style of the hydrophilic patterns surrounded by the hydrophobic barriers. The designed platform can cultivate and screen various types of cells in individual nanoliter droplets as miniaturized TMEs [99]. Through the modulation of the size and distance between the hydrophilic spots on the DMA, PROgrammable Merging of Adjacent Droplets (proMAD) can be employed to generate 3D multicellular spheroids by fusing multiple neighboring droplets of single spheroids. Thus, the proMAD method can be applied to the pharmacokinetic field where high-throughput screening and generation of hetero-type spheroids are required. Currently, graphene and its derivatives have shown promise in improving cell adhesion due to rapid absorption of ECM materials [100,101]. Namely, Kim et al., (2020) developed a graphene-oxide (GO) microarray platform that exhibited efficient cancer spheroid formation [102]. The HepG2 cells were cultured on this vertically coated GO platform, forming spheroids that grew from outside to inside. By treating with different anticancer drugs, the spheroid sizes could be quantitively monitored at various concentrations of the drugs.

Recently, researchers envisioned a novel platform to guide facile and highly reproducible fabrication method to reliably generate MCTs. In numerous attempts to overcome unwanted irregular spheroid growth, a customized microarray-based platform has been fabricated, with its size and shape well-suited to multicellular spheroid growth.

3.3. Matrix-Based Platform

In the regenerative and tissue engineering field, bio-mimicking scaffolds are routinely used to provide mechanical support for cell growth and tissue repair [103,104]. Biomaterials are derived from both natural materials and synthetic polymers with biocompatible properties [105]. Some synthetic polymers and polysaccharides such as HA, chitosan, alginate, poly(lactic-co-glycolic acid), and PEG have excellent physicochemical abilities and can be fabricated with minimum variability [106,107]. To engineer tumor ECMs suitable for cell adhesion, these biomaterials require further modification with integrin-binding domains to guarantee MCT formation [108]. In contrast to synthetic polymers, natural polymers such as collagen and Matrigel have been used in numerous approaches owing to their inherent cytocompatibility [109]. Earlier versions of scaffolds were limited to low cell numbers with slow processes and specific type of cells. At present, versatile techniques for handling scaffolds with diverse shapes and various types of cells co-cultured or compatible with tumor cell microenvironment have been developed [110]. In a previous study, a platform with rapid self-assembly of cells and matrix material of various shapes using microfabricated molds was introduced (Figure 5) [43]. Researchers have demonstrated that various molds with dumbbell, cross-like, spherical, and cuboidal shaped cell morphologies could be generated on this platform. Using this approach, different cell lines, such as breast cancer, osteosarcoma, and endothelial cell lines, were cultured with a range of cell seeding amounts. Non-spherical molds like dumbbell and cuboidal shapes retained their shape even after elimination from the molds and during long periods of culturing. Additionally, the shape of molds could be patterned to position numerous cell types in an accurate and controlled way, indicating a significant role in tissue or organ implantation.

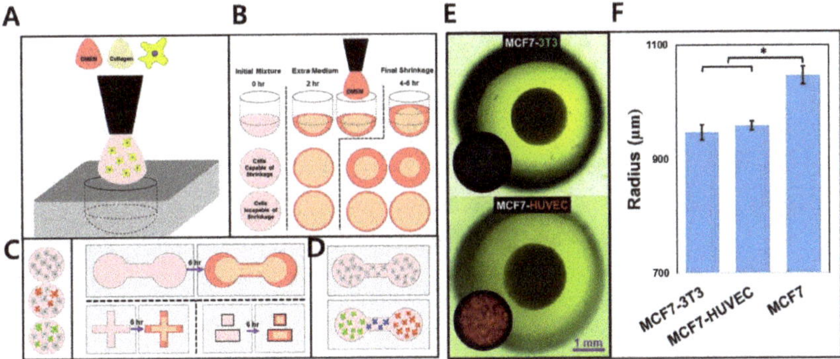

Figure 5. (**A**) Fabrication process of the collagen matrix platform. (**B**) Cell consolidation and shrinkage due to structural transformation and medium addition. (**C**) Creation of homogeneous co-culture structure with different morphologies (dumbbells, cuboids, and crosses). (**D**) Heterogeneous components formed by different cells at specific location of the molds. (**E**) Bright field and fluorescence microscopy of MCF-3T3 (green) and HUVEC (red) co-cultured with MCF7. (**F**) Calculated radius of a spheroid depending on cell type. (*p*-values: * < 0.01). Reprinted with permission from [43]. Copyright 2019, Elsevier.

Matrix-based scaffolds could also be applied to multiple biomaterials to better mimic a 3D in-vivo tumor. For instance, polyurethane (PU) was used in the scaffolds with enhanced long-term multicellular incubation involving cancer and endothelial cells [44]. The purpose of the study was to mimic pancreatic ductal adenocarcinoma (PDAC), which is a deadly disease. Two different compartments were organized: an inner tumor zone coated with fibronectin (FN) for cancer cell growth and a surrounding stromal zone treated with collagen I (COL) facilitating stellate and endothelial cell adhesion. Three types of cells were successfully generated in vitro with the shape of the scaffolds varying according to their usage and target cell lines. Zhang et al., (2020) illustrated cross-linked nanofibrous-type scaffolds with the usage of alkaline phosphatase (ALP) and carboxylesterase (CES) [111]. Two biomaterials were guided in a designated manner to induce coexistence of nanofibrils and vesicles, followed by the generation of nanoaggregates, resulting in the cross-linked scaffolds.

Overall, synthetic polymers have certain limitations in recapitulating precise structure and composition of ECM, which leads to inflammatory responses induced by the implantation of the fabricated materials. Native biopolymers derived from ECM have excellent biocompatibility and generally facilitate regenerative response after engraftment. However, they have limitations such that the recreation of the nanostructure of native ECM using single or multiple biopolymers is difficult.

4. Biosensing Methods to Assess Drug Efficacy in Multicellular Spheroids

Multicellular spheroids are used in drug screening tools as an in-vitro spheroid model for the selection and identification of drug candidates [112]. Because MCTs are superior to other 2D culture cells in producing physiological conditions of tumors, such as oxygen mobility, nutrients, and drugs, they have been subjected to a more sophisticated model of in-vivo drug testing assessments. Furthermore, drug delivery could be administered to MCTs models with a facile clinical evaluation [113,114]. Since drug uptake and diffusion were accurately replicated in 3D multicellular models, MCTs could be applied to drug penetration analysis. Different approaches have been developed to assess drug-screening platforms, such as immunofluorescence, fluorescence activated cell sorter (FACS), absorbance assay, electrochemical detection (ECD), and optical coherence tomography. In this study, the abovementioned tools are classified into two types: invasive and non-invasive measurements, depending on the characteristics of each measurement. In this section, various biosensing tools that have been used to assess the efficacy of drugs in multicellular cancer spheroids are discussed (Table 2).

Table 2. Drug-screening methods for multicellular cancer spheroids.

Cell Line	Anticancer Drug	Screening Tools	Ref.
MCF-7, HeLa, Caco-2	5-fluorouracil, cetuximab, panitumumab	Immunofluorescence	[115]
HT-29	Perifosine	Immunofluorescence	[116]
HT-29	Carboxyl-modified polystyrene nanoparticle	Immunofluorescence	[117]
Human breast cancer cell line (MCF-7 and MDA-MB-231)	Copper(II)-tropolone complex	Cell viability assay	[118]
MCF7, MDA-MB-231, SKBR3, MCF12A	Preussin	Cell viability assay	[119]
MCF-7	PLGA-MnO2 nanoparticles	Cell viability assay	[120]
HCT-116	Doxorubicin	Electrical impedance spectroscopy (EIS)	[121]
Human neuroblastoma and glioblastoma (SH-SY5Y, U-87MG)	Curcumin	Electrochemical detection (Cyclic voltammetry, Differential Pulse Voltammetry)	[122]
HCT-116	Immersion media (glycerol and ScaleView-A2)	Optical coherence tomography	[123]
MCF7	Paclitaxel	Optical coherence tomography	[110]

4.1. Invasive Sensing Methods

In invasive screening, conventional analyses have been used for drug-screening measurements to evaluate drug toxicity. For instance, invasive screening could be introduced to MCTs monitoring in two approaches: immunofluorescence and cell availability assays. Inevitably, damage arises in these drug-screening assessments, influencing the overall results of drug-screening experiments. Additionally, invasive screening necessitates fixation of cells, addition of toxic reagents, and the breakdown of cell populations into separate cells.

4.1.1. Immunofluorescence

Immunofluorescence is a worldwide tool that facilitates vivid and cellular monitoring in biological studies [124,125]. The monitoring mechanism involves using specific antibodies that are chemically compatible with fluorescent dyes. Once conjugated to antigen–dye complexes, these labeled antibodies bind to cellular antigens, which can be visualized in a fluorescence image. This technique is useful to demonstrate target antigens in tissues or circulating fluids, assisting diagnosis and monitoring of life-threatening disorders. Because cancers are chronic diseases worldwide, several studies using an immunofluorescence assay have been reported. In a previous study, an acoustic droplet-based microarray platform was engineered to facilitate screening of patient-derived spheroids [115]. For rapid and precise screening of cultivated human samples, MCF-7, HeLa, and Caco-2 cells were stained with calcein and PI. Additionally, fluorescence 3D modeling was performed for clinical demonstration of patient-derived spheroids. Through immunofluorescence analysis, protein expression was visually provided comparing before and after drug treatments. In drug-screening analysis, the drug penetration assessment is an indispensable component to confirm the efficacy of anticancer drugs. Machálková et al., (2019) performed drug penetration analysis on spheroids by laser scanning confocal microscopy and matrix-assisted laser desorption/ionization mass spectrometry imaging [116]. Human colorectal carcinoma (HT-29) cells were cultured, forming multicellular spheroids. Thereafter, the cells were treated with perifosine drug. d, drug distribution was confirmed, and the spheroid regions were colocalized with apoptosis, proliferation, and metastatic features. In modeling NP penetration for drug delivery, it is necessary to mimic tumor microenvironments, allowing specific interactions between the ligands on the surface of carrier and tumor cells [66]. Cutrona et al., (2019) demonstrated uptake and transport of NPs in spheroids using an automated confocal microscopy (Figure 6A,B) [117]. Profiling HT-29 multicellular spheroids, quantitative analysis was performed to observe penetration of synthetic NPs. The morphology of 4 days of HT-29 spheroids was observed through β-catenin, TGN46, actin, and nuclei staining, followed by Z-stack confocal imaging. Above all, the penetration studies of drug screening illustrated therapeutic target across tumor cells in vivo. Immunofluorescence analysis allows fast monitoring and precise quantification for drug screening at cellular and molecular levels. However, immunofluorescence assays have limitations such as staining compact spheroids with a dye unable to penetrate to the inner cells. Basically, staining protocols necessitate fixation steps that results in unavoidable death of target spheroids.

4.1.2. Cell Viability Assay

When cancer cells proliferate or die, they emit specific biomarkers [126–128]. In drug screening, cells treated with toxic compounds undergo two phenomena: proliferation stops or necrosis arises, and apoptosis, which leads to cell death [129]. Basically, necrosis refers to swelling cells and bursting membranes emitting inner cellular components, caused by toxic chemicals or sudden physical stress [130]. In case of apoptosis, cells contract, DNA breaks down into specific fragments, and cells eventually die after a series of processes such as breakdown by blood cells [131]. Various cell-based assays, such as tetrazolium reduction assays (MTT, MTS, XTT, and WST-1), lactate dehydrogenase assay (LDH), Cell Counting Kit-8 (CCK-8), hematoxylin and eosin (H&E) assay, and terminal deoxynucleotidyl-transferase-mediated dUTP nick-end labeling apoptosis assays [132–137], have been reported. Fundamentally, cell viability assays are based on detecting cellular components using reagents, dyes, and a series of reduction-mediating electrons. All of cell viability assays require incubation of a proper reagent with an estimated number of viable cells to label fluorescent byproducts that can be visualized through a plate reader. In one study, anticancer activity of copper (II)–tropolone complex was investigated in breast multicellular spheroids [118]. MCF7 (breast adenocarcinoma) and MDA-MB-231 (triple-negative breast adenocarcinoma) multicellular spheroids were generated, and a cytotoxicity assay, the MTT assay, was performed to determine the anticancer effect of Cu(trp)$_2$ in comparison with the conventional drugs of cisplatin (CDDP) and doxorubicin. The IC$_{50}$ values of Cu(trp)$_2$ were four- and sevenfold lower than IC$_{50}$ values of CDDP on MCF7 and MDA-MB-231 cells, indicating the anticancer effect of Cu(trp)$_2$. LDH assays have also been used in breast cancer studies profiling a cytotoxic effect of preussin drug derived from the marine-sponge-associated fungus (Figure 6C–F) [119]. The anticancer effect of preussin at concentrations of 50 and 100 μM on MCF7, SKBR3, and MCF12A cells resulted in approximately 100% LDH release compared to control groups. Since LDH released from damaging cells, the amount of LDH release correlated to drug efficacy of preussin. In addition, Murphy et al., (2021) suggested a drug delivery model incorporating with manganese dioxide nanoparticles (MnO$_2$ NPs) to design controlled oxygen production and promote nature killer cell function [120]. To increase biocompatibility, MnO2 nanoparticles were encapsulated to poly(lactic-co-glytic), forming PLGA-MnO$_2$ NPs. The cytotoxicity of proposed NPs was analyzed using the MTS assay on MCF-7 multicellular spheroids. In this assay, PLGA-MnO2 NPs exhibited significantly enhanced biocompatibility compared to PEG-MnO2 NPs. In diverse approaches, cell viability assays successfully quantified drug efficacy in a detailed and precise manner.

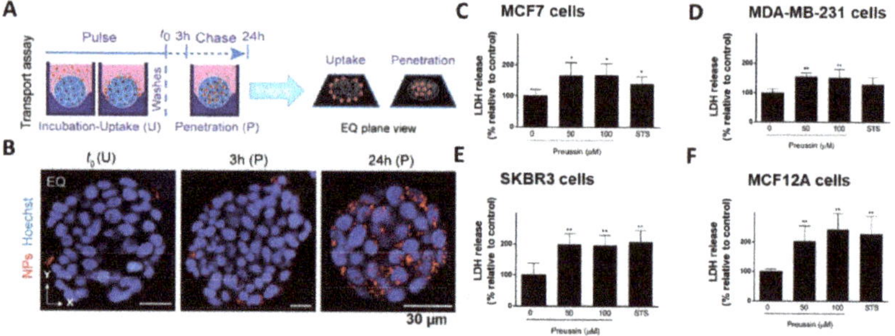

Figure 6. (**A**) Schematic illustration of transport assay for sensing NP drug uptake and penetration. (**B**) Confocal microscopic images showing spheroids at different stages of the NP uptake wave at Z-stacks. LDH cell viability assay for evaluating cytotoxic effect of preussin at different concentrations of 0 μM, 50 μM, 100 μM, and STS (1 μM) on (**C**) MCF7, (**D**) MDA-MB-231, (**E**) SKBR3, and (**F**) MCF-12A cells. (* $p < 0.05$; ** $p < 0.01$). Reprinted with permission from [117]. Copyright 2019, Wiley Online Library; Reprinted with permission from [119]. Copyright 2019, MDPI.

4.2. Non-Invasive Sensing Methods

Recently, cytotoxic assessments of drug candidates in a non-destructive manner were designed to exhibit precise drug screening results [138]. Non-invasive measures should be cell-friendly and non-toxic to cells, making them ideal for determining anticancer drugs' effects. Likewise, measuring spheroid viability via non-invasive screening has been considered as an alternative to conventional colorimetric assays. In general, non-invasive methods do not affect the internal or external cellular environment and allow normal physiological metabolism; thus, it is possible to monitor and evaluate the original appearance of the cell. Ultimately, methods that omit the destructive pretreatment steps can be an excellent index that can clearly reflect the cells in the living organism.

4.2.1. Electrochemical Biosensing

Recently, an ECD method with a fast and sensitive technique was considered as a drug-screening tool [139–141]. When an ECD system is electrically stimulated, the chemical response of the stimulus can be observed, which measures an analyte electrochemically. The electrochemical reaction involves oxidation/reduction via movement of electrons. Depending on the type of electrical signals and regulation factors, various electrochemical measurements exist, such as linear sweep voltammetry, cyclic voltammetry, differential pulse voltammetry (DPV), chronoamperometry, chronopotentiometry, electrochemical impedance spectroscopy (EIS), and electrical impedance tomography. The electrochemical signal intensity is proportional to the cell population, indicating cell viability. Dong et al., (2020) demonstrated monitoring of MCTs in a parallelized wireless sensing system (Figure 7D–F) [121]. An EIS-based platform was devised to continuously measure the size and viability of cancer microtissues for 90 h. The cancer microtissues were treated with different concentrations of doxorubicin (0.1, 1, and 10 µM), which is widely known as a chemotherapeutic agent. The diameter and cell viability were decreased with increasing concentration of doxorubicin in HCT-116. Some researchers have applied the hybrid function platform to MCTs by increasing the electrochemical signal intensity. In one study, a gold-nanostructure-based platform was designed to detect curcumin, which is known as a natural anticancer compound in a brain cancer model (Figure 7A–C) [122]. On this platform, a neuroblastoma (SH-SY5Y) and glioblastoma (U87-MG) were co-cultured to form spontaneous multicellular spheroids without any treatment. Comparing the toxicity assessments of DPV and CCK-8, the electrochemical method proved to be more sensitive (29.4%) with a low concentration of curcumin (30 µM). Spheroids were also monitored for a long time, enabling real-time, non-invasive analysis of potential drug candidates.

Electrochemical detection analysis proved its precision and rapidness without damaging the cellular components, making it an excellent method for monitoring spheroid viability. Preparation steps are minimized such that only the electrochemical device and a live sample are required for the analysis. Above all, live multicellular spheroids can be preserved before and after electrochemical detection. Since the redox reaction is a sudden response, optimization of external components is required.

4.2.2. Optical Coherence Tomography-Based Biosensing

Optical coherence tomography (OCT) refers to a technique capable of imaging an entire spheroid with a high resolution and millimeter-scale penetration depth, in addition to providing physiological and morphological cues about the spheroid [142,143]. It allows label-free imaging, using the intrinsic contrast lens to analyze a sample in a non-destructive manner, thereby enabling clinical cancer research. The OCT technology can be used to analyze the pharmacokinetic response of drug candidates and can be applied to determine the microenvironment and vascular interactions of tumors [144,145]. OCT provides tissue morphology at much higher resolutions than other imaging models such as MRI and ultrasound. The impermeable region of inner spheroid biology can be investigated through the OCT analysis. Confocal microscopy can be used to examine multicellular spheroids through greater depth of imaging. Hari et al., (2019) used this method to conduct refractive index measurements of human colon cancer (HCT116) spheroids [123]. The aim of this research was to study the presence of hypoxia in oxygen-poor cores of the spheroids, which have been known to play an important role in tumor proliferation. The core of the HCT116 spheroids showed debris of stained cell components after sufficient maturity. This study demonstrated the spheroid imaging of hypoxia that could be a future target for the drug-screening analysis. In other cell lines, ovarian cancer cell (OVCAR-8) was applied to visualize 3D structures and monitor necrotic zones within the MCTs (Figure 7F) [146]. In this study, a swept-source optical coherence tomography platform was fabricated, and OCT images were obtained. Mathematical models were employed to calculate growth kinetics of the spheroid size and necrotic tissues. The root-mean-square error (d)used an indicator to evaluate mathematical models, and lower scores showed better performance.

Remarkably, the Boltzmann model proved facile performance as it had the lowest RMSE and AIC score for the growth kinetics of multicellular ovarian tumor spheroids. This result demonstrated OCT for potential drug-screening development for cancer research. Recently, the OCT has been employed not only to determine the tissue structure, but also tissue dynamics. It can be used to assess the necrotic tissue dynamics in multicellular spheroids, which is crucial to investigate the drug efficacy response [147,148]. Two OCT-based methods were reported, such as logarithmic intensity variance (LIV) and OCT correlation decay speed analysis (OCDS), a technique that quantifies and visualizes tissue dynamics (Figure 7G) [110]. Further, human breast adenocarcinoma (MCF7) multicellular spheroids were visualized using OCT, and two methods were used to evaluate the tissue dynamics. First, the LIV methods indicated that necrosis occurred at the center due to lack of nutrient uptake, which is consistent with the necrotic process of a solid tumor. The gradual increase in the OCDS plots demonstrated that the necrotic response became more intense. After paclitaxel drug treatment, the indicators of the necrotic process occurred at the outer region of the spheroids, which was investigated through following analysis.

Overall, OCT allows non-invasive and in-depth analysis tumor imaging. However, some limitations exist in that measurable depth is limited and application samples are restricted.

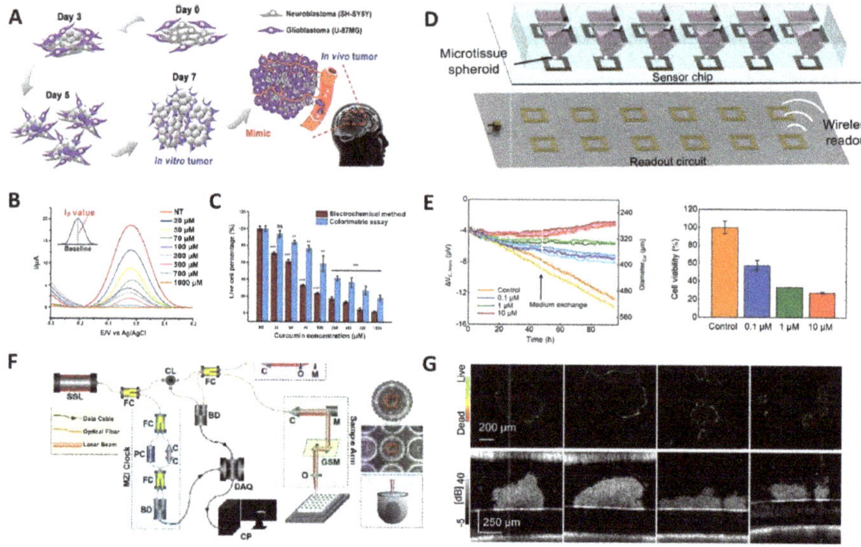

Figure 7. (**A**) Schematic diagram of a real tumor-mimicking Co-culture spheroid model. (**B**) DPV signal graph of the brain tumor model treated with varying concentrations of curcumin. (**C**) Calculated live cell percentage derived from the electrochemical signal peak (Ip_a) from (**B**) and CCK-8 results indicated in a bar graph. (Student's *t*-test, $N = 3$, * $p < 0.05$, ** $p < 0.01$, *** $p < 0.001$). (**D**) An overview of the wireless impedance-based sensor platform that was connected to a PDMS microfluidic device. (**E**) Normalized $\Delta V_{C,Norm}$ values of 0.1, 1, and 10 µM doxorubicin samples and standard medium conditions (control). ΔV_C indicates evaporation-compensated voltage shift. (**F**) Validated ATP measurements of cancer spheroids after 96 h of incubation. The cell viability values were based on the control samples (the error bar indicating mean absolute errors, $N = 2$). (**F**) Schematic of the OCT system for 3D imaging of the spheroid. (**G**) Cross-sectional drug toxicity evaluation of the MCF7 spheroids. The upper panel shows the fluorescence images and the lower panel shows the results obtained through the OCT intensity microscopy. Reprinted with permission from [122]. Copyright 2020, Wiley Online Library; Reprinted with permission from [121]. Copyright 2020, American Chemical Society; Reprinted with permission from [146]. Copyright 2021, Optical Society of America; Reprinted with permission from [110]. Copyright 2020, Optical Society of America.

5. Conclusions

Through the studies discussed above, we categorized 3D multicellular spheroid formation into two main categories: "3D cell-culture technology" and "multicellular cancer spheroid formation platform." Three-dimensional cell culture refers to cells aggregated in a fluidic external cue or attached to an ECM-mimicking microenvironment. Compared to conventional 2D cell culture methods, it exhibits better tissue structure complexity and maintains numerous types of cell interactions. The 3D cell culture can be achieved through two approaches: anchorage-independent and anchorage-type approaches. The former allows effective nutrient uptake through high surface, and the latter enables stable cell growth in ECM-mimicking biomaterials. To fabricate real tumors, multicellular spheroids are adopted in the culture technology and platforms are employed to evaluate anticancer drug efficiency.

With the advent of 3D multicellular screening platforms, pharmacokinetic analysis of the MCTs can be realized. These platforms offer biologically similar in vivo structures of a solid tumor that can be conjugated to investigate complex mechanisms of tumor response to drugs that are not compatible with the 2D culture platform. Furthermore, they could be economically and ethically superior to 2D platforms as they simultaneously enable screening in bulk and reduce the frequency of animal model usage. Advancements in material science and drug-screening methods facilitate in-depth examination of the 3D culture platform. It is predicted that in the future, advanced biosensing tools will be developed to better evaluate drug efficacy and toxicity. Moreover, various applications are expected to be available for drug discovery, oncogenesis research, tissue engineering, and cell physiology, and to augment biomedical industry productivity.

Through the development of MCT generation platform and biosensing techniques, it enables accurate simulation of physiological response of tumor cells, allowing for in-depth analysis of oncology study. With highly simulated cancer models, it is accessible to track cancer cells and biological changes that regulate the behavior of tumors and responsiveness to clinical drugs. Recently, patient-derived cancer organoid (PDO) models, which are elicited from patient tumor tissues, have been an alternative to in vitro models in the biomedical field. Combining three-dimensional cell culture techniques and PDO models, patient-specific drug therapy will be secured and allow safe and efficient treatment, increasing the possibility of survival in the future.

Author Contributions: K.-H.L. and T.-H.K. wrote and revised the manuscript. All authors have read and agreed to the published version of the manuscript.

Funding: This research was supported by the Chung-Ang University Graduate Research Scholarship in 2020 and by a grant from the National Research Foundation of Korea (NRF) (Grant Nos. NRF-2019M3A9H2031820 and NRF-2019R1A4A1028700).

Institutional Review Board Statement: Not applicable.

Informed Consent Statement: Not applicable.

Data Availability Statement: Not applicable. No new data were created or analyzed in this study.

Conflicts of Interest: The authors declare no conflict of interests.

References

1. Jahanban-Esfahlan, R.; Seidi, K.; Banimohamad-Shotorbani, B.; Jahanban-Esfahlan, A.; Yousefi, B. Combination of nanotechnology with vascular targeting agents for effective cancer therapy. *J. Cell. Physiol.* **2018**, *233*, 2982–2992. [CrossRef]
2. Jahanban-Esfahlan, R.; Seidi, K.; Zarghami, N. Tumor vascular infarction: Prospects and challenges. *Int. J. Hematol.* **2017**, *105*, 244–256. [CrossRef] [PubMed]
3. Whiteside, T.J.O. The tumor microenvironment and its role in promoting tumor growth. *Oncogene* **2008**, *27*, 5904–5912. [CrossRef]
4. Balkwill, F.R.; Capasso, M.; Hagemann, T. The tumor microenvironment at a glance. *J. Cell Sci.* **2012**, *125*, 5591–5596. [CrossRef]
5. Arneth, B.J.M. Tumor microenvironment. *Medicina* **2020**, *56*, 15. [CrossRef]
6. Bae, J.; Han, S.; Park, S. Recent Advances in 3D Bioprinted Tumor Microenvironment. *Biochip J.* **2020**, *14*, 137–147. [CrossRef]

7. Szakács, G.; Paterson, J.K.; Ludwig, J.A.; Booth-Genthe, C.; Gottesman, M.M. Targeting multidrug resistance in cancer. *Nat. Rev. Drug Discov.* **2006**, *5*, 219–234. [CrossRef] [PubMed]
8. Gillet, J.-P.; Gottesman, M.M. Mechanisms of multidrug resistance in cancer. In *Multi-Drug Resistance in Cancer*; Springer: Berlin, Germany, 2010; pp. 47–76.
9. Hanahan, D.; Coussens, L.M. Accessories to the crime: Functions of cells recruited to the tumor microenvironment. *Cancer Cell* **2012**, *21*, 309–322. [CrossRef] [PubMed]
10. Frisch, J.; Angenendt, A.; Hoth, M.; Prates Roma, L.; Lis, A.J.C. STIM-Orai channels and reactive oxygen species in the tumor microenvironment. *Cancers* **2019**, *11*, 457. [CrossRef] [PubMed]
11. Patil, A.A.; Rhee, W.J. Exosomes: Biogenesis, Composition, Functions, and Their Role in Pre-metastatic Niche Formation. *Biotechnol. Bioprocess Eng.* **2019**, *24*, 689–701. [CrossRef]
12. Gupta, G.P.; Massagué, J. Cancer metastasis: Building a framework. *Cell* **2006**, *127*, 679–695. [CrossRef]
13. Joyce, J.A.; Pollard, J.W. Microenvironmental regulation of metastasis. *Nat. Rev. Cancer* **2009**, *9*, 239–252. [CrossRef]
14. Ha, N.-H.; Faraji, F.; Hunter, K.W. Mechanisms of metastasis. In *Cancer Target Drug Delivery*; Springer: Berlin, Germany, 2013; pp. 435–458.
15. Lovitt, C.J.; Shelper, T.B.; Avery, V.M. Advanced cell culture techniques for cancer drug discovery. *Biology* **2014**, *3*, 345–367. [CrossRef] [PubMed]
16. Loessner, D.; Stok, K.S.; Lutolf, M.P.; Hutmacher, D.W.; Clements, J.A.; Rizzi, S.C. Bioengineered 3D platform to explore cell–ECM interactions and drug resistance of epithelial ovarian cancer cells. *Biomaterials* **2010**, *31*, 8494–8506. [CrossRef]
17. Cukierman, E.; Pankov, R.; Stevens, D.R.; Yamada, K.M. Taking cell-matrix adhesions to the third dimension. *Science* **2001**, *294*, 1708–1712. [CrossRef] [PubMed]
18. Duval, K.; Grover, H.; Han, L.-H.; Mou, Y.; Pegoraro, A.F.; Fredberg, J.; Chen, Z. Modeling physiological events in 2D vs. 3D cell culture. *Physiology* **2017**, *32*, 266–277. [CrossRef] [PubMed]
19. Li, Y.; Kilian, K.A. Bridging the gap: From 2D cell culture to 3D microengineered extracellular matrices. *Adv. Healthc. Mater.* **2015**, *4*, 2780–2796. [CrossRef] [PubMed]
20. Vinci, M.; Gowan, S.; Boxall, F.; Patterson, L.; Zimmermann, M.; Lomas, C.; Mendiola, M.; Hardisson, D.; Eccles, S.A. Advances in establishment and analysis of three-dimensional tumor spheroid-based functional assays for target validation and drug evaluation. *BMC Biol.* **2012**, *10*, 1–21. [CrossRef] [PubMed]
21. Navin, N.; Kendall, J.; Troge, J.; Andrews, P.; Rodgers, L.; McIndoo, J.; Cook, K.; Stepansky, A.; Levy, D.; Esposito, D.; et al. Tumour evolution inferred by single-cell sequencing. *Nature* **2011**, *472*, 90–94. [CrossRef]
22. Jang, M.; Koh, I.; Lee, S.J.; Cheong, J.-H.; Kim, P. Droplet-based microtumor model to assess cell-ECM interactions and drug resistance of gastric cancer cells. *Sci. Rep.* **2017**, *7*, 1–10. [CrossRef]
23. Hirschhaeuser, F.; Menne, H.; Dittfeld, C.; West, J.; Mueller-Klieser, W.; Kunz-Schughart, L.A. Multicellular tumor spheroids: An underestimated tool is catching up again. *J. Biotechnol.* **2010**, *148*, 3–15. [CrossRef]
24. Sant, S.; Johnston, P.A. The production of 3D tumor spheroids for cancer drug discovery. *Drug Discov. Today Technol.* **2017**, *23*, 27–36. [CrossRef] [PubMed]
25. Knight, E.; Przyborski, S. Advances in 3D cell culture technologies enabling tissue-like structures to be created in vitro. *J. Anat.* **2015**, *227*, 746–756. [CrossRef] [PubMed]
26. Maltman, D.J.; Przyborski, S.A. Developments in three-dimensional cell culture technology aimed at improving the accuracy of in vitro analyses. *Biochem. Soc. Trans.* **2010**, *38*, 1072–1075. [CrossRef] [PubMed]
27. Haycock, J.W. 3D cell culture: A review of current approaches and techniques. In *3D Cell Culture*; Springer: Berlin, Germany, 2011; pp. 1–15.
28. Zhao, L.; Xiu, J.; Liu, Y.; Zhang, T.; Pan, W.; Zheng, X.; Zhang, X. A 3D printed hanging drop dripper for tumor spheroids analysis without recovery. *Sci. Rep.* **2019**, *9*, 1–14. [CrossRef] [PubMed]
29. Park, J.; Kim, H.; Park, J.-K. Microfluidic channel-integrated hanging drop array chip operated by pushbuttons for spheroid culture and analysis. *Analyst* **2020**, *145*, 6974–6980. [CrossRef] [PubMed]
30. RA, V.; Kumari, S.; Poddar, P.; Dhara, D.; Maiti, S. Poly (N-isopropylacrylamide)-Based Polymers as Additive for Rapid Generation of Spheroid via Hanging Drop Method. *Macromol. Biosci.* **2020**, *20*, 2000180. [CrossRef]
31. Jiang, N.; Chen, Z.; Li, B.; Guo, S.; Li, A.; Zhang, T.; Fu, X.; Si, S.; Cui, Y. Effects of rotary cell culture system-simulated microgravity on the ultrastructure and biological behavior of human MDA-MB-231 breast cancer cells. *Precis. Radiat. Oncol.* **2019**, *3*, 87–93. [CrossRef]
32. Chen, Z.Y.; Jiang, N.; Guo, S.; Li, B.B.; Yang, J.Q.; Chai, S.B.; Yan, H.F.; Sun, P.M.; Zhang, T.; Sun, H.W. Effect of simulated microgravity on metabolism of HGC-27 gastric cancer cells. *Oncol. Lett.* **2020**, *19*, 3439–3450. [CrossRef] [PubMed]
33. Zhang, H.-W.; Hu, X.-B.; Qin, Y.; Jin, Z.-H.; Zhang, X.-W.; Liu, Y.-L.; Huang, W.-H. Conductive polymer coated scaffold to integrate 3D cell culture with electrochemical sensing. *Anal. Chem.* **2019**, *91*, 4838–4844. [CrossRef] [PubMed]
34. Or, T.; Saem, S.; Esteve, A.; Osorio, D.A.; De France, K.J.; Vapaavuori, J.; Hoare, T.; Cerf, A.; Cranston, E.D.; Moran-Mirabal, J.M. Patterned cellulose nanocrystal aerogel films with tunable dimensions and morphologies as ultra-porous scaffolds for cell culture. *ACS Appl. Nano Mater.* **2019**, *2*, 4169–4179. [CrossRef]
35. Fu, J.; Li, X.B.; Wang, L.X.; Lv, X.H.; Lu, Z.; Wang, F.; Xia, Q.; Yu, L.; Li, C.M. One-Step Dip-Coating-Fabricated Core–Shell Silk Fibroin Rice Paper Fibrous Scaffolds for 3D Tumor Spheroid Formation. *ACS Appl. Bio Mater.* **2020**, *3*, 7462–7471. [CrossRef]

36. Hainline, K.M.; Gu, F.; Handley, J.F.; Tian, Y.F.; Wu, Y.; de Wet, L.; Vander Griend, D.J.; Collier, J.H. Self-Assembling Peptide Gels for 3D Prostate Cancer Spheroid Culture. *Macromol. Biosci.* **2019**, *19*, 1800249. [CrossRef]

37. Chen, B.; Wu, Y.; Ao, Z.; Cai, H.; Nunez, A.; Liu, Y.; Foley, J.; Nephew, K.; Lu, X.; Guo, F. High-throughput acoustofluidic fabrication of tumor spheroids. *Lab Chip* **2019**, *19*, 1755–1763. [CrossRef]

38. Hai, J.; Zeng, X.; Zhu, Y.; Wang, B. Anions reversibly responsive luminescent nanocellulose hydrogels for cancer spheroids culture and release. *Biomaterials* **2019**, *194*, 161–170. [CrossRef]

39. Ramadhan, W.; Kagawa, G.; Moriyama, K.; Wakabayashi, R.; Minamihata, K.; Goto, M.; Kamiya, N. Construction of higher-order cellular microstructures by a self-wrapping co-culture strategy using a redox-responsive hydrogel. *Sci. Rep.* **2020**, *10*, 1–13. [CrossRef]

40. Wu, K.W.; Kuo, C.T.; Tu, T.Y. A Highly Reproducible Micro U-Well Array Plate Facilitating High-Throughput Tumor Spheroid Culture and Drug Assessment. *Glob. Chall.* **2021**, *5*, 2000056. [CrossRef]

41. Dhamecha, D.; Le, D.; Chakravarty, T.; Perera, K.; Dutta, A.; Menon, J.U. Fabrication of PNIPAm-based thermoresponsive hydrogel microwell arrays for tumor spheroid formation. *Mater. Sci. Eng. C* **2021**, *125*, 112100. [CrossRef]

42. Cui, H.; Wang, X.; Wesslowski, J.; Tronser, T.; Rosenbauer, J.; Schug, A.; Davidson, G.; Popova, A.A.; Levkin, P.A. Assembly of Multi-Spheroid Cellular Architectures by Programmable Droplet Merging. *Adv. Mater.* **2021**, *33*, 2006434. [CrossRef]

43. Shahin-Shamsabadi, A.; Selvaganapathy, P.R. A rapid biofabrication technique for self-assembled collagen-based multicellular and heterogeneous 3D tissue constructs. *Acta Biomater.* **2019**, *92*, 172–183. [CrossRef] [PubMed]

44. Gupta, P.; Pérez-Mancera, P.A.; Kocher, H.; Nisbet, A.; Schettino, G.; Velliou, E.G. A novel scaffold-based hybrid multicellular model for pancreatic ductal adenocarcinoma—Toward a better mimicry of the in vivo tumor microenvironment. *Front. Bioeng. Biotechnol.* **2020**, *8*, 290. [CrossRef] [PubMed]

45. Ham, S.L.; Joshi, R.; Thakuri, P.S.; Tavana, H. Liquid-based three-dimensional tumor models for cancer research and drug discovery. *Exp. Biol. Med.* **2016**, *241*, 939–954. [CrossRef]

46. Kelm, J.M.; Timmins, N.E.; Brown, C.J.; Fussenegger, M.; Nielsen, L.K. Method for generation of homogeneous multicellular tumor spheroids applicable to a wide variety of cell types. *Biotechnol. Bioeng.* **2003**, *83*, 173–180. [CrossRef]

47. Mitteregger, R.; Vogt, G.; Rossmanith, E.; Falkenhagen, D. Rotary cell culture system (RCCS): A new method for cultivating hepatocytes on microcarriers. *Int. J. Artif. Organs* **1999**, *22*, 816–822. [CrossRef]

48. Foty, R. A simple hanging drop cell culture protocol for generation of 3D spheroids. *J. Vis. Exp.* **2011**, 2720. [CrossRef] [PubMed]

49. Tung, Y.-C.; Hsiao, A.Y.; Allen, S.G.; Torisawa, Y.-S.; Ho, M.; Takayama, S. High-throughput 3D spheroid culture and drug testing using a 384 hanging drop array. *Analyst* **2011**, *136*, 473–478. [CrossRef] [PubMed]

50. Ivascu, A.; Kubbies, M. Rapid generation of single-tumor spheroids for high-throughput cell function and toxicity analysis. *J. Biomol. Screen.* **2006**, *11*, 922–932. [CrossRef]

51. Hsiao, A.Y.; Tung, Y.-C.; Kuo, C.-H.; Mosadegh, B.; Bedenis, R.; Pienta, K.J.; Takayama, S. Micro-ring structures stabilize microdroplets to enable long term spheroid culture in 384 hanging drop array plates. *Biomed. Microdevices* **2012**, *14*, 313–323. [CrossRef]

52. Lin, R.Z.; Chang, H.Y. Recent advances in three-dimensional multicellular spheroid culture for biomedical research. *Biotechnol. J.* **2008**, *3*, 1172–1184. [CrossRef]

53. Lv, D.; Hu, Z.; Lu, L.; Lu, H.; Xu, X. Three-dimensional cell culture: A powerful tool in tumor research and drug discovery. *Oncol. Lett.* **2017**, *14*, 6999–7010. [CrossRef]

54. Nyberg, S.L.; Hardin, J.; Amiot, B.; Argikar, U.A.; Remmel, R.P.; Rinaldo, P. Rapid, large-scale formation of porcine hepatocyte spheroids in a novel spheroid reservoir bioartificial liver. *Liver Transplant.* **2005**, *11*, 901–910. [CrossRef] [PubMed]

55. Fischbach, C.; Chen, R.; Matsumoto, T.; Schmelzle, T.; Brugge, J.S.; Polverini, P.J.; Mooney, D.J. Engineering tumors with 3D scaffolds. *Nat. Methods* **2007**, *4*, 855–860. [CrossRef]

56. Dhiman, H.K.; Ray, A.R.; Panda, A.K. Three-dimensional chitosan scaffold-based MCF-7 cell culture for the determination of the cytotoxicity of tamoxifen. *Biomaterials* **2005**, *26*, 979–986. [CrossRef]

57. Weaver, V.M.; Lelièvre, S.; Lakins, J.N.; Chrenek, M.A.; Jones, J.C.; Giancotti, F.; Werb, Z.; Bissell, M.J. β4 integrin-dependent formation of polarized three-dimensional architecture confers resistance to apoptosis in normal and malignant mammary epithelium. *Cancer Cell* **2002**, *2*, 205–216. [CrossRef]

58. Zhang, X.; Wang, W.; Yu, W.; Xie, Y.; Zhang, X.; Zhang, Y.; Ma, X. Development of an in vitro multicellular tumor spheroid model using microencapsulation and its application in anticancer drug screening and testing. *Biotechnol. Prog.* **2005**, *21*, 1289–1296. [CrossRef]

59. Ohtaka, K.; Watanabe, S.; Iwazaki, R.; Hirose, M.; Sato, N. Role of extracellular matrix on colonic cancer cell migration and proliferation. *Biochem. Biophys. Res. Commun.* **1996**, *220*, 346–352. [CrossRef]

60. Hollister, S.J. Porous scaffold design for tissue engineering. *Nat. Mater.* **2005**, *4*, 518–524. [CrossRef] [PubMed]

61. Zhang, M.; Boughton, P.; Rose, B.; Lee, C.S.; Hong, A.M. The use of porous scaffold as a tumor model. *Int. J. Biomater.* **2013**, *2013*, 396056. [CrossRef]

62. Nemati, S.; Kim, S.-J.; Shin, Y.M.; Shin, H. Current progress in application of polymeric nanofibers to tissue engineering. *Nano Converg.* **2019**, *6*, 36. [CrossRef]

63. Park, M.; Lee, G.; Ryu, K.; Lim, W. Improvement of Bone Formation in Rats with Calvarial Defects by Modulating the Pore Size of Tricalcium Phosphate Scaffolds. *Biotechnol. Bioprocess Eng.* **2019**, *24*, 885–892. [CrossRef]

64. Sun, M.T.; O'Connor, A.J.; Milne, I.; Biswas, D.; Casson, R.; Wood, J.; Selva, D. Development of Macroporous Chitosan Scaffolds for Eyelid Tarsus Tissue Engineering. *J. Tissue Eng. Regen. Med.* **2019**, *16*, 595–604. [CrossRef]
65. Girard, Y.K.; Wang, C.; Ravi, S.; Howell, M.C.; Mallela, J.; Alibrahim, M.; Green, R.; Hellermann, G.; Mohapatra, S.S.; Mohapatra, S. A 3D fibrous scaffold inducing tumoroids: A platform for anticancer drug development. *PLoS ONE* **2013**, *8*, e75345. [CrossRef] [PubMed]
66. Bae, C.-S.; Lee, C.-M.; Ahn, T. Encapsulation of Apoptotic Proteins in Lipid Nanoparticles to Induce Death of Cancer Cells. *Biotechnol. Bioprocess Eng.* **2020**, *25*, 264–271. [CrossRef]
67. Feng, S.; Duan, X.; Lo, P.-K.; Liu, S.; Liu, X.; Chen, H.; Wang, Q. Expansion of breast cancer stem cells with fibrous scaffolds. *Integr. Biol.* **2013**, *5*, 768–777. [CrossRef]
68. Shang, M.; Soon, R.H.; Lim, C.T.; Khoo, B.L.; Han, J. Microfluidic modelling of the tumor microenvironment for anti-cancer drug development. *Lab Chip* **2019**, *19*, 369–386. [CrossRef]
69. Caliari, S.R.; Burdick, J.A. A practical guide to hydrogels for cell culture. *Nat. Methods* **2016**, *13*, 405–414. [CrossRef]
70. Ashworth, J.C.; Morgan, R.L.; Lis-Slimak, K.; Meade, K.A.; Jones, S.; Spence, K.; Slater, C.E.; Thompson, J.L.; Grabowska, A.M.; Clarke, R.B.; et al. Preparation of a User-Defined Peptide Gel for Controlled 3D Culture Models of Cancer and Disease. *J. Vis. Exp.* **2020**, e61710. [CrossRef] [PubMed]
71. Jung, J.P.; Moyano, J.V.; Collier, J.H. Multifactorial optimization of endothelial cell growth using modular synthetic extracellular matrices. *Integr. Biol.* **2011**, *3*, 185–196. [CrossRef] [PubMed]
72. Oh, S.; Ryu, H.; Tahk, D.; Ko, J.; Chung, Y.; Lee, H.K.; Lee, T.R.; Jeon, N.L. "Open-top" microfluidic device for in vitro three-dimensional capillary beds. *Lab Chip* **2017**, *17*, 3405–3414. [CrossRef] [PubMed]
73. Bersini, S.; Jeon, J.S.; Dubini, G.; Arrigoni, C.; Chung, S.; Charest, J.L.; Moretti, M.; Kamm, R.D. A microfluidic 3D in vitro model for specificity of breast cancer metastasis to bone. *Biomaterials* **2014**, *35*, 2454–2461. [CrossRef] [PubMed]
74. Truong, D.; Puleo, J.; Llave, A.; Mouneimne, G.; Kamm, R.D.; Nikkhah, M. Breast cancer cell invasion into a three dimensional tumor-stroma microenvironment. *Sci. Rep.* **2016**, *6*, 1–18.
75. Rodríguez-Enríquez, S.; Gallardo-Pérez, J.C.; Avilés-Salas, A.; Marín-Hernández, A.; Carreño-Fuentes, L.; Maldonado-Lagunas, V.; Moreno-Sánchez, R. Energy metabolism transition in multi-cellular human tumor spheroids. *J. Cell. Physiol.* **2008**, *216*, 189–197. [CrossRef] [PubMed]
76. De Witt Hamer, P.C.; Leenstra, S.; Van Noorden, C.J.; Zwinderman, A.H. Organotypic glioma spheroids for screening of experimental therapies: How many spheroids and sections are required? *Cytom. Part A* **2009**, *75*, 528–534. [CrossRef] [PubMed]
77. Jiang, Y.; Chen, J.; Deng, C.; Suuronen, E.J.; Zhong, Z. Click hydrogels, microgels and nanogels: Emerging platforms for drug delivery and tissue engineering. *Biomaterials* **2014**, *35*, 4969–4985. [CrossRef]
78. Zhu, D.; Tong, X.; Trinh, P.; Yang, F. Mimicking cartilage tissue zonal organization by engineering tissue-scale gradient hydrogels as 3D cell niche. *Tissue Eng. Part A* **2018**, *24*, 1–10. [CrossRef]
79. Tibbitt, M.W.; Anseth, K.S. Hydrogels as extracellular matrix mimics for 3D cell culture. *Biotechnol. Bioeng.* **2009**, *103*, 655–663. [CrossRef]
80. Geckil, H.; Xu, F.; Zhang, X.; Moon, S.; Demirci, U. Engineering hydrogels as extracellular matrix mimics. *Nanomedicine* **2010**, *5*, 469–484. [CrossRef]
81. Mun, S.G.; Choi, H.W.; Lee, J.M.; Lim, J.H.; Ha, J.H.; Kang, M.-J.; Kim, E.-J.; Kang, L.; Chung, B.G. rGO nanomaterial-mediated cancer targeting and photothermal therapy in a microfluidic co-culture platform. *Nano Converg.* **2020**, *7*, 10. [CrossRef] [PubMed]
82. Hauck, N.; Seixas, N.; Centeno, S.P.; Schlüßler, R.; Cojoc, G.; Müller, P.; Guck, J.; Wöll, D.; Wessjohann, L.A.; Thiele, J. Droplet-assisted microfluidic fabrication and characterization of multifunctional polysaccharide microgels formed by multicomponent reactions. *Polymers* **2018**, *10*, 1055. [CrossRef]
83. Hann, S.D.; Niepa, T.H.; Stebe, K.J.; Lee, D. One-step generation of cell-encapsulating compartments via polyelectrolyte complexation in an aqueous two phase system. *ACS Appl. Mater. Interfaces* **2016**, *8*, 25603–25611. [CrossRef] [PubMed]
84. Song, Y.; Chan, Y.K.; Ma, Q.; Liu, Z.; Shum, H.C. All-aqueous electrosprayed emulsion for templated fabrication of cytocompatible microcapsules. *ACS Appl. Mater. Interfaces* **2015**, *7*, 13925–13933. [CrossRef]
85. Iqbal, M.; Tao, Y.; Xie, S.; Zhu, Y.; Chen, D.; Wang, X.; Huang, L.; Peng, D.; Sattar, A.; Shabbir, M.A.B.; et al. Aqueous two-phase system (ATPS): An overview and advances in its applications. *Biol. Proced. Online* **2016**, *18*, 1–18. [CrossRef] [PubMed]
86. Kim, D.; Park, S.; Yoo, H.; Park, S.; Kim, J.; Yum, K.; Kim, K.; Kim, H. Overcoming anticancer resistance by photodynamic therapy-related efflux pump deactivation and ultrasound-mediated improved drug delivery efficiency. *Nano Converg.* **2020**, *7*, 30. [CrossRef]
87. Griffin, D.R.; Kasko, A.M. Photodegradable macromers and hydrogels for live cell encapsulation and release. *J. Am. Chem. Soc.* **2012**, *134*, 13103–13107. [CrossRef]
88. Steinhilber, D.; Rossow, T.; Wedepohl, S.; Paulus, F.; Seiffert, S.; Haag, R. A microgel construction kit for bioorthogonal encapsulation and pH-controlled release of living cells. *Angew. Chem. Int.* **2013**, *52*, 13538–13543. [CrossRef] [PubMed]
89. Teramura, Y.; Iwata, H. Cell surface modification with polymers for biomedical studies. *Soft Matter.* **2010**, *6*, 1081–1091. [CrossRef]
90. Yarmush, M.L.; King, K.R. Living-cell microarrays. *Annu. Rev. Biomed. Eng.* **2009**, *11*, 235–257. [CrossRef]
91. Starkuviene, V.; Pepperkok, R.; Erfle, H. Transfected cell microarrays: An efficient tool for high-throughput functional analysis. *Expert Rev. Proteom.* **2007**, *4*, 479–489. [CrossRef]

92. Gidrol, X.; Fouqué, B.; Ghenim, L.; Haguet, V.; Picollet-D'hahan, N.; Schaack, B. 2D and 3D cell microarrays in pharmacology. *Curr. Opin. Pharmacol.* **2009**, *9*, 664–668. [CrossRef]
93. Roach, K.L.; King, K.R.; Uygun, B.E.; Kohane, I.S.; Yarmush, M.L.; Toner, M. High throughput single cell bioinformatics. *Biotechnol. Prog.* **2009**, *25*, 1772–1779. [CrossRef] [PubMed]
94. Hook, A.L.; Thissen, H.; Voelcker, N.H. Advanced substrate fabrication for cell microarrays. *Biomacromolecules* **2009**, *10*, 573–579. [CrossRef]
95. Fernandes, T.G.; Diogo, M.M.; Clark, D.S.; Dordick, J.S.; Cabral, J.M. High-throughput cellular microarray platforms: Applications in drug discovery, toxicology and stem cell research. *Trends Biotechnol.* **2009**, *27*, 342–349. [CrossRef] [PubMed]
96. Tu, T.Y.; Wang, Z.; Bai, J.; Sun, W.; Peng, W.K.; Huang, R.Y.J.; Thiery, J.P.; Kamm, R.D. Rapid prototyping of concave microwells for the formation of 3D multicellular cancer aggregates for drug screening. *Adv. Healthc. Mater.* **2014**, *3*, 609–616. [CrossRef] [PubMed]
97. Albritton, J.L.; Roybal, J.D.; Paulsen, S.J.; Calafat, N.J.; Flores-Zaher, J.A.; Farach-Carson, M.C.; Gibbons, D.L.; Miller, J.S. Ultrahigh-throughput generation and characterization of cellular aggregates in laser-ablated microwells of poly (dimethylsiloxane). *RSC Adv.* **2016**, *6*, 8980–8991. [CrossRef]
98. Chiu, C.-Y.; Chen, Y.-C.; Wu, K.-W.; Hsu, W.-C.; Lin, H.-P.; Chang, H.-C.; Lee, Y.-C.; Wang, Y.-K.; Tu, T.-Y. Simple in-house fabrication of microwells for generating uniform hepatic multicellular cancer aggregates and discovering novel therapeutics. *Materials* **2019**, *12*, 3308. [CrossRef] [PubMed]
99. Meckenstock, R.U.; von Netzer, F.; Stumpp, C.; Lueders, T.; Himmelberg, A.M.; Hertkorn, N.; Schmitt-Kopplin, P.; Harir, M.; Hosein, R.; Haque, S. Water droplets in oil are microhabitats for microbial life. *Science* **2014**, *345*, 673–676. [CrossRef]
100. Kang, E.-S.; Kim, D.-S.; Suhito, I.R.; Choo, S.-S.; Kim, S.-J.; Song, I.; Kim, T.-H. Guiding osteogenesis of mesenchymal stem cells using carbon-based nanomaterials. *Nano Converg.* **2017**, *4*, 1–14. [CrossRef] [PubMed]
101. Koh, I.; Kim, P. In vitro reconstruction of brain tumor microenvironment. *Biochip J.* **2019**, *13*, 1–7. [CrossRef]
102. Kim, C.H.; Suhito, I.R.; Angeline, N.; Han, Y.; Son, H.; Luo, Z.; Kim, T.H. Vertically Coated Graphene Oxide Micro-Well Arrays for Highly Efficient Cancer Spheroid Formation and Drug Screening. *Adv. Healthc. Mater.* **2020**, *9*, 1901751. [CrossRef]
103. Hussey, G.S.; Dziki, J.L.; Badylak, S.F. Extracellular matrix-based materials for regenerative medicine. *Nat. Rev. Mater.* **2018**, *3*, 159–173. [CrossRef]
104. Goodarzi, H.; Hashemi-Najafabadi, S.; Baheiraei, N.; Bagheri, F. Preparation and Characterization of Nanocomposite Scaffolds (Collagen/β-TCP/SrO) for Bone Tissue Engineering. *J. Tissue Eng. Regen. Med.* **2019**, *16*, 237–251. [CrossRef] [PubMed]
105. Badylak, S.F. The extracellular matrix as a biologic scaffold material. *Biomaterials* **2007**, *28*, 3587–3593. [CrossRef] [PubMed]
106. Tian, H.; Tang, Z.; Zhuang, X.; Chen, X.; Jing, X. Biodegradable synthetic polymers: Preparation, functionalization and biomedical application. *Prog. Polym. Sci.* **2012**, *37*, 237–280. [CrossRef]
107. Sionkowska, A. Current research on the blends of natural and synthetic polymers as new biomaterials. *Prog. Polym. Sci.* **2011**, *36*, 1254–1276. [CrossRef]
108. Baker, A.E.; Tam, R.Y.; Shoichet, M.S. Independently tuning the biochemical and mechanical properties of 3D hyaluronan-based hydrogels with oxime and Diels–Alder chemistry to culture breast cancer spheroids. *Biomacromolecules* **2017**, *18*, 4373–4384. [CrossRef] [PubMed]
109. Gu, L.; Mooney, D.J. Biomaterials and emerging anticancer therapeutics: Engineering the microenvironment. *Nat. Rev. Cancer* **2016**, *16*, 56–66. [CrossRef] [PubMed]
110. Abd El-Sadek, I.; Miyazawa, A.; Shen, L.T.-W.; Makita, S.; Fukuda, S.; Yamashita, T.; Oka, Y.; Mukherjee, P.; Matsusaka, S.; Oshika, T. Optical coherence tomography-based tissue dynamics imaging for longitudinal and drug response evaluation of tumor spheroids. *Biomed. Opt. Express* **2020**, *11*, 6231–6248. [CrossRef]
111. Zhang, S.; Cortes, W.; Zhang, Y. Constructing Cross-Linked Nanofibrous Scaffold via Dual-Enzyme-Instructed Hierarchical Assembly. *Langmuir* **2020**, *36*, 6261–6267. [CrossRef] [PubMed]
112. Zanoni, M.; Piccinini, F.; Arienti, C.; Zamagni, A.; Santi, S.; Polico, R.; Bevilacqua, A.; Tesei, A. 3D tumor spheroid models for in vitro. therapeutic screening: A systematic approach to enhance the biological relevance of data obtained. *Sci. Rep.* **2016**, *6*, 1–11. [CrossRef] [PubMed]
113. Torchilin, V.P. Recent advances with liposomes as pharmaceutical carriers. *Nat. Rev. Drug Discov.* **2005**, *4*, 145–160. [CrossRef] [PubMed]
114. Torchilin, V.P. Passive and active drug targeting: Drug delivery to tumors as an example. *Drug Deliv.* **2010**, 3–53. [CrossRef]
115. Xia, Y.; Chen, H.; Li, J.; Hu, H.; Qian, Q.; He, R.-X.; Ding, Z.; Guo, S.-S. Acoustic Droplet-Assisted Superhydrophilic–Superhydrophobic Microarray Platform for High-Throughput Screening of Patient-Derived Tumor Spheroids. *ACS Appl. Mater. Interfaces* **2021**, *13*, 23489–23501. [CrossRef] [PubMed]
116. Machálková, M.; Pavlatovská, B.; Michálek, J.; Pruška, A.; Štěpka, K.; Nečasová, T.; Radaszkiewicz, K.A.; Kozubek, M.; Šmarda, J.; Preisler, J.; et al. Drug penetration analysis in 3D cell cultures using fiducial-based semiautomatic coregistration of MALDI MSI and immunofluorescence images. *Anal. Chem.* **2019**, *91*, 13475–13484. [CrossRef]
117. Cutrona, M.B.; Simpson, J.C. A High-Throughput Automated Confocal Microscopy Platform for Quantitative Phenotyping of Nanoparticle Uptake and Transport in Spheroids. *Small* **2019**, *15*, 1902033. [CrossRef] [PubMed]
118. Balsa, L.M.; Ruiz, M.C.; de la Parra, L.S.M.; Baran, E.J.; León, I.E. Anticancer and antimetastatic activity of copper (II)-tropolone complex against human breast cancer cells, breast multicellular spheroids and mammospheres. *J. Inorg. Biochem.* **2020**, *204*, 110975. [CrossRef]

119. Malhão, F.; Ramos, A.A.; Buttachon, S.; Dethoup, T.; Kijjoa, A.; Rocha, E. Cytotoxic and antiproliferative effects of Preussin, a hydroxypyrrolidine derivative from the marine sponge-associated Fungus Aspergillus candidus KUFA 0062, in a panel of breast cancer cell lines and using 2D and 3D cultures. *Mar. Drugs.* **2019**, *17*, 448. [CrossRef]
120. Murphy, D.A.; Cheng, H.; Yang, T.; Yan, X.; Adjei, I.M. Reversing Hypoxia with PLGA-Encapsulated Manganese Dioxide Nanoparticles Improves Natural Killer Cell Response to Tumor Spheroids. *Mol. Pharm.* **2021**, *18*, 2935–2946. [CrossRef] [PubMed]
121. Dong, L.; Ravaynia, P.S.; Huang, Q.-A.; Hierlemann, A.; Modena, M.M. Parallelized wireless sensing system for continuous monitoring of microtissue spheroids. *ACS Sens.* **2020**, *5*, 2036–2043. [CrossRef]
122. Suhito, I.R.; Angeline, N.; Lee, K.H.; Kim, H.; Park, C.G.; Luo, Z.; Kim, T.H. A Spheroid-Forming Hybrid Gold Nanostructure Platform That Electrochemically Detects Anticancer Effects of Curcumin in a Multicellular Brain Cancer Model. *Small* **2021**, *17*, 2002436. [CrossRef]
123. Hari, N.; Patel, P.; Ross, J.; Hicks, K.; Vanholsbeeck, F. Optical coherence tomography complements confocal microscopy for investigation of multicellular tumour spheroids. *Sci. Rep.* **2019**, *9*, 1–11. [CrossRef]
124. Miller, D.M.; Shakes, D.C. Immunofluorescence microscopy. *Methods Cell Biol.* **1995**, *48*, 365–394. [PubMed]
125. Odell, I.D.; Cook, D. Immunofluorescence techniques. *J. Investig. Dermatol.* **2013**, *133*, e4. [CrossRef]
126. Sawyers, C.L. The cancer biomarker problem. *Nature* **2008**, *452*, 548–552. [CrossRef] [PubMed]
127. Kim, D.-H.; Paek, S.-H.; Choi, D.-Y.; Lee, M.-K.; Park, J.-N.; Cho, H.-M.; Paek, S.-H. Real-time Monitoring of Biomarkers in Serum for Early Diagnosis of Target Disease. *Biochip J.* **2020**, *14*, 2–17. [CrossRef]
128. Paek, S.-H. Real-time Monitoring of Biomarkers: Current Status and Future Perspectives. *Biochip J.* **2020**, *14*. [CrossRef]
129. Frankfurt, O.S.; Krishan, A. Apoptosis-based drug screening and detection of selective toxicity to cancer cells. *Anticancer Drugs* **2003**, *14*, 555–561. [CrossRef] [PubMed]
130. Balkwill, F. Tumour necrosis factor and cancer. *Nat. Rev. Cancer* **2009**, *9*, 361–371. [CrossRef] [PubMed]
131. Lowe, S.W.; Lin, A.W. Apoptosis in cancer. *J. Heterocycl. Chem.* **2000**, *21*, 485–495. [CrossRef] [PubMed]
132. Denton, D.; Kumar, S. Terminal deoxynucleotidyl transferase (TdT)-mediated dUTP nick-end labeling (TUNEL) for detection of apoptotic cells in Drosophila. *Cold Spring Harb. Protoc.* **2015**, *2015*, 568–571. [CrossRef] [PubMed]
133. Fallone, C.A.; Loo, V.G.; Lough, J.; Barkun, A.N. Hematoxylin and eosin staining of gastric tissue for the detection of Helicobacter pylori. *Helicobacter* **1997**, *2*, 32–35. [CrossRef] [PubMed]
134. Smith, S.M.; Wunder, M.B.; Norris, D.A.; Shellman, Y.G. A simple protocol for using a LDH-based cytotoxicity assay to assess the effects of death and growth inhibition at the same time. *PLoS ONE* **2011**, *6*, e26908. [CrossRef]
135. Van Meerloo, J.; Kaspers, G.J.; Cloos, J. Cell sensitivity assays: The MTT assay. In *Cancer Cell Culture*; Springer: Berlin, Germany, 2011; pp. 237–245.
136. Peskin, A.V.; Winterbourn, C.C. A microtiter plate assay for superoxide dismutase using a water-soluble tetrazolium salt (WST-1). *Clin. Chim. Acta* **2000**, *293*, 157–166. [CrossRef]
137. Roehm, N.W.; Rodgers, G.H.; Hatfield, S.M.; Glasebrook, A.L. An improved colorimetric assay for cell proliferation and viability utilizing the tetrazolium salt XTT. *J. Immunol. Methods* **1991**, *142*, 257–265. [CrossRef]
138. Imperiale, T.F. Noninvasive screening tests for colorectal cancer. *J. Dig. Dis.* **2012**, *30*, 16–26. [CrossRef] [PubMed]
139. Park, I.-H.; Hong, Y.; Jun, H.-S.; Cho, E.-S.; Cho, S. DAQ based impedance measurement system for low cost and portable electrical cell-substrate impedance sensing. *Biochip J.* **2018**, *12*, 18–24. [CrossRef]
140. Simoska, O.; Sans, M.; Fitzpatrick, M.D.; Crittenden, C.M.; Eberlin, L.S.; Shear, J.B.; Stevenson, K.J. Real-time electrochemical detection of pseudomonas aeruginosa phenazine metabolites using transparent carbon ultramicroelectrode arrays. *ACS Sens.* **2018**, *4*, 170–179. [CrossRef] [PubMed]
141. Suhito, I.R.; Lee, W.; Baek, S.; Lee, D.; Min, J.; Kim, T.-H. Rapid and sensitive electrochemical detection of anticancer effects of curcumin on human glioblastoma cells. *Sens. Actuators B Chem.* **2019**, *288*, 527–534. [CrossRef]
142. Wojtkowski, M. High-speed optical coherence tomography: Basics and applications. *Appl. Opt.* **2010**, *49*, D30–D61. [CrossRef] [PubMed]
143. Drexler, W. Ultrahigh-resolution optical coherence tomography. *J. Biomed. Opt.* **2004**, *9*, 47–74. [CrossRef] [PubMed]
144. Assayag, O.; Grieve, K.; Devaux, B.; Harms, F.; Pallud, J.; Chretien, F.; Boccara, C.; Varlet, P. Imaging of non-tumorous and tumorous human brain tissues with full-field optical coherence tomography. *Neuroimage Clin.* **2013**, *2*, 549–557. [CrossRef] [PubMed]
145. Kut, C.; Chaichana, K.L.; Xi, J.; Raza, S.M.; Ye, X.; McVeigh, E.R.; Rodriguez, F.J.; Quiñones-Hinojosa, A.; Li, X. Detection of human brain cancer infiltration ex vivo and in vivo using quantitative optical coherence tomography. *Sci. Transl. Med.* **2015**, *7*, 292ra100. [CrossRef]
146. Yan, F.; Gunay, G.; Valerio, T.I.; Wang, C.; Wilson, J.A.; Haddad, M.S.; Watson, M.; Connell, M.O.; Davidson, N.; Fung, K.-M. Characterization and quantification of necrotic tissues and morphology in multicellular ovarian cancer tumor spheroids using optical coherence tomography. *Biomed. Opt. Express* **2021**, *12*, 3352–3371. [CrossRef] [PubMed]
147. Huang, Y.; Wang, S.; Guo, Q.; Kessel, S.; Rubinoff, I.; Chan, L.L.-Y.; Li, P.; Liu, Y.; Qiu, J.; Zhou, C. Optical coherence tomography detects necrotic regions and volumetrically quantifies multicellular tumor spheroids. *Cancer Res.* **2017**, *77*, 6011–6020. [CrossRef] [PubMed]
148. Farhat, G.; Mariampillai, A.; Yang, V.X.; Czarnota, G.J.; Kolios, M.C. Optical coherence tomography speckle decorrelation for detecting cell death. In *Proceedings of the Biomedical Applications of Light Scattering V*; SPIE: Washington, DC, USA, 2011; pp. 1–15.

 *biosensors*

Review

Recent Trends in Exhaled Breath Diagnosis Using an Artificial Olfactory System

Chuntae Kim [1,†], Iruthayapandi Selestin Raja [1,†], Jong-Min Lee [2], Jong Ho Lee [3], Moon Sung Kang [4], Seok Hyun Lee [4], Jin-Woo Oh [1,5,*] and Dong-Wook Han [1,4,*]

[1] BIO-IT Foundry Technology Institute, Pusan National University, Busan 46241, Korea; chuntae1122@gmail.com (C.K.); rajaselestin@gmail.com (I.S.R.)
[2] School of Nano Convergence Technology, Hallym University, Chuncheon 24252, Korea; jongminlee1984@gmail.com
[3] Daan Korea Corporation, Seoul 06252, Korea; nunssob@gmail.com
[4] Department of Cogno-Mechatronics Engineering, Pusan National University, Busan 46241, Korea; mskang7909@gmail.com (M.S.K.); seokhyun2285@gmail.com (S.H.L.)
[5] Department of Nanoenergy Engineering, Pusan National University, Busan 46241, Korea
* Correspondence: ojw@pusan.ac.kr (J.-W.O.); nanohan@pusan.ac.kr (D.-W.H.)
† These authors contributed equally to this work.

Abstract: Artificial olfactory systems are needed in various fields that require real-time monitoring, such as healthcare. This review introduces cases of detection of specific volatile organic compounds (VOCs) in a patient's exhaled breath and discusses trends in disease diagnosis technology development using artificial olfactory technology that analyzes exhaled human breath. We briefly introduce algorithms that classify patterns of odors (VOC profiles) and describe artificial olfactory systems based on nanosensors. On the basis of recently published research results, we describe the development trend of artificial olfactory systems based on the pattern-recognition gas sensor array technology and the prospects of application of this technology to disease diagnostic devices. Medical technologies that enable early monitoring of health conditions and early diagnosis of diseases are crucial in modern healthcare. By regularly monitoring health status, diseases can be prevented or treated at an early stage, thus increasing the human survival rate and reducing the overall treatment costs. This review introduces several promising technical fields with the aim of developing technologies that can monitor health conditions and diagnose diseases early by analyzing exhaled human breath in real time.

Keywords: artificial olfactory system; health monitoring; exhaled breath diagnosis; volatile organic compounds; gas sensor; electronic nose

Citation: Kim, C.; Raja, I.S.; Lee, J.-M.; Lee, J.H.; Kang, M.S.; Lee, S.H.; Oh, J.-W.; Han, D.-W. Recent Trends in Exhaled Breath Diagnosis Using an Artificial Olfactory System. *Biosensors* **2021**, *11*, 337. https://doi.org/10.3390/bios11090337

Received: 5 August 2021
Accepted: 10 September 2021
Published: 14 September 2021

1. Introduction

The olfactory sense is the oldest human sense. Therefore, humans tend to analyze the environment by sniffing [1]. Various studies have been conducted to mimic the sensory cognitive mechanism, and research on sensor systems focusing on olfactory cognitive models has attracted significant attention [2]. Sniffing is a complicated process as there are over 400 species of olfactory receptor gene, and signals through various receptors are comprehensively determined in the brain to provide information about smell and human cognition [3]. The concept of an "artificial nose" is based on a technology that grasps information about odors and uses them as data. In other words, it is an electronic system technology that analyzes the state and composition of a substance through smell. Figure 1 shows the concept of an artificial nose system inspired by the olfactory perception pathway. A biomimetic gas detection platform can be designed by constructing a nanosensor array based on the olfactory receptor tissue and by combining a data processing technology through pattern analysis of signal data.

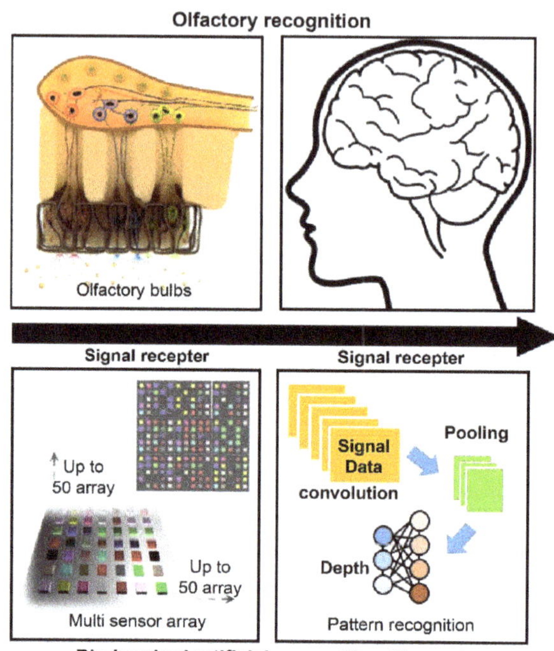

Figure 1. Concept of an artificial nose (e-nose) system based on the olfactory perception pathway.

Persaud et al. proposed the concept of using different types of sensors as arrays and applying unique signal patterns to specific odors to enhance the selectivity, reliability, and accuracy of gas detection systems [4]. The developed system was called an electronic nose (e-nose). All corresponding technologies associated to this system are called e-noses. The principle of the e-nose is similar to that of the human olfactory mechanism. That is, the odor recognition is based on the reaction of the smell factor and olfactory receptors. The e-nose reacts with odor substances using sensor unit devices instead of olfactory receptor cells as receptors, and analyzes patterns using a computer instead of the brain [5].

In the 21st century, the e-nose technology has evolved significantly along with the development of NT/IT technologies. Low-cost, high-performance sensor devices have been developed on the basis of various nanobiosensor technologies, and the data processing analysis technologies have been improved owing to the advent of artificial intelligence (AI)-integrated technologies. The evolution of AI and big data processing technologies has subsequently led to the development of high-level e-nose technologies that utilize a large number of sensor arrays [6]. The current artificial-nose technology can sniff a smell that cannot be smelled by humans using a sensor unit with a sensitivity of parts-per-billion level [7]. Recently, this technology has been used in various fields such as chemistry, medical care, food quality management, and military industry [8–11]. In the case of conventional gas sensors, the information on the gas concentration level is acquired by quantitatively analyzing specific chemicals. The e-nose sensors do not require such an extensive process and have been increasingly used because of their simple structure that recognizes patterns for only specific odors through a database.

This paper introduces several research cases that analyze the smell of breath exhaled by humans using an artificial olfactory system; in many cases, an e-nose is used. When a person is diagnosed with a certain disease, various volatile organic compounds (VOCs) are produced in vivo owing to metabolic disorders or free radicals [12–14]. Such VOCs serve as unique biometric data depending on the specificity, stage, and condition of the disease. In other words, as the conditions under which VOCs occur depend on the particular disease and its various stages, they are classified as "fingerprints" of the disease conditions. Therefore, the analysis of the exhaled gas can provide information on the physiological and health status of an individual. This information can be used for the early diagnosis of many diseases during the induction stage [15]. In this study, we describe the overall concept of disease diagnosis using human respiratory gas, which has gained significant attention in recent years. Furthermore, we describe the trend of human respiratory gas analysis technologies used for diagnosing such diseases and analyze future scenarios of human respiratory gas disease analysis.

2. Exhaled Breath Diagnosis

In conventional approaches, diseases were distinguished on the basis of the breath of patients. Recent reports indicate that dogs that are trained to detect various drugs and explosives by smell can also distinguish cancer patients from healthy people. The olfactory analysis method is a natural approach that is characterized by many possibilities. The current drunk-driving control technologies are based on the blood alcohol concentration measurement, and human breath gas analysis methods are being actively researched. A majority of this research is directed toward the analysis of respiratory diseases such as lung cancer and asthma [16,17]. Once the relationship between a disease and breath gas components is identified, the disease diagnosis based on the exhaled breath analysis is expected to emerge as a future disease diagnostic method [18].

In general, the diagnostic methods using expensive analytical equipment by collecting samples of patients' blood, tissues, etc. require a time-consuming sample collection process and a skilled operator owing to complex protocols. In addition, such methods can only be performed in a limited number of places such as hospitals. In contrast, disease diagnosis using the analysis of breathing gas can be easily performed by operators and users, and the identification process of the analysis results can be verified in real time. Disease diagnosis by analyzing breath gas is considered to be an innovative non-invasive diagnostic method. Breath gas diagnosis has potential advantages over other diagnostic methods such as blood sampling, urine sampling, biopsy, endoscopy, and imaging. First, it is a completely non-invasive approach that allows the development of a user-friendly, simple, and intuitive diagnostic platform [19]. Second, the sample collection in this method is advantageous because it has superior processing when compared to serum or urine sampling [20]. Finally, it is the most convenient method as it does not pose the problem of bio-hazardous specimens within the current regulations.

Breath gas diagnosis is based on the physiological phenomenon of gas exchange occurring in the alveoli. Human blood contains chemicals that reflect physiological phenomena and metabolic conditions in the human body. Low-molecular organic compounds are contained in the human exhaled breath through the lungs during the respiratory process and are released out of the body [21]. The health conditions of the human body are reflected by various VOCs contained in breathing gases that undergo various biological reactions [22]. Blake et al. [23] reported that various metabolic diseases and pathological conditions could be observed in relation to the concentration increase of the aforementioned VOCs in the exhaled breath gas. These are associated with the respiratory diseases, such as cardiovascular disease (CVD), diabetes, asthma, bacterial infertility and inflammatory disease, cancer, COPD, and Alzheimer's disease, as well as with other diseases, according to the results of a recent breath analysis. In addition, a number of studies have been conducted on the applicability of diagnostic technologies using these VOC gases.

Table 1 summarizes the biomarker VOCs for each disease that can be used for its diagnosis and the currently used corresponding diagnostic methods. Although it cannot be used as a replacement for the conventional diagnostic method, it has the potential to serve as a portable diagnostic device capable of periodic self-diagnosis, which solves the problems of location and equipment. Through an accurate and efficient analysis of specific biomarker VOCs present in the breath, we demonstrate the possibility of developing a new selective self-diagnosis platform.

Table 1. Conventional measurements with the candidate group for biomarker volatile organic compounds (VOCs) in exhaled breath gas by a diagnosable disease. A simple and efficient e-nose platform that can distinguish VOCs in exhaled gas can be used as a selective self-diagnostic technology.

Disease	Conventional Measurement	Biomarker VOCs	Ref.
Diabetes	Glucose level Clinical biomarkers	Acetone	[24]
Bacterial infection	Computed tomography (CT) Gram stain Microorganism culture Morphological analysis High isoprene	Ammonia Hydrogen cyanide Nitric oxide Ethane Pentane	[18,25–28]
Asthma	Spirometry Peak expiratory flow Lung function testing Bronchoprovocation test	Acetone Nitric oxide Isoprene Ammonia	[29,30]
COPD	Spirometry X-Ray, CT Peak expiratory flow Lung function testing	Acetone Ethane	[31,32]
Cardiovascular disease (CVD)	HDL & LDL cholesterol High blood pressure Clinical biomarkers Obesity	Acetone Pentane Isoprene	[33,34]
Cancer	Clinical biomarkers Biopsy CT, X-ray, MRI	Acetone Formaldehyde Ethane Pentane Isoprene Ethanol	[35–38]

2.1. Diabetes

Diabetes is an abnormal carbohydrate metabolic disease that is characterized by hyperglycemic symptoms [39]. It is classified as either type 2 or type 1 depending on the problems with insulin secretion or various degrees of peripheral nerve resistance to insulin action [40,41]. Currently, the most widely applied diabetes diagnostic methods include the fasting plasma glucose, oral glucose tolerance, and glycated hemoglobin [42]. Diabetes is a condition of abnormal blood sugar in patients, and the complications that follow place a considerable burden on clinical and public health. Accordingly, an effective intervention that detects the glucose abnormalities early and prevents progression from prediabetes to diabetes is of utmost importance. Diabetes should be thoroughly managed by the patient through a self-diagnosis. The primary self-diagnosis method is based on glucose level measurement using existing blood collection methods [43]. Although the low-cost diagnostic technology is currently in use, users' reluctance to collect blood remains to be addressed. If low-cost and efficient non-invasive diagnostic methods are developed in the near future, it is expected to be an innovation in the health care market related to diabetes.

Galassetti et al. [23] presented a correlation between ethanol and acetone in exhaled breath, which is related to serum glucose levels. In the case of patients suffering from diabetic ketoacidosis, a study showed that the acetone concentration in the exhaled gas increases to hundreds of ppmv [44]. When the blood sugar levels remain high over long periods of time, the fatty and amino acids are burned to produce energy. The ketone body produced in the body is a 3-carbon ketone body derived from the oxidation of non-esterified fatty acids, and is found in the state of hydroxyacetone (1-hydroxyacetone) and 1,2-propanediol (PPD) through acetoacetate decarboxylase [45]. The ketone bodies are stored in the blood and thus, lower the pH level. Therefore, glucose cannot be used as an energy source for untreated diabetic patients. As a result, the ketone bodies are produced as by-products and energy sources when fat is broken down instead of glucose. High levels of ketones are produced as a result of low insulin levels in diabetic ketoacidosis [46]. Therefore, the exhaled acetone can be detected in patients with diabetic ketoacidosis and a high-fat diet [25]. These research results can be used for developing a diabetes diagnosis technology for acetone level monitoring in exhaled gas. The introduction of a simple and low-cost measurement technology could enable its use in a wide range of selective testing methods and reduce patient discomfort or pain during blood collection.

2.2. Various VOCs Derived from Inflammatory Diseases

In addition to diabetes, a few other diseases are expected to be diagnosed through the analysis of VOCs in exhaled breath gas. In vivo immune responses that are caused by bacterial infections or inflammatory reactions produce various types of VOC in the body. The pathological problems can interfere with normal metabolism, and abnormal chemical reactions can cause the detection of some VOCs in the exhaled gas [44]. By detecting volatile chemicals such as ammonia, nitrogen oxides, and hydrogen sulfide, the cause of pathological reactions can be analyzed and implemented in the diagnostic technology [47]. Mathew et al. [46] reviewed various types of VOC caused by metabolic processes in the body and metabolic disorders caused by various pathological reactions, and suggested the possibility of the implementation of a breath diagnosis technology. The level of ammonia (NH3) in breath gas can be used as an important biomarker. In the human body, ammonia is produced during protein metabolism and is regulated through the urea circuit owing to its toxicity [48]. In the case of liver and kidney problems, ammonia levels increase [49]. In addition, bacteria such as H. pylori, which cause peptic ulcers from the oral cavity to the duodenum, excrete ammonia gas and hydrogen sulfide through metabolic processes [50]. Nitrogen monoxide and nitrogen oxide are important signaling substances produced in the human body. Within the respiratory system, NO regulates the tension of blood vessels and bronchi (promotes the expansion of blood vessels and airways), promotes the coordinated beating of ciliated epithelial cells, and acts as an important neurotransmitter of non-adrenergic, non-cholinergic neurons running in the bronchi. Diseases related to inflammation can be mainly analyzed through the concentration of NO in the breath [51–58]. Currently, a technology that can measure NO in the respiratory tract using laser analysis sensors is being developed and used as a reference for diagnosing inflammatory diseases [59].

Isoprene is a unit molecule that forms cholesterol and is involved in cholesterol metabolism. Its concentration can be used as a sensitive and non-invasive metric for analyzing various metabolic effects in the human body [60]. Among the cholesterol metabolic diseases, CVDs and hypertension are categorized as representative diseases with a very high risk, and the patient needs constant management through periodic and voluntary diagnosis [61,62]. The currently used self-diagnosis method is the patient family history analysis and periodic monitoring of the blood pressure level [63,64]. Research on the classification of exhaled breath components of patients based on various levels of VOC other than hydrocarbons in the breath is being actively conducted [35]. The following information is from a review by Mathew et. al. [46] for journal diagnostics in 2015. In the case of hydrocarbons such as ethane and pentane, it is caused by oxidation of lipid components in cells [65]. This component is found when a problem occurs in the metabolism of lipid components, and is advantageously released through breath gas owing to its low solubility. It is mainly associated with respiratory obstructive diseases caused by inflammation, such as asthma, COPD, obstructive sleep apnea, and ARDS [66–70]. The hydrocarbon molecules are biomarkers of oxidative stress, and pentane and ethane concentrations increase owing to physical and mental stress [71–73]. It shows a significant difference in concentration in sepsis or SIRS patients [70] and can be used in the diagnosis of inflammatory bowel disease, sleep apnea, cancer, and ischemic heart disease [74–77].

Asthma [78] and COPD [79], which are obstructive respiratory diseases, should be detected early and prevented from worsening through rapid response. According to the currently widely known medical manual, patients with active asthma need periodic checkups every one to six months, depending on the severity of asthma. Asthma symptoms are diagnosed through equipment that detects lung function, such as spirometry, in hospitals [80]. In addition, lung capacity indicators are periodically managed at home using a personal peak flow meter (which approximately costs $20) [81]. On the basis of pathological research results on the levels of various VOC biomarker substances, such as NO, in the patient's exhaled strain and hydrocarbons, research on inflammatory asthma patients breathing diagnosis using breath gas analysis is actively underway [82–84].

2.3. Cancer

Conventionally, cancer diagnostics include genetic, epigenetic, proteomic, and glycomic biomarker screening, as well as some non-invasively collected biofluids [85–89]. In the case of cancer, early detection may increase the chance of complete recovery [90]. If a self-diagnosis technology enabling easy breathing gas analysis is developed, health and medical expenses for cancer treatment can be reduced, and average life expectancy can be increased. A malignant tumor, commonly known as cancer, is a disease wherein cell mutations are caused by several risk factors to identify the cause, which leads to abnormal cell growth and metastasis. It is reported that approximately 100 types of cancer affect the human body.

Because the reactions of cancer cells in the body are complex, various types of VOCs can be used as exhaled breath biomarkers. The first study on cancer diagnosis was conducted for lung cancer, which is a malignant tumor of the lung tissue through direct gas exchange. O'Neil et al. [91] conducted a study to select candidate groups of biomarker VOCs through GC/MS by collecting breath samples from eight lung cancer patients. It was found that among the 386 component gases that were detected with this technology, 45 components were at >75% occurrence level and 28 components were at >90% occurrence level. The research results were significant because they could distinguish between normal samples and patient respiration gas using a classification process via a computer program. Since then, many research results related to the analysis of biomarker VOCs that are common in lung cancer patients have been published [92,93].

Phillips et al. published a study on biomarker VOCs produced by the intracellular oxidative stress caused by breast cancer [94]. Oxidative stress is the process of oxidizing biologically important molecular substances, including DNA and proteins, when an increased amount of reactive oxygen species (ROS) is leaked into the cytoplasm in mitochondria [95]. This causes the decomposition of fatty acids and the peroxidation of lipids by abundant oxygen radicals [96,97]. Breast cancer patients and control patients were classified using SPSS treatment, and higher negative value (NPV) and lower positive value (PPP) were derived, respectively, as compared with the results of the screening mammograms. Kumar et al. published a selected ion flow tube mass spectrometry (SIFT-MS) analysis of exhaled breath for VOC profiling of esophagogastric cancer [98]. In approximately 17 VOC species, the concentrations of hexanoic acid, phenol, methyl phenol, and ethyl phenol differed statistically from those in the positive control group.

Until recently, studies on exhaled breath gas analysis for various cancer disease models and classified patient and control groups have been published steadily. If the integrated analysis sensor array technology that can classify breathing gas samples according to the composition of VOCs is used, a new concept diagnosis technology can be developed for various disease models described above. Furthermore, if breath diagnosis technology is deployed as a low-cost artificial-nose platform consisting of simple chemical sensor units, it will be possible to design self-diagnostic devices that can be used by individuals in real time at home. This technology could be an excellent innovation in health and medical industry.

3. Nanosensor Array E-Nose for Exhaled Breath Diagnosis

The sensor array technology has recently been widely applied in disease diagnosis for exhaled breath gas analysis because it is efficient for analyzing multiple VOCs, including human breath gases [99]. Because exhaled breath gas samples have different compositions of the compounds, their individual analysis using conventional devices, such as existing analytical instrument GC-MS, is limited. The sample component profile can be patterned and recognized using the nanosensor array technology. The artificial nose, which can analyze the components of real-time breathing based on a simple system, is a novel technology that can be used for disease self-diagnosis in healthcare. In terms of the practical application and commercialization of e-nose sensors, low cost, ease of use, and miniaturization are key factors. To meet these requirements, sensor technologies are being developed based on the mechanisms derived from the unique characteristics of various materials [100]. In this section, we discuss electrochemical sensors (metal oxide (MO) nanomaterial-based e-nose sensors) and colorimetric sensors (metal-containing dye sensor arrays and functional phage sensor arrays).

Table 2 summarizes representative nanosensor technologies that can be used for diagnosis based on recent gas detection methods. We aim to introduce an electrochemical sensor method and a technology to detect a small amount of gas mixture using a color sensor. To detect various VOCs present in exhaled breath, artificial olfactory models have been developed using a variety of nanosensor technologies that distinguish specific gas molecules. These technologies are expected to be applicable to disease diagnosis in the future.

Table 2. Gas sensor technology applicable to future disease diagnosis.

Measurement Target	Sensor Type	Sensing Materials	Ref.
Lung cancer	Electrochemical sensor	Undoped SnO_2, $Co\text{-}SnO_2$, and $Ni\text{-}SnO_2$ nanoparticles with cyclic voltammetry and electrochemical impedance spectroscopy/screen-printed electrode	[101]
	Colorimetric sensor	Colorimetric sensor array containing Lewis acid/base dyes (metal–organic complex dye)	[102,103]
Diabetes	Electrochemical sensor	Co_3O_4 thin film with a cubic spinel phase with AC impedance analyses/gold interdigitated electrode pattern	[104]
		Pristine SnO_2 nanofiber (undoped) and Eu-doped SnO_2 nanofibers (1, 2, and 3 mol% of Eu^{3+}) with gold electrodes and Pt wires	[105]
Ethanol in a VOC mixture	Electrochemical sensor	CeO_2–TiO_2 core shell nanorods with Pt electrodes	[106]
		Pristine SnO_2 and Yb-doped SnO_2 hollow nanofiber (0.5, 1.0, and 1.5 wt% Yb) with an Au electrode and a Pt wire	[107]
Cancer cell culture	Colorimetric sensor	Functional M13 bacteriophage-based colorimetric sensor array	[108,109]

3.1. Metal Oxide-Based Electrochemical Sensor Array for Disease Diagnosis

Metal oxide (MO) nanoparticles are promising candidates for sensor element design owing to their remarkable physicochemical properties, adjustable surface properties, and good stability [110]. These nanoparticles have a high density of trapped charged oxygen species (O^{2-}, O^-, and O^{2-}), creating a surface charged layer in the sensor element. When the reacting gaseous molecules adsorb oxygen ions on an MO surface, they alter the surface-trapped charge density [111,112]. The number of oxidation states used for gas sensing at the parts-per-billion level can be controlled by the nanoparticle size, shape, and composition. Many transition metal elements such as Fe, Co, Ni, Mn, Al, and Cu have been used as dopants for improving the electrical and optical characteristics of MOs and for enhancing their sensitivity to gases [113].

Khatoon et al. [101] doped Co and Ni with tin oxide (SnO_2) using a sol–gel method and investigated it as a sensor material for e-nose development. (Figure 2a) They applied an MO-based screen-printed electrode as the working electrode to determine the levels of 1-propanol and isopropyl alcohol in cyclic voltammetry. Furthermore, Ni-SnO_2 and Co-SnO_2 were found selective to 1-propanol and isopropyl alcohol, respectively, among other investigated VOCs (acetone, toluene, formaldehyde, 2-butanol, and ethyl acetate). Liu et al. [114,115] developed various CeO_2-based gas sensors attached with different $MMnO_3$ (M: Sr, Ca, La, and Sm) sensing electrodes and conducted a comparative study for detecting acetone gas. CeO_2-$MMnO_3$ compounds were prepared using a simple sol–gel method.

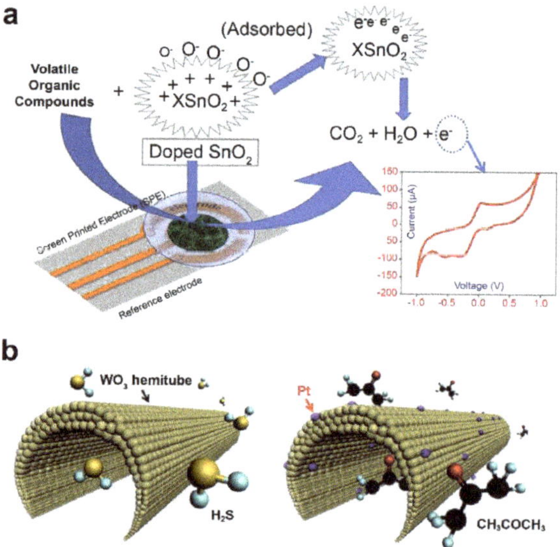

Figure 2. Metal–oxide nanomaterial-based electrochemical sensors. (**a**) Doped SnO_2 nanomaterial sensor for lung cancer diagnosis. Data reproduced from Ref. [101]. Copyrights ACS 2020. (**b**) Diabetes diagnosis model using polycrystalline WO_3 hemi-tube for H_2S, selective acetone detecting sensors. Data reproduced from Ref. [116]. Copyrights ACS 2013.

Nanostructured materials with various morphologies, including nanorods, nanowires, nanosheets, and nanofibers, have been developed for e-nose applications because their large surface area-to-volume greatly facilitates the conversion of gas response into electric signals. Srinivasan et al. [104] developed an acetone sensor using nanostructured Co_3O_4 thin films for the detection of diabetic ketoacidosis (DKA). The presence of acetone at a trace level (1.8 ppm) in human exhaled breath signifies the presence of DKA from a diagnostic perspective.

The acetone level of the exhaled breath air of type-2 diabetes mellitus patients exceeded 1.71 ppm, whereas that of type-1 diabetes patients was 2.19 ppm. It is known that the concentration of acetone is directly proportional to metabolic disorders in humans. The Occupational Safety and Health Administration states that the allowable human exposure level of acetone is 1000 ppm in industries. Different nanostructured cobalt oxide sensing elements were synthesized using a spray pyrolysis method at different deposition temperatures (473 to 773 K) [117]. It was found that the nanostructured material was more sensitive to acetone when analyzing different solvents such as acetone, EtOH, NH_3, xylene, toluene, and acetaldehyde. The sensor fabricated at 773 K exhibited a response of 235 toward acetone (50 ppm) at room temperature.

Furthermore, the sensor showed a limit of detection of 1 ppm, which is lower than the minimum threshold level of DKA. Ren et al. synthesized four different Fe-doped

TiO$_2$ thin films on Ti plates using the microarc oxidation technique to measure the level of ethanol gas [118]. The Fe-doped TiO$_2$ thin films fabricated by introducing 0.5 mM K$_4$(FeCN)$_6\cdot$3H$_2$O into 0.5 M Na$_3$PO$_4$ showed a better sensitivity to ethanol gas with a response of 7.9. This value was significantly larger than the responses of other samples (less than 5.2, at 275 °C) prepared with different formulations of K$_4$(FeCN)$_6\cdot$3H$_2$O and Na$_3$PO$_4$. Li et al. synthesized pristine SnO$_2$ and Er-SnO$_2$ nanobelts using the thermal evaporation method. When analyzing 100 ppm of various gases including formaldehyde, ethanediol, ethanol, and acetone at temperature ranging from 150 °C to 260 °C, it was found that the Er-SnO$_2$ nanobelt was more sensitive to formaldehyde gas than other gases. The experimental results revealed that the gas response of a single Er-SnO$_2$ nanobelt device was 9, with response and recovery times of 17 s and 25 s, respectively [119].

Wang et al. [120] also prepared SnO$_2$ and SnO$_2$/NiO electrospun nanofibers, which were subsequently subjected to thermocompression and calcination processes. A fabricated SnO$_2$/NiO sensor was more sensitive to ethanol vapor than to other gases such as H$_2$S, CO, NH$_3$, and acetone. A SnO$_2$/NiO nanofiber exhibited a higher Ra/Rg value (27.5) than a pristine SnO$_2$ nanofiber (2.4) on sensing 100 ppm of ethanol and showed average response and recovery times of 2.9 s and 4.7 s, respectively. Li et al. [121] synthesized porous Nb$_2$O$_5$-TiO$_2$ n–n junction nanofibers with different Nb molar ratios by electrospinning.

Choi et al. used detection sensors to demonstrate promising clinical applications for diagnostic purposes through correlation analysis between exhalation components and specific diseases [116]. They utilized 1D fibers with uniformly applied platinum nanoparticle catalysts on a porous tin oxide (SnO$_2$) sensor material surrounded by layers of thin shells (Figure 2b). When acetone gas was adsorbed on the surface of the material under study, it was applied to a sensor for detecting acetone concentration at approximately 120 ppb, which resulted in a change in the electrical resistance value. The developed nanofiber sensor increased the resistance of the material by up to six times at a concentration of acetone of 1000 ppb, allowing the diagnosis of diabetes. According to another study, wherein trained dogs diagnosed lung cancer, on average, toluene was detected approximately at an 80% accuracy [122]. This sensor was found to detect toluene with an accuracy of approximately 70%. Active research is being conducted on using multisensor arrays for various-disease diagnosis, such as lung cancer and diabetes. An increased number of research studies related to the detection of VOCs based on electrochemical sensors via the change in the resistance of MOs are being published.

3.2. Colorimetric Sensor Array for an Artificial Nose System

The pattern information for specific reactions can be stored in the fingerprint form using metal-containing dye arrays that react at gas concentrations of hundreds of parts-per-billion. This technique is called smell seeding, and a number of studies focusing on the development of a personal chemical dosimeter for the detection, identification, and quantification of environmental and workplace VOCs have recently been published [123–128]. Furthermore, studies have been published on the development of medical diagnosis tools based on breath analysis. Rakow et al. fabricated a sensor array using the color transfer phenomenon of metal-containing dyes such as metalloporphyrin, which is sensitive to gas, and presented an artificial olfactory sensor model [129]. When there is a specific odor, the structure of metalloporphyrins and the color change (Figure 3a) [130]. Using 2D-displayed array metalloporphyrins, a pattern-recognition e-nose that detects a wide range of olfactants (including alcohols, amines, ethers, phosphines, phosphites, thioethers, and thiols) and weakly ligating solvent vapors (arenes, halocarbons, and ketones) was developed (Figure 3b) [130].

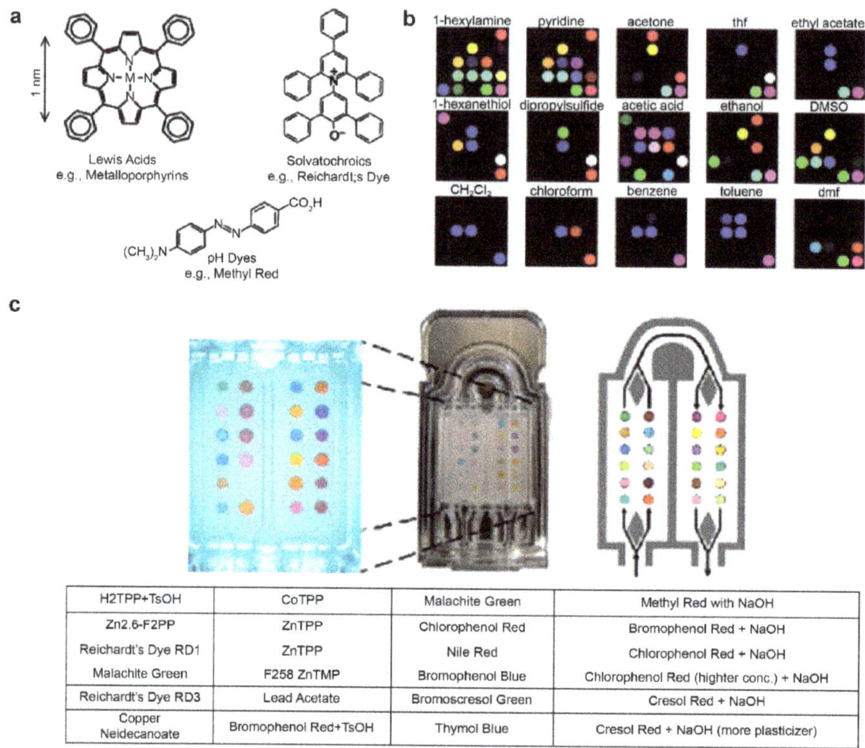

H2TPP+TsOH	CoTPP	Malachite Green	Methyl Red with NaOH
Zn2.6-F2PP	ZnTPP	Chlorophenol Red	Bromophenol Red + NaOH
Reichardt's Dye RD1	ZnTPP	Nile Red	Chlorophenol Red + NaOH
Malachite Green	F258 ZnTMP	Bromophenol Blue	Chlorophenol Red (highter conc.) + NaOH
Reichardt's Dye RD3	Lead Acetate	Bromoscresol Green	Cresol Red + NaOH
Copper Neidecanoate	Bromophenol Red+TsOH	Thymol Blue	Cresol Red + NaOH (more plasticizer)

Figure 3. Metal-containing dye sensor array-type e-nose. (**a**) Suslick used strong chemical reactions such as "Lewis acid/base dyes (i.e., metal ion-containing dyes)," "Brønsted acidic or basic dyes (i.e., pH indicators)," and "dyes with large permanent dipoles (i.e., zwitterionic solvatochromic dyes)." Various types of sensor arrays were fabricated using the base dye (**b**) Electronic nose model that classifies various types of VOCs using the manufactured sensor array. (**c**) Exhalation analysis of lung cancer patients using the e-nose technology. The images (**a,b**) were adapted with permission from [130]. The image (**c**) was adapted with permission from [103].

Mazzone et al. [103] analyzed exhaled breath samples from lung cancer patients using a colorimetric sensor array. The exhaled breath of 92 lung cancer patients and 229 control patients was obtained via the chromaticity sensor array. (Figure 3c) The technique is further expected to evaluate specific histologies of patients and to optimize them by incorporating clinical risk factors [103]. Other technologies for pathogenic fungal identification and rapid detection of bacteria have also been reported [131,132].

Kim et al. developed a superior chemical gas detection layer by simultaneously controlling nanostructures and catalytic functionalization [133]. The nanostructures derived from electrospinning act as highly dispersed ultrasmall catalysts. Electrospinning-derived 1D-dimensional MOs are nanomaterials with excellent advantages such as large surface-to-volume ratios, high porosity, and high gas permeability, which are beneficial for building chemical gas detection platforms. These materials are expanded to a variety of nanoarchitectures derived from metal organic frameworks, graphene oxide, and polymer templates, and they are used as a sensing sensor material [134–136].

Filamentous bacteriophage material is a functional bioreceptor with directed evolution (DE) properties (the Nobel Prize in Chemistry 2018) [137]. Kim et al. developed a multi-array sensor using a functional bacteriophage colorimetric sensor [138]. Phage-based colorimetric sensors classify various VOCs that can be used as an e-nose platform.

Phage display technology, which is based on the principle of natural selection and proliferation by mutation, can discover functional bio-reporter peptide sequences with selectivity for specific molecules [139]. Through a simple genetic manipulation, it is possible to produce a functional phage material for a specific bio-reporter peptide discovered through phage display screening. The phage has ssDNA, which contains genetic information. It is synthesized through biological reactions and can be mass-produced as a bioreporter material with high-purity specific reactions. Oh et al. analyzed the bioreceptor function of the functional bacteriophage and presented the results of optical colorimetric sensor devices using the phage as a building unit (Figure 4) [110]. This single phage unit has a uniform fiber shape of less than 1 µm and has liquid crystal characteristics. Thus, a self-assembled nanostructure can be fabricated. This self-assembled phage structure forms a microstructure with quasi-ordered pitches and scattered reflected light [140]. The color of the phage structure changes according to the size and arrangement of the phage bundles. The external stimuli caused by chemicals change the arrangement of the bundles and change the color of the phage structure. These are used as a chemical gas sensor for VOC classification [109], food origin analysis [141], environment monitoring [142], and breath diagnosis [143]. Extensive research on the pattern-recognition integrated analysis technology utilizing various color sensor arrays is currently underway. Various studies have been published for the following five cell types: human hepatocellular adenocarcinoma (SK-Hep-1), cervical cancer (HeLa), human colon cancer (HCT116), human non-small lung cancer (NCI-H1299), and normal human embryonic kidney (HEK293); cells of these types were incubated in minimum essential media (MEM) containing 10% fetal serum culture [108]. Because each cell type produces a unique composition of VOCs, the three-band optoelectronic sensors produce a unique color. The results of the unique color change were obtained with 99.8% reliable data via 2D-linear discriminant analysis. The results of this study suggest the possibility of developing an effective diagnosis technology through further research.

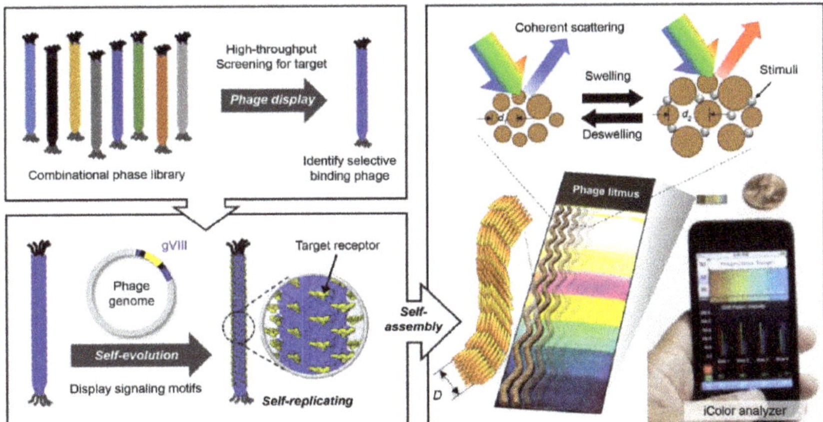

Figure 4. Colorimetric sensor system using an M13 bacteriophage as a functional biomaterial [109]. Peptide-based bioreceptor materials can secure various peptide sequences with desired functional groups using the phage display technology. Filamentous bacteriophage material with a scale less than 1 µm has approximately 2700 pairs of functional proteins on its surface. Hence, it can be used as a highly sensitive bioreceptor material. The synthesis process is based on internal DNA genetic information, and high-purity mass production is possible with simple genetic modifications. Phage units produced by bacteria have the same structure of a certain size and, thus have liquid crystal properties. A self-assembled structure can be produced using a bacteriophage as a unit, and it has liquid crystal properties; therefore, a color matrix can be produced by creating light scattering at regular distances formed by the structure. Based on the principle that phage self-assembled structures change the surface structure in response to external stimuli, a highly sensitive colorimetric sensor can be manufactured. This technique is referred to as the phage litmus.

4. Signal Processing Technology Based on Olfactory Cognitive Mechanisms

An artificial olfactory sensor is composed of a gas sensor array, which consists of multiple independent channels and an AI, which is an e-nose system that can detect odors and quantitatively measure their types and concentrations. Fundamentally, signal detection and data processing are required in the olfactory recognition mechanism. By leveraging various sensor arrays to construct multichannel reporters, signal data processing is required. An increasing number of signal receptors results in varying patterns of sensor responses, and a systematic data analysis algorithm using a multisensor array allows for trending classification. By utilizing the pattern analysis of the measurement data using the sensor array, various diseases can be detected, and the influence of external environmental factors (smoking, gender, age, etc.) can be minimized using the accumulated database. The improved analysis algorithm for the accumulated data results in a high commercialization potential of the artificial nose system. Table 3 summarizes various disease diagnosis research cases through data processing using existing sensor arrays.

Table 3. Classification of disease groups through a data analysis algorithm using a sensor array.

Disease	Sensor	Data Process	Ref.
Lung cancer	Gold nanoparticle-based electrochemical sensor	PCA	[144]
Lung cancer cell culture	Cyranose® 320	LDA, PNN, KNN	[145]
Lung cancer	Cyranose® 320	SVM	[93,146]
Lung cancer and COPD	QCM sensor array	PLS-DA	[147]
Lung, breast, colorectal, and prostate cancers	Electrochemical sensor single array	PCA	[148]
Pulmonary disease	GC-MS/Chemo-nanoarray	DFA	[149]
Tuberculosis	BH114-Bloodhound	ANN	[150]
Urinary tract infections	BH114-Bloodhound	ANN, PCA	[151]
Brain cancer organoids	Polymer-carbon black based electro-chemical sensor array	Normalized pattern	[152]
Lung and gastric cancer, asthma and COPD	FET sensor	ANN, DFA	[153]
Renal dysfunction	Electrochemical sensor array	PCA	[154]
Gastric cancer	Aeonose	ANN	[155]
Gastric cancer	Metal–organic ligand-based nanosensor array	DFA	[156]
Pneumonia	Cyranose® 320	PLS-DA	[157]
Ear, nose, and throat infection	Cyranose® 320	PCA	[158]
Parkinson's disease	Nanosensor array	KNN	[159]
Head and neck cancer	GC-MS	PCA, SVM	[160]
Human armpit body odor classification	Tagushi gas sensors	PCA	[161]
Colorectal cancer	GC-MS	DFA	[162]
Ovarian cancer	GC-MS	DFA	[163]
Seventeen types of diseases	Gold nanoparticle-based nanosensor array	ANN, hierarchal clustering analysis	[164]

PCA; Principal component analysis; LDA; Linear discriminant analysis; PNN; Probabilistic neural network; ANN; Artificial neural network; KNN; k-Neural network; SVM; Support vector machine; DFA; Deterministic finite automaton; QCM; Quartz microbalance; PLS-DA; Partial least squares discriminant analysis; Cyranose® 320; a commercialized e-nose device, Cyrano Science, Pasadena, CA, USA; BH114-Bloodhound; a commercialized gas sensor array, Leeds, UK; Aenose; a commercialized e-nose device, The eNose Company, Zutphen, The Netherlands; FET; Field effect transistor.

4.1. Artificial Intelligence Data Processing-Based Multisensor Pattern Recognition

Multivariate analysis of gas mixture data is required for the analysis of sensor array data. Principal component analysis (PCA) is a method that is often used for visually distinguishing between the same sample groups in a plot. A PCA plot is a two-dimensional picture of data from which the data maximum variance can be obtained. The intelligent olfactory sensor can be implemented through advanced data processing techniques using AI and machine learning (ML) with pre-processed data. The supervised learning methods are mainly used to establish a functional relationship between the measurement space and classification elements [165]. In the past decades, many learning methods have been developed, for example, least squares regression (PLS), support vector machine (SVM), artificial neural network (ANN), decision tree (DT), and K-nearest neighbor (KNN). Among these, neural networks such as multilayer perceptrons (MLPs) have been widely used. Recently, the deep neural network (DNN) has been developed as part of a broader family of ML methods based on artificial neural networks. The ML is currently the most common application of AI and its principle is based on the automatic detection of patterns in data and these patterns can be used for future pattern recognition. It has become possible to derive new information by predicting or classifying collected data or by extracting information from appropriate data [166,167].

Deep learning (DL) is also a branch of ML, and it is achieved through the ML based on neural networks. A neural network is inspired by biology, as shown in Figure 5, and it operates in the same way a biological brain solves problems with large units of axons connected to neurons. With the introduction of the concept of dropout, the over-fitting problem of neural networks was resolved, the accuracy was increased significantly, and the DL technology was newly highlighted [168]. This significantly reduced the computational time owing to the advancement of GPU hardware. In addition, the accuracy was considerably enhanced because accurate conclusions could be drawn through very fast iterative learning with big data. Recently, DL algorithms based on convolutional neural networks (CNNs) have shown excellent performance in various fields such as computer vision [169]. In particular, the performance of DL was further improved by the newly introduced ReLU activation function. The signal processing functions include sigmoid, tanh, and ReLU. The ReLU is a function that returns 0 when the value less than 0 is returned, and returns the same value when a value greater than 0 is found. The desired result can be outputted by applying ReLU in the internal hidden layer and sigmoid returning 1 if the value is greater than 0 in the output layer. This factor could be essential for the successful implementation of an intelligent olfactory sensor intended for the application in medical, environmental, and safety fields that require high performance based on big data [170–172]. Recently, several studies on the application of DL to intelligent olfactory sensors have been reported [173].

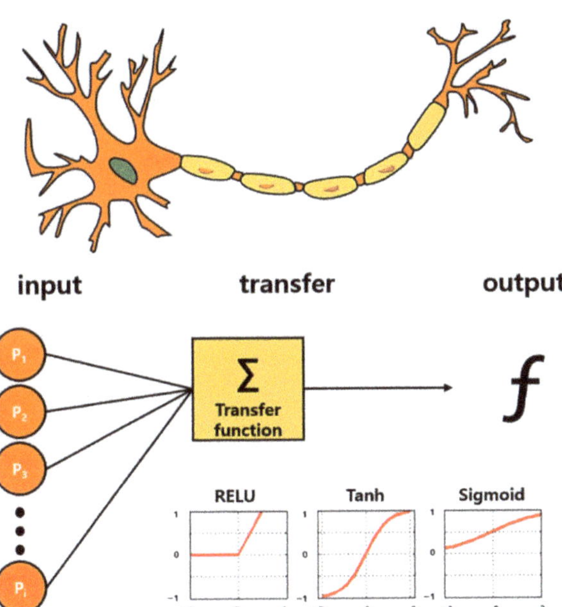

Figure 5. Bioinspired from neuronal pathway: Artificial intelligence algorithms. Deep learning technology based on the neuronal signal process. A structure that draws correct conclusions via iterative learning through activation functions called Sigmoid, Tanh, and ReLU. By applying ReLU to the inner hidden layer and by applying the sigmoid function in the last output layer, the accuracy is significantly increased.

4.2. Multimodal (MM) Analysis

To analyze the concentration of a specific gas accurately, the influence of other gas components should be minimized. Several technologies are being developed for each sensor to minimize the effects of gases other than the gas component under analysis. However, few gases possess the same physical (mass, absorbance, etc.) and chemical (adsorption degree, reaction, etc.) properties. The gas components with the same chemical or physical properties cannot be separated using only a chemical or physical sensor. When a gas is analyzed based on several chemical or physical sensors, some measurement errors occur, and measurement sensitivity and reliability are not guaranteed.

By simultaneously analyzing measurement information of heterogeneous multimodal (MM) sensors, such as sensors that analyze chemical or physical properties, gas components that cannot be separated by the same type of the sensor method can now be separated. As a result, the gas analysis specificity can be improved. The ultralow-concentration (ppt or less) gases can be distinguished using MM information analysis based on measurement information of various kinds provided by pattern recognition-based sensors and AI technologies. The sensor-array that we introduced is also the multimodality concept. By arranging various sensors with different characteristics, each single sensor unit can complement each other against an external variable that cannot be distinguished. By extending the concept of multimodality, multi-array sensors of physical, chemical sensors, biosensors, etc. enables the identification of advanced information based on advanced discrimination.

The olfactory sensor technology is expected to develop into olfactory intelligence using pattern recognition, AI, and MM analysis. Consequently, the measurement sensitivity, accuracy, and reliability can be improved. Moreover, it is expected to develop into a future technology that provides new functions, such as early diagnosis/warning.

5. Conclusions

Clinical diagnosis and post-treatment monitoring technologies based on respiratory gas analysis have several advantages such as non-invasiveness, patient convenience, low cost, and real-time analysis. For successful implementation and use as a self-diagnostic device, this technology should be developed in the direction of miniaturization and simple sampling.

Organic compounds derived from the human body are indicators that assist with disease diagnosis. Body odor, sputum, urine, sweat, breath, etc. are the sources of odor required to analyze these organic compounds. Conductive polymer compounds with different physicochemical properties can be used to analyze the channel-specific pattern of electrical conductivity that changes depending on the type and concentration of gas molecules. Alternatively, olfactory receptors can be immobilized directly on each channel to provide specificity to gas molecules. In order for respiratory gas analysis to be established as a clinical examination tool, an in-depth and extensive clinical study of respiratory gas components and diseases should be conducted.

Olfactory sensors can be miniaturized and integrated using nanosensor technology, and they possess the advantage of low production costs. Currently, research is being carried out on the disease diagnosis analysis using various nanosensors. If a low-cost, easy-to-use, and portable artificial olfactory sensor is implemented, it can be extended and applied to all fields that require continuous monitoring. For the future development of diagnostic sensor technology in the form of an artificial olfactory perception model, multichannelling, miniaturization, weight reduction, low manufacturing cost, and sensor network construction are essential. Additionally, for the development of clinically reliable sensor array technology, a sensor element with high sensitivity as well as high selectivity performance should be developed, and a consistent breathing gas sampling method should be secured.

Advanced AI analysis methods such as DL are attracting significant attention because of their widespread use to improve the accuracy of signal analysis and result prediction of multichannel sensors in the future. Although the disease diagnosis using human breathing gas is still under development, a simple and convenient disease screening method using breathing gas has been designed through continuous gas analyzer development, sensor array technology development, and research on the relationship between human breathing gas and diseases. It is expected that this technology will be applied to disease screening in various forms, starting with the diseases.

Intelligent olfactory sensor technology holds potential for applications and services in various industries including medical, environmental, and safety fields that were previously impossible. In particular, the clinical equivalence and efficacy evaluation with existing medical devices for early disease monitoring using intelligent olfactory sensors in the medical field have been reported. Therefore, for the successful implementation of this technology, it is essential to develop a new technology that combines MM sensors that can provide high sensitivity, selectivity, and reliability with high-performance ML using big data.

Author Contributions: C.K., J.-W.O. and D.-W.H. proposed the research objective and the structure of the review article. C.K. and I.S.R. wrote the paper using materials supplied by J.-M.L., J.H.L., M.S.K. and S.H.L. J.-W.O. and D.-W.H. revised and improved the manuscript. J.-W.O. and D.-W.H. supervised the study. All authors have read and agreed to the published version of the manuscript.

Funding: This research was supported by the Basic Science Research Program through the National Research Foundation of Korea (NRF) funded by the Ministry of Education (No. 2020R1I1A1A01072566 and 2020R1I1A1A01069244) and by the Korea Medical Device Development Fund grant funded by the Korea government (the Ministry of Science and ICT, the Minis-try of Trade, Industry and Energy, the Ministry of Health and Welfare, the Ministry of Food and Drug Safety) (NTIS Number: 9991006781).

Institutional Review Board Statement: Not applicable.

Informed Consent Statement: Not applicable.

Data Availability Statement: Not applicable. No new data were created or analyzed in this study.

Conflicts of Interest: The authors declare no conflict of interest.

References

1. Sarafoleanu, C.; Mella, C.; Georgescu, M.; Perederco, C. The importance of the olfactory sense in the human behavior and evolution. *J. Med. Life* **2009**, *2*, 196.
2. Gardner, J.W. *Electronic Noses and Olfaction 2000*; IOP Publishing: Bristol, UK, 2001.
3. Young, J.M.; Shykind, B.M.; Lane, R.P.; Tonnes-Priddy, L.; Ross, J.A.; Walker, M.; Williams, E.M.; Trask, B.J. Odorant receptor expressed sequence tags demonstrate olfactory expression of over 400 genes, extensive alternate splicing and unequal expression levels. *Genome Biol.* **2003**, *4*, 1–15. [CrossRef]
4. Persaud, K.; Dodd, G. Analysis of discrimination mechanisms in the mammalian olfactory system using a model nose. *Nature* **1982**, *299*, 352–355. [CrossRef] [PubMed]
5. Gardner, J.W.; Bartlett, P.N. A brief history of electronic noses. *Sens. Actuators B Chem.* **1994**, *18*, 210–211. [CrossRef]
6. Aleixandre, M.; Lozano, J.; Gutiérrez, J.; Sayago, I.; Fernández, M.; Horrillo, M. Portable e-nose to classify different kinds of wine. *Sens. Actuators B Chem.* **2008**, *131*, 71–76. [CrossRef]
7. Sysoev, V.V.; Goschnick, J.; Schneider, T.; Strelcov, E.; Kolmakov, A. A gradient microarray electronic nose based on percolating SnO_2 nanowire sensing elements. *Nano Lett.* **2007**, *7*, 3182–3188. [CrossRef]
8. Wilson, A.D. Review of electronic-nose technologies and algorithms to detect hazardous chemicals in the environment. *Procedia Technol.* **2012**, *1*, 453–463. [CrossRef]
9. Kou, L.; Zhang, D.; Liu, D. A novel medical e-nose signal analysis system. *Sensors* **2017**, *17*, 402. [CrossRef]
10. Baldwin, E.A.; Bai, J.; Plotto, A.; Dea, S. Electronic noses and tongues: Applications for the food and pharmaceutical industries. *Sensors* **2011**, *11*, 4744–4766. [CrossRef] [PubMed]
11. Raj, V.B.; Singh, H.; Nimal, A.; Sharma, M.; Gupta, V. Oxide thin films (ZnO, TeO_2, SnO_2 and TiO_2) based surface acoustic wave (SAW) E-nose for the detection of chemical warfare agents. *Sens. Actuators B Chem.* **2013**, *178*, 636–647. [CrossRef]
12. Incalza, M.A.; D'Oria, R.; Natalicchio, A.; Perrini, S.; Laviola, L.; Giorgino, F. Oxidative stress and reactive oxygen species in endothelial dysfunction associated with cardiovascular and metabolic diseases. *Vascul. Pharmacol.* **2018**, *100*, 1–19. [CrossRef]
13. Rani, V.; Deep, G.; Singh, R.K.; Palle, K.; Yadav, U.C. Oxidative stress and metabolic disorders: Pathogenesis and therapeutic strategies. *Life Sci.* **2016**, *148*, 183–193. [CrossRef]
14. Hoamisligil, G.S. Inflammation and metabolic disorders. *Nature* **2006**, *444*, 860–867. [CrossRef]

15. Boots, A.W.; Bos, L.D.; van der Schee, M.P.; van Schooten, F.-J.; Sterk, P.J. Exhaled molecular fingerprinting in diagnosis and monitoring: Validating volatile promises. *Trends Mol. Med.* **2015**, *21*, 633–644. [CrossRef]
16. Phillips, M.; Cataneo, R.N.; Cummin, A.R.; Gagliardi, A.J.; Gleeson, K.; Greenberg, J.; Maxfield, R.A.; Rom, W.N. Detection of lung cancer with volatile markers in the breath. *Chest* **2003**, *123*, 2115–2123. [CrossRef] [PubMed]
17. Dweik, R.; Boggs, P.; Erzurum, S.; Irvin, C.; Leigh, M.; Lundberg, J.; Olin, A.; Plummer, A.; Taylor, D. An official ATS clinical practice guideline: Interpretation of exhaled nitric oxide levels (FENO) for clinical applications. *Am. J. Respir. Crit. Care Med.* **2011**, *184*, 602–615. [CrossRef] [PubMed]
18. Cao, W.; Duan, Y. Breath analysis: Potential for clinical diagnosis and exposure assessment. *Clin. Chem.* **2006**, *52*, 800–811. [CrossRef] [PubMed]
19. Cheng, W.-H.; Lee, W.-J. Technology development in breath microanalysis for clinical diagnosis. *J. Lab. Clin. Med.* **1999**, *133*, 218–228. [CrossRef]
20. Braun, P.X.; Gmachl, C.F.; Dweik, R.A. Bridging the collaborative gap: Realizing the clinical potential of breath analysis for disease diagnosis and monitoring–tutorial. *IEEE Sens. J.* **2012**, *12*, 3258–3270. [CrossRef]
21. Das, S.; Pal, M. Non-invasive monitoring of human health by exhaled breath analysis: A comprehensive review. *J. Electrochem. Soc.* **2020**, *167*, 037562. [CrossRef]
22. Buszewski, B.; Kęsy, M.; Ligor, T.; Amann, A. Human exhaled air analytics: Biomarkers of diseases. *Biomed. Chromatogr.* **2007**, *21*, 553–566. [CrossRef]
23. Galassetti, P.R.; Novak, B.; Nemet, D.; Rose-Gottron, C.; Cooper, D.M.; Meinardi, S.; Newcomb, R.; Zaldivar, F.; Blake, D.R. Breath ethanol and acetone as indicators of serum glucose levels: An initial report. *Diabetes Technol. Ther.* **2005**, *7*, 115–123. [CrossRef]
24. Wang, Z.; Wang, C. Is breath acetone a biomarker of diabetes? A historical review on breath acetone measurements. *J. Breath Res.* **2013**, *7*, 037109.
25. Hibbard, T.; Killard, A.J. Breath ammonia analysis: Clinical application and measurement. *Crit. Rev. Anal. Chem.* **2011**, *41*, 21–35. [CrossRef]
26. Enderby, B.; Smith, D.; Carroll, W.; Lenney, W. Hydrogen cyanide as a biomarker for Pseudomonas aeruginosa in the breath of children with cystic fibrosis. *Pediatr. Pulmonol.* **2009**, *44*, 142–147. [CrossRef] [PubMed]
27. Marteus, H.; Törnberg, D.; Weitzberg, E.; Schedin, U.; Alving, K. Origin of nitrite and nitrate in nasal and exhaled breath condensate and relation to nitric oxide formation. *Thorax* **2005**, *60*, 219–225. [CrossRef] [PubMed]
28. McCurdy, M.R.; Bakhirkin, Y.; Wysocki, G.; Lewicki, R.; Tittel, F.K. Recent advances of laser-spectroscopy-based techniques for applications in breath analysis. *J. Breath Res.* **2007**, *1*, 014001. [CrossRef] [PubMed]
29. Van der Schee, M.P.; Fens, N.; Brinkman, P.; Bos, L.D.J.; Angelo, M.D.; Nijsen, T.M.E.; Raabe, R.; Knobel, H.H.; Vink, T.J.; Sterk, P.J. Effect of transportation and storage using sorbent tubes of exhaled breath samples on diagnostic accuracy of electronic nose analysis. *J. Breath Res.* **2012**, *7*, 016002. [CrossRef]
30. Dweik, R.A.; Amann, A. Exhaled breath analysis: The new frontier in medical testing. *J. Breath Res.* **2008**, *2*, 030301. [CrossRef]
31. Cazzola, M.; Segreti, A.; Capuano, R.; Bergamini, A.; Martinelli, E.; Calzetta, L.; Rogliani, P.; Ciaprini, C.; Ora, J.; Paolesse, R. Analysis of exhaled breath fingerprints and volatile organic compounds in COPD. *COPD Res. Pract.* **2015**, *1*, 1–8. [CrossRef]
32. Maniscalco, M.; Paris, D.; Melck, D.J.; Molino, A.; Carone, M.; Ruggeri, P.; Caramori, G.; Motta, A. Differential diagnosis between newly diagnosed asthma and COPD using exhaled breath condensate metabolomics: A pilot study. *Eur. Respir. J.* **2018**, *51*, 1701825. [CrossRef] [PubMed]
33. Pabst, F.; Miekisch, W.; Fuchs, P.; Kischkel, S.; Schubert, J.K. Monitoring of oxidative and metabolic stress during cardiac surgery by means of breath biomarkers: An observational study. *J. Cardiothorac. Surg.* **2007**, *2*, 1–8. [CrossRef] [PubMed]
34. Cikach Jr, F.S.; Dweik, R.A. Cardiovascular biomarkers in exhaled breath. *Prog. Cardiovasc. Dis.* **2012**, *55*, 34–43. [CrossRef]
35. Bajtarevic, A.; Ager, C.; Pienz, M.; Klieber, M.; Schwarz, K.; Ligor, M.; Ligor, T.; Filipiak, W.; Denz, H.; Fiegl, M. Noninvasive detection of lung cancer by analysis of exhaled breath. *BMC Cancer* **2009**, *9*, 348. [CrossRef]
36. Guntner, A.T.; Koren, V.; Chikkadi, K.; Righettoni, M.; Pratsinis, S.E. E-nose sensing of low-ppb formaldehyde in gas mixtures at high relative humidity for breath screening of lung cancer? *ACS Sens.* **2016**, *1*, 528–535. [CrossRef]
37. Fuchs, P.; Loeseken, C.; Schubert, J.K.; Miekisch, W. Breath gas aldehydes as biomarkers of lung cancer. *Int. J. Cancer* **2010**, *126*, 2663–2670. [CrossRef]
38. Filipiak, W.; Filipiak, A.; Sponring, A.; Schmid, T.; Zelger, B.; Ager, C.; Klodzinska, E.; Denz, H.; Pizzini, A.; Lucciarini, P. Comparative analyses of volatile organic compounds (VOCs) from patients, tumors and transformed cell lines for the validation of lung cancer-derived breath markers. *J. Breath Res.* **2014**, *8*, 027111. [CrossRef] [PubMed]
39. McCulloch, D.K. *Classification of Diabetes Mellitus and Genetic Diabetic Syndromes*; UpToDate: Waltham, MA, USA, 2007.
40. Harris, M.I.; Robbins, D.C. Prevalence of adult-onset IDDM in the US population. *Diabetes Care* **1994**, *17*, 1337–1340. [CrossRef] [PubMed]
41. Landin-Olsson, M.; Nilsson, K.; Lernmark, Å.; Sundkvist, G. Islet cell antibodies and fasting C-peptide predict insulin requirement at diagnosis of diabetes mellitus. *Diabetologia* **1990**, *33*, 561–568. [CrossRef]
42. Niskanen, L.K.; Tuomi, T.; Karjalainen, J.; Groop, L.C.; Uusitupa, M.I. GAD antibodies in NIDDM: Ten-year follow-up from the diagnosis. *Diabetes Care* **1995**, *18*, 1557–1565. [CrossRef] [PubMed]
43. Pasquale, L.R.; Kang, J.H.; Manson, J.E.; Willett, W.C.; Rosner, B.A.; Hankinson, S.E. Prospective study of type 2 diabetes mellitus and risk of primary open-angle glaucoma in women. *Ophthalmology* **2006**, *113*, 1081–1086. [CrossRef] [PubMed]

44. Massick, S. Portable breath acetone measurements combine chemistry and spectroscopy. In Proceedings of the SPIE, San Jose, CA, USA, 28 January–1 February 2007.

45. Reichard, G.; Skutches, C.; Hoeldtke, R.; Owen, O. Acetone metabolism in humans during diabetic ketoacidosis. *Diabetes* **1986**, *35*, 668–674. [CrossRef] [PubMed]

46. Mathew, T.L.; Pownraj, P.; Abdulla, S.; Pullithadathil, B. Technologies for clinical diagnosis using expired human breath analysis. *Diagnostics* **2015**, *5*, 27–60. [CrossRef]

47. Perez-Guaita, D.; Kokoric, V.; Wilk, A.; Garrigues, S.; Mizaikoff, B. Towards the determination of isoprene in human breath using substrate-integrated hollow waveguide mid-infrared sensors. *J. Breath Res.* **2014**, *8*, 026003. [CrossRef]

48. Cox, M.M.; Nelson, D.L. *Lehninger Principles of Biochemistry*; WH Freeman: New York, NY, USA, 2013.

49. Iorio, R.; Sepe, A.; Giannattasio, A.; Cirillo, F.; Vegnente, A. Hypertransaminasemia in childhood as a marker of genetic liver disorders. *J. Gastroenterol.* **2005**, *40*, 820–826. [CrossRef]

50. Linden, D.R. Hydrogen sulfide signaling in the gastrointestinal tract. *Antioxid. Redox Signal.* **2014**, *20*, 818–830. [CrossRef]

51. Belvisi, M.; Ward, J.; Mitchell, J.; Barnes, P. Nitric oxide as a neurotransmitter in human airways. *Arch. Int. Pharmacodyn. Ther.* **1995**, *329*, 97–110. [PubMed]

52. Jain, B.; Rubinstein, I.; Robbins, R.A.; Leise, K.L.; Sisson, J.H. Modulation of airway epithelial cell ciliary beat frequency by nitric oxide. *Biochem. Biophys. Res. Commun.* **1993**, *191*, 83–88. [CrossRef] [PubMed]

53. Belvisi, M.G.; Stretton, C.D.; Yacoub, M.; Barnes, P.J. Nitric oxide is the endogenous neurotransmitter of bronchodilator nerves in humans. *Eur. J. Pharmacol.* **1992**, *210*, 221–222. [CrossRef]

54. Blitzer, M.L.; Loh, E.; Roddy, M.-A.; Stamler, J.S.; Creager, M.A. Endothelium-derived nitric oxide regulates systemic and pulmonary vascular resistance during acute hypoxia in humans. *Am. J. Cardiol.* **1996**, *28*, 591–596. [CrossRef]

55. Combes, X.; Mazmanian, M.; Gourlain, H.; Herve, P. Effect of 48 hours of nitric oxide inhalation on pulmonary vasoreactivity in rats. *Am. J. Respir. Crit. Care Med.* **1997**, *156*, 473–477. [CrossRef] [PubMed]

56. Melot, C.; Vermeulen, F.; Maggiorini, M.; Gilbert, E.; Naeije, R. Site of pulmonary vasodilation by inhaled nitric oxide in microembolic lung injury. *Am. J. Respir. Crit. Care Med.* **1997**, *156*, 75–85. [CrossRef]

57. Buga, G.M.; Ignarro, L.J. Electrical field stimulation causes endothelium-dependent and nitric oxide-mediated relaxation of pulmonary artery. *Am. J. Physiol. Heart Circ. Physiol.* **1992**, *262*, H973–H979. [CrossRef]

58. Que, L.G.; Yang, Z.; Stamler, J.S.; Lugogo, N.L.; Kraft, M. S-nitrosoglutathione reductase: An important regulator in human asthma. *Am. J. Respir. Crit. Care Med.* **2009**, *180*, 226–231. [CrossRef] [PubMed]

59. McManus, J.B.; Shorter, J.H.; Nelson, D.D.; Zahniser, M.S.; Glenn, D.E.; McGovern, R.M. Pulsed quantum cascade laser instrument with compact design for rapid, high sensitivity measurements of trace gases in air. *Appl. Phys. B* **2008**, *92*, 387–392. [CrossRef]

60. Stone, B.G.; Besse, T.J.; Duane, W.C.; Dean Evans, C.; DeMaster, E.G. Effect of regulating cholesterol biosynthesis on breath isoprene excretion in men. *Lipids* **1993**, *28*, 705–708. [CrossRef]

61. Wilson, P.W.F. Overview of Established Risk Factors for Cardiovascular Disease. Available online: https://www.uptodate.com/contents/overview-of-established-risk-factors-for-cardiovascular-disease (accessed on 5 August 2021).

62. Wilson, P.W.F.; Givens, J. *Cardiovascular Disease Risk Assessment for Primary Prevention in Adults: Our Approach*; UpToDate: Wellesley, MA, USA, 2020.

63. Canto, J.G.; Kiefe, C.I.; Rogers, W.J.; Peterson, E.D.; Frederick, P.D.; French, W.J.; Gibson, C.M.; Pollack, C.V.; Ornato, J.P.; Zalenski, R.J. Number of coronary heart disease risk factors and mortality in patients with first myocardial infarction. *JAMA* **2011**, *306*, 2120–2127. [CrossRef]

64. Rapsomaniki, E.; Timmis, A.; George, J.; Pujades-Rodriguez, M.; Shah, A.D.; Denaxas, S.; White, I.R.; Caulfield, M.J.; Deanfield, J.E.; Smeeth, L. Blood pressure and incidence of twelve cardiovascular diseases: Lifetime risks, healthy life-years lost, and age-specific associations in 1·25 million people. *Lancet* **2014**, *383*, 1899–1911. [CrossRef]

65. Timmis, K.N.; McGenity, T.; Van Der Meer, J.R.; de Lorenzo, V. *Handbook of Hydrocarbon and Lipid Microbiology*; Springer: Berlin/Heidelberg, Germany, 2010.

66. Efron, B. Censored data and the bootstrap. *J. Am. Stat. Assoc.* **1981**, *76*, 312–319. [CrossRef]

67. Mazzone, P.J.; Hammel, J.; Dweik, R.; Na, J.; Czich, C.; Laskowski, D.; Mekhail, T. Diagnosis of lung cancer by the analysis of exhaled breath with a colorimetric sensor array. *Thorax* **2007**, *62*, 565–568. [CrossRef]

68. Shawe-Taylor, J.; Cristianini, N. *Kernel Methods for Pattern Analysis*; Cambridge University Press: Cambridge, UK, 2004.

69. Guyon, I.; Weston, J.; Barnhill, S.; Vapnik, V. Gene selection for cancer classification using support vector machines. *Mach. Learn.* **2002**, *46*, 389–422. [CrossRef]

70. Paredi, P.; Kharitonov, S.A.; Leak, D.; Ward, S.; Cramer, D.; Barnes, P.J. Exhaled ethane, a marker of lipid peroxidation, is elevated in chronic obstructive pulmonary disease. *Am. J. Respir. Crit. Care Med.* **2000**, *162*, 369–373. [CrossRef]

71. Scotter, J.; Allardyce, R.; Langford, V.; Hill, A.; Murdoch, D. The rapid evaluation of bacterial growth in blood cultures by selected ion flow tube–mass spectrometry (SIFT-MS) and comparison with the BacT/ALERT automated blood culture system. *J. Microbiol. Methods* **2006**, *65*, 628–631. [CrossRef]

72. Thorn, R.M.S.; Reynolds, D.M.; Greenman, J. Multivariate analysis of bacterial volatile compound profiles for discrimination between selected species and strains in vitro. *J. Microbiol. Methods* **2011**, *84*, 258–264. [CrossRef] [PubMed]

73. Allardyce, R.A.; Hill, A.L.; Murdoch, D.R. The rapid evaluation of bacterial growth and antibiotic susceptibility in blood cultures by selected ion flow tube mass spectrometry. *Diagn. Microbiol. Infect. Dis.* **2006**, *55*, 255–261. [CrossRef] [PubMed]

74. Wendland, B.E.; Aghdassi, E.; Tam, C.; Carrrier, J.; Steinhart, A.H.; Wolman, S.L.; Baron, D.; Allard, J.P. Lipid peroxidation and plasma antioxidant micronutrients in Crohn disease. *Am. J. Clin. Nutr.* **2001**, *74*, 259–264. [CrossRef] [PubMed]

75. Olopade, C.O.; Christon, J.A.; Zakkar, M.; Swedler, W.I.; Rubinstein, I.; Hua, C.-w.; Scheff, P.A. Exhaled pentane and nitric oxide levels in patients with obstructive sleep apnea. *Chest* **1997**, *111*, 1500–1504. [CrossRef] [PubMed]

76. Hietanen, E.; Bartsch, H.; Bereziat, J.; Camus, A.; McClinton, S.; Eremin, O.; Davidson, L.; Boyle, P. Diet and oxidative stress in breast, colon and prostate cancer patients: A case-control study. *Eur. J. Clin. Nutr.* **1994**, *48*, 575–586.

77. Mendis, S.; Sobotka, P.A.; Leia, F.L.; Euler, D.E. Breath pentane and plasma lipid peroxides in ischemic heart disease. *Free Radic. Biol. Med.* **1995**, *19*, 679–684. [CrossRef]

78. Fanta, C.; Fletcher, S.; Wood, R.; Bochner, B.; Hollingsworth, H. *An Overview of Asthma Management*; UpToDate: Wellesley, MA, USA, 2009.

79. Rennard, S.I.; Stolel, J.; Wilson, K. *Chronic Obstructive Pulmonary Disease: Definition, Clinical Manifestations, Diagnosis, and Staging*; UpToDate: Waltham, MA, USA, 2009.

80. Wang, P.; Zhang, G.; Wondimu, T.; Ross, B.; Wang, R. Hydrogen Sulfide and Asthma. *Exp. Physiol.* **2011**, *96*, 847–852. [CrossRef]

81. Gerald, L.B.; Carr, T.F. *Patient Education: How to Use a Peak Flow Meter (beyond the Basics)*; UpToDate: Wellesley, MA, USA, 2020.

82. Wenzel, S. *Treatment of Severe Asthma in Adolescents and Adults*; Bochner, B.S.T.W., Hollingsworth, H., Eds.; (online giriş Aralık 2014.); UpToDate: Wellesley, MA, USA, 2009.

83. Shelhamer, J.H.; Levine, S.J.; Wu, T.; Jacoby, D.B.; Kaliner, M.A.; Rennard, S.I. Airway inflammation. *Ann. Intern. Med.* **1995**, *123*, 288–304. [CrossRef]

84. Dweik, R.A. *Exhaled Nitric Oxide Analysis and Applications*; UpToDate: Wellesley, MA, USA, 2019.

85. Calzone, K.A. Genetic biomarkers of cancer risk. In *Seminars in Oncology Nursing*; Elsevier: Amsterdam, The Netherlands, 2012; p. 28.

86. Herceg, Z.; Hainaut, P. Genetic and epigenetic alterations as biomarkers for cancer detection, diagnosis and prognosis. *Mol. Oncol.* **2007**, *1*, 26–41. [CrossRef] [PubMed]

87. Li, D.; Chan, D.W. Proteomic cancer biomarkers from discovery to approval: It's worth the effort. *Expert Rev. Proteom.* **2014**, *11*, 135–136. [CrossRef] [PubMed]

88. Aizpurua-Olaizola, O.; Toraño, J.S.; Falcon-Perez, J.M.; Williams, C.; Reichardt, N.; Boons, G.-J. Mass spectrometry for glycan biomarker discovery. *Trends Anal. Chem.* **2018**, *100*, 7–14. [CrossRef]

89. Mishra, A.; Verma, M. Cancer biomarkers: Are we ready for the prime time? *Cancers* **2010**, *2*, 190–208. [CrossRef]

90. Hamblin, M.R. Shining light on the head: Photobiomodulation for brain disorders. *BBA Clin.* **2016**, *6*, 113–124. [CrossRef]

91. O'neill, H.; Gordon, S.; O'neill, M.; Gibbons, R.; Szidon, J. A computerized classification technique for screening for the presence of breath biomarkers in lung cancer. *Clin. Chem.* **1988**, *34*, 1613–1618. [CrossRef]

92. Phillips, M.; Gleeson, K.; Hughes, J.M.B.; Greenberg, J.; Cataneo, R.N.; Baker, L.; McVay, W.P. Volatile organic compounds in breath as markers of lung cancer: A cross-sectional study. *Lancet* **1999**, *353*, 1930–1933. [CrossRef]

93. Machado, R.F.; Laskowski, D.; Deffenderfer, O.; Burch, T.; Zheng, S.; Mazzone, P.J.; Mekhail, T.; Jennings, C.; Stoller, J.K.; Pyle, J. Detection of lung cancer by sensor array analyses of exhaled breath. *Am. J. Respir. Crit. Care Med.* **2005**, *171*, 1286–1291. [CrossRef]

94. Phillips, M.; Cataneo, R.N.; Ditkoff, B.A.; Fisher, P.; Greenberg, J.; Gunawardena, R.; Kwon, C.S.; Rahbari-Oskoui, F.; Wong, C. Volatile markers of breast cancer in the breath. *Breast J.* **2003**, *9*, 184–191. [CrossRef]

95. Knight, J.A. Free radicals: Their history and current status in aging and disease. *Ann. Clin. Lab. Sci.* **1998**, *28*, 331–346. [PubMed]

96. Kneepkens, C.F.; Lepage, G.; Roy, C.C. The potential of the hydrocarbon breath test as a measure of lipid peroxidation. *Free Radic. Biol. Med.* **1994**, *17*, 127–160. [CrossRef]

97. Phillips, M.; Cataneo, R.N.; Greenberg, J.; Gunawardena, R.; Naidu, A.; Rahbari-Oskoui, F. Effect of age on the breath methylated alkane contour, a display of apparent new markers of oxidative stress. *J. Lab. Clin. Med.* **2000**, *136*, 243–249. [CrossRef] [PubMed]

98. Kumar, S.; Huang, J.; Abbassi-Ghadi, N.; Španěl, P.; Smith, D.; Hanna, G.B. Selected ion flow tube mass spectrometry analysis of exhaled breath for volatile organic compound profiling of esophago-gastric cancer. *Anal. Chem.* **2013**, *85*, 6121–6128. [CrossRef] [PubMed]

99. Zhou, X.; Xue, Z.; Chen, X.; Huang, C.; Bai, W.; Lu, Z.; Wang, T. Nanomaterial-based gas sensors used for breath diagnosis. *J. Mater. Chem. B* **2020**, *8*, 3231–3248. [CrossRef] [PubMed]

100. Bogue, R. Nanosensors: A review of recent progress. *Sens. Rev.* **2008**, *28*, 12–17. [CrossRef]

101. Khatoon, Z.; Fouad, H.; Alothman, O.Y.; Hashem, M.; Ansari, Z.A.; Ansari, S.A. Doped SnO_2 nanomaterials for e-nose based electrochemical sensing of biomarkers of lung cancer. *ACS Omega* **2020**, *5*, 27645–27654. [CrossRef]

102. Janzen, M.C.; Ponder, J.B.; Bailey, D.P.; Ingison, C.K.; Suslick, K.S. Colorimetric sensor arrays for volatile organic compounds. *Anal. Chem.* **2006**, *78*, 3591–3600. [CrossRef]

103. Mazzone, P.J.; Wang, X.-F.; Xu, Y.; Mekhail, T.; Beukemann, M.C.; Na, J.; Kemling, J.W.; Suslick, K.S.; Sasidhar, M. Exhaled breath analysis with a colorimetric sensor array for the identification and characterization of lung cancer. *J. Thorac. Oncol.* **2012**, *7*, 137–142. [CrossRef] [PubMed]

104. Srinivasan, P.; Prakalya, D.; Jeyaprakash, B. UV-activated ZnO/CdO nn isotype heterostructure as breath sensor. *J. Alloys Compd.* **2020**, *819*, 152985. [CrossRef]

105. Jiang, Z.; Zhao, R.; Sun, B.; Nie, G.; Ji, H.; Lei, J.; Wang, C. Highly sensitive acetone sensor based on Eu-doped SnO_2 electrospun nanofibers. *Ceram. Int.* **2016**, *42*, 15881–15888. [CrossRef]

106. Chen, Y.-J.; Xiao, G.; Wang, T.-S.; Zhang, F.; Ma, Y.; Gao, P.; Zhu, C.-L.; Zhang, E.; Xu, Z.; Li, Q.-h. Synthesis and enhanced gas sensing properties of crystalline CeO_2/TiO_2 core/shell nanorods. *Sens. Actuators B Chem.* **2011**, *156*, 867–874. [CrossRef]
107. Wang, T.; Ma, S.; Cheng, L.; Luo, J.; Jiang, X.; Jin, W. Preparation of Yb-doped SnO_2 hollow nanofibers with an enhanced ethanol–gas sensing performance by electrospinning. *Sens. Actuators B Chem.* **2015**, *216*, 212–220. [CrossRef]
108. Moon, J.-S.; Kim, W.-G.; Shin, D.-M.; Lee, S.-Y.; Kim, C.; Lee, Y.; Han, J.; Kim, K.; Yoo, S.Y.; Oh, J.-W. Bioinspired M-13 bacteriophage-based photonic nose for differential cell recognition. *Chem. Sci.* **2017**, *8*, 921–927. [CrossRef]
109. Oh, J.-W.; Chung, W.-J.; Heo, K.; Jin, H.-E.; Lee, B.Y.; Wang, E.; Zueger, C.; Wong, W.; Meyer, J.; Kim, C. Biomimetic virus-based colourimetric sensors. *Nat. Commun.* **2014**, *5*, 1–8. [CrossRef]
110. Barsan, N.; Koziej, D.; Weimar, U. Metal oxide-based gas sensor research: How to? *Sens. Actuators B Chem.* **2007**, *121*, 18–35. [CrossRef]
111. Wang, C.; Yin, L.; Zhang, L.; Xiang, D.; Gao, R. Metal oxide gas sensors: Sensitivity and influencing factors. *Sensors* **2010**, *10*, 2088–2106. [CrossRef]
112. Rothschild, A.; Komem, Y. The effect of grain size on the sensitivity of nanocrystalline metal-oxide gas sensors. *J. Appl. Phys.* **2004**, *95*, 6374–6380. [CrossRef]
113. Thirupathi, B.; Smirniotis, P.G. Co-doping a metal (Cr, Fe, Co, Ni, Cu, Zn, Ce, and Zr) on Mn/TiO_2 catalyst and its effect on the selective reduction of NO with NH_3 at low-temperatures. *Appl. Catal. B* **2011**, *110*, 195–206. [CrossRef]
114. Liu, T.; Zhang, Y.; Yang, X.; Hao, X.; Liang, X.; Liu, F.; Liu, F.; Yan, X.; Ouyang, J.; Lu, G. CeO_2-based mixed potential type acetone sensor using $MFeO_3$ (M: Bi, La and Sm) sensing electrode. *Sens. Actuators B Chem.* **2018**, *276*, 489–498. [CrossRef]
115. Liu, T.; Yang, X.; Ma, C.; Hao, X.; Liang, X.; Liu, F.; Liu, F.; Yang, C.; Zhu, H.; Lu, G. CeO_2-based mixed potential type acetone sensor using $MMnO_3$ (M: Sr, Ca, La and Sm) sensing electrode. *Solid State Ion.* **2018**, *317*, 53–59. [CrossRef]
116. Choi, S.-J.; Lee, I.; Jang, B.-H.; Youn, D.-Y.; Ryu, W.-H.; Park, C.O.; Kim, I.-D. Selective diagnosis of diabetes using Pt-functionalized WO_3 hemitube networks as a sensing layer of acetone in exhaled breath. *Anal. Chem.* **2013**, *85*, 1792–1796. [CrossRef]
117. Srinivasan, P.; Kulandaisamy, A.J.; Mani, G.K.; Babu, K.J.; Tsuchiya, K.; Rayappan, J.B.B. Development of an acetone sensor using nanostructured Co_3O_4 thin films for exhaled breath analysis. *RSC Adv.* **2019**, *9*, 30226–30239. [CrossRef]
118. Ren, F.-J.; Yu, X.-b.; Ling, Y.-h.; Feng, J.-y. Micro-arc oxidization fabrication and ethanol sensing performance of Fe-doped TiO_2 thin films. *Int. J. Miner. Metall. Mater.* **2012**, *19*, 461–466. [CrossRef]
119. Li, S.; Liu, Y.; Wu, Y.; Chen, W.; Qin, Z.; Gong, N.; Yu, D. Highly sensitive formaldehyde resistive sensor based on a single Er-doped SnO_2 nanobelt. *Phys. B Condens. Matter* **2016**, *489*, 33–38. [CrossRef]
120. Wang, Y.; Zhang, H.; Sun, X. Electrospun nanowebs of NiO/SnO_2 pn heterojunctions for enhanced gas sensing. *Appl. Surf. Sci.* **2016**, *389*, 514–520. [CrossRef]
121. Li, G.; Zhang, X.; Lu, H.; Yan, C.; Chen, K.; Lu, H.; Gao, J.; Yang, Z.; Zhu, G.; Wang, C. Ethanol sensing properties and reduced sensor resistance using porous Nb_2O_5-TiO_2 nn junction nanofibers. *Sens. Actuators B Chem.* **2019**, *283*, 602–612. [CrossRef]
122. Rudnicka, J.; Walczak, M.; Kowalkowski, T.; Jezierski, T.; Buszewski, B. Determination of volatile organic compounds as potential markers of lung cancer by gas chromatography–mass spectrometry versus trained dogs. *Sens. Actuators B Chem.* **2014**, *202*, 615–621. [CrossRef]
123. Zhang, C.; Suslick, K.S. Colorimetric sensor array for soft drink analysis. *J. Agric. Food Chem.* **2007**, *55*, 237–242. [CrossRef] [PubMed]
124. Lim, S.H.; Feng, L.; Kemling, J.W.; Musto, C.J.; Suslick, K.S. An optoelectronic nose for the detection of toxic gases. *Nat. Chem.* **2009**, *1*, 562–567. [CrossRef] [PubMed]
125. Musto, C.J.; Lim, S.H.; Suslick, K.S. Colorimetric detection and identification of natural and artificial sweeteners. *Anal. Chem.* **2009**, *81*, 6526–6533. [CrossRef]
126. Suslick, B.A.; Feng, L.; Suslick, K.S. Discrimination of complex mixtures by a colorimetric sensor array: Coffee aromas. *Anal. Chem.* **2010**, *82*, 2067–2073. [CrossRef]
127. Feng, L.; Musto, C.J.; Kemling, J.W.; Lim, S.H.; Suslick, K.S. A colorimetric sensor array for identification of toxic gases below permissible exposure limits. *Chem. Commun.* **2010**, *46*, 2037–2039. [CrossRef]
128. Lin, H.; Jang, M.; Suslick, K.S. Preoxidation for colorimetric sensor array detection of VOCs. *J. Am. Chem. Soc.* **2011**, *133*, 16786–16789. [CrossRef]
129. Rakow, N.A.; Suslick, K.S. A colorimetric sensor array for odour visualization. *Nature* **2000**, *406*, 710–713. [CrossRef]
130. Suslick, K.S.; Rakow, N.A.; Sen, A. Colorimetric sensor arrays for molecular recognition. *Tetrahedron* **2004**, *60*, 11133–11138. [CrossRef]
131. Carey, J.R.; Suslick, K.S.; Hulkower, K.I.; Imlay, J.A.; Imlay, K.R.; Ingison, C.K.; Ponder, J.B.; Sen, A.; Wittrig, A.E. Rapid identification of bacteria with a disposable colorimetric sensing array. *J. Am. Chem. Soc.* **2011**, *133*, 7571–7576. [CrossRef]
132. Zhang, Y.; Askim, J.R.; Zhong, W.; Orlean, P.; Suslick, K.S. Identification of pathogenic fungi with an optoelectronic nose. *Analyst* **2014**, *139*, 1922–1928. [CrossRef]
133. Kim, S.-J.; Choi, S.-J.; Jang, J.-S.; Cho, H.-J.; Kim, I.-D. Innovative nanosensor for disease diagnosis. *Acc. Chem. Res.* **2017**, *50*, 1587–1596. [CrossRef]
134. Koo, W.-T.; Jang, J.-S.; Kim, I.-D. Metal-organic frameworks for chemiresistive sensors. *Chem* **2019**, *5*, 1938–1963. [CrossRef]

135. Choi, S.-J.; Jang, B.-H.; Lee, S.-J.; Min, B.K.; Rothschild, A.; Kim, I.-D. Selective detection of acetone and hydrogen sulfide for the diagnosis of diabetes and halitosis using SnO_2 nanofibers functionalized with reduced graphene oxide nanosheets. *ACS Appl. Mater. Interfaces* **2014**, *6*, 2588–2597. [CrossRef]

136. Choi, S.-H.; Ankonina, G.; Youn, D.-Y.; Oh, S.-G.; Hong, J.-M.; Rothschild, A.; Kim, I.-D. Hollow ZnO nanofibers fabricated using electrospun polymer templates and their electronic transport properties. *ACS Nano* **2009**, *3*, 2623–2631. [CrossRef]

137. Smith, G.P.; Petrenko, V.A. Phage display. *Chem. Rev.* **1997**, *97*, 391–410. [CrossRef]

138. Kim, C.; Lee, H.; Devaraj, V.; Kim, W.-G.; Lee, Y.; Kim, Y.; Jeong, N.-N.; Choi, E.J.; Baek, S.H.; Han, D.-W. Hierarchical cluster analysis of medical chemicals detected by a bacteriophage-based colorimetric sensor array. *Nanomaterials* **2020**, *10*, 121. [CrossRef] [PubMed]

139. Winter, G.; Griffiths, A.D.; Hawkins, R.E.; Hoogenboom, H.R. Making antibodies by phage display technology. *Annu. Rev. Immunol.* **1994**, *12*, 433–455. [CrossRef]

140. Chung, W.-J.; Oh, J.-W.; Kwak, K.; Lee, B.Y.; Meyer, J.; Wang, E.; Hexemer, A.; Lee, S.-W. Biomimetic self-templating supramolecular structures. *Nature* **2011**, *478*, 364–368. [CrossRef] [PubMed]

141. Seol, D.; Moon, J.-S.; Lee, Y.; Han, J.; Jang, D.; Kang, D.-J.; Moon, J.; Jang, E.; Oh, J.-W.; Chung, H. Feasibility of using a bacteriophage-based structural color sensor for screening the geographical origins of agricultural products. *Spectrochim. Acta A Mol. Biomol. Spectrosc.* **2018**, *197*, 159–165. [CrossRef] [PubMed]

142. Moon, J.S.; Lee, Y.; Shin, D.M.; Kim, C.; Kim, W.G.; Park, M.; Han, J.; Song, H.; Kim, K.; Oh, J.W. Identification of Endocrine Disrupting Chemicals using a Virus-Based Colorimetric Sensor. *Chem. Asian J.* **2016**, *11*, 3097–3101. [CrossRef] [PubMed]

143. Lee, J.-M.; Lee, Y.; Devaraj, V.; Nguyen, T.M.; Kim, Y.-J.; Kim, Y.H.; Kim, C.; Choi, E.J.; Han, D.-W.; Oh, J.-W. Investigation of bioelectronic nose based on programable surface chemistry of M13 bacteriophages for volatile organic compound detection: From basic properties of the biosensor to practical application. *Biosens. Bioelectron.* **2021**, *188*, 113339. [CrossRef] [PubMed]

144. Peng, G.; Tisch, U.; Adams, O.; Hakim, M.; Shehada, N.; Broza, Y.Y.; Billan, S.; Abdah-Bortnyak, R.; Kuten, A.; Haick, H. Diagnosing lung cancer in exhaled breath using gold nanoparticles. *Nat. Nanotechnol.* **2009**, *4*, 669–673. [CrossRef]

145. Thriumani, R.; Zakaria, A.; Hashim, Y.Z.H.-Y.; Jeffree, A.I.; Helmy, K.M.; Kamarudin, L.M.; Omar, M.I.; Shakaff, A.Y.M.; Adom, A.H.; Persaud, K.C. A study on volatile organic compounds emitted by in-vitro lung cancer cultured cells using gas sensor array and SPME-GCMS. *BMC Cancer* **2018**, *18*, 362. [CrossRef]

146. Tirzīte, M.; Bukovskis, M.; Strazda, G.; Jurka, N.; Taivans, I. Detection of lung cancer in exhaled breath with an electronic nose using support vector machine analysis. *J. Breath Res.* **2017**, *11*, 036009. [CrossRef] [PubMed]

147. Rocco, G. Every breath you take: The value of the electronic nose (e-nose) technology in the early detection of lung cancer. *J. Thorac. Cardiovasc. Surg.* **2018**, *155*, 2622–2625. [CrossRef] [PubMed]

148. Peng, G.; Hakim, M.; Broza, Y.; Billan, S.; Abdah-Bortnyak, R.; Kuten, A.; Tisch, U.; Haick, H. Detection of lung, breast, colorectal, and prostate cancers from exhaled breath using a single array of nanosensors. *Br. J. Cancer* **2010**, *103*, 542–551. [CrossRef] [PubMed]

149. Peled, N.; Hakim, M.; Bunn Jr, P.A.; Miller, Y.E.; Kennedy, T.C.; Mattei, J.; Mitchell, J.D.; Hirsch, F.R.; Haick, H. Non-invasive breath analysis of pulmonary nodules. *J. Thorac. Oncol.* **2012**, *7*, 1528–1533. [CrossRef]

150. Pavlou, A.K.; Magan, N.; Jones, J.M.; Brown, J.; Klatser, P.; Turner, A.P. Detection of Mycobacterium tuberculosis (TB) in vitro and in situ using an electronic nose in combination with a neural network system. *Biosens. Bioelectron.* **2004**, *20*, 538–544. [CrossRef]

151. Pavlou, A.K.; Magan, N.; McNulty, C.; Jones, J.M.; Sharp, D.; Brown, J.; Turner, A.P. Use of an electronic nose system for diagnoses of urinary tract infections. *Biosens. Bioelectron.* **2002**, *17*, 893–899. [CrossRef]

152. Kateb, B.; Ryan, M.; Homer, M.; Lara, L.; Yin, Y.; Higa, K.; Chen, M.Y. Sniffing out cancer using the JPL electronic nose: A pilot study of a novel approach to detection and differentiation of brain cancer. *NeuroImage* **2009**, *47*, T5–T9. [CrossRef] [PubMed]

153. Shehada, N.; Cancilla, J.C.; Torrecilla, J.S.; Pariente, E.S.; Brönstrup, G.; Christiansen, S.; Johnson, D.W.; Leja, M.; Davies, M.P.; Liran, O. Silicon nanowire sensors enable diagnosis of patients via exhaled breath. *ACS Nano* **2016**, *10*, 7047–7057. [CrossRef]

154. Voss, A.; Baier, V.; Reisch, R.; von Roda, K.; Elsner, P.; Ahlers, H.; Stein, G. Smelling renal dysfunction via electronic nose. *Ann. Biomed. Eng.* **2005**, *33*, 656–660. [CrossRef]

155. Schuermans, V.N.; Li, Z.; Jongen, A.C.; Wu, Z.; Shi, J.; Ji, J.; Bouvy, N.D. Pilot study: Detection of gastric cancer from exhaled air analyzed with an electronic nose in Chinese patients. *Surg. Innov.* **2018**, *25*, 429–434. [CrossRef]

156. Xu, Z.; Broza, Y.; Ionsecu, R.; Tisch, U.; Ding, L.; Liu, H.; Song, Q.; Pan, Y.; Xiong, F.; Gu, K. A nanomaterial-based breath test for distinguishing gastric cancer from benign gastric conditions. *Br. J. Cancer* **2013**, *108*, 941–950. [CrossRef] [PubMed]

157. Hanson, C.W.; Thaler, E.R. Electronic nose prediction of a clinical pneumonia score: Biosensors and microbes. *J. Am. Soc. Anesthesiol.* **2005**, *102*, 63–68. [CrossRef] [PubMed]

158. Shykhon, M.; Morgan, D.; Dutta, R.; Hines, E.; Gardner, J. Clinical evaluation of the electronic nose in the diagnosis of ear, nose and throat infection: A preliminary study. *J. Laryngol. Otol.* **2004**, *118*, 706–709. [CrossRef] [PubMed]

159. Nakhleh, M.; Badarny, S.; Winer, R.; Jeries, R.; Finberg, J.; Haick, H. Distinguishing idiopathic Parkinson's disease from other parkinsonian syndromes by breath test. *Parkinsonism Relat. Disord.* **2015**, *21*, 150–153. [CrossRef] [PubMed]

160. Hakim, M.; Billan, S.; Tisch, U.; Peng, G.; Dvrokind, I.; Marom, O.; Abdah-Bortnyak, R.; Kuten, A.; Haick, H. Diagnosis of head-and-neck cancer from exhaled breath. *Br. J. Cancer* **2011**, *104*, 1649–1655. [CrossRef]

161. Wongchoosuk, C.; Lutz, M.; Kerdcharoen, T. Detection and classification of human body odor using an electronic nose. *Sensors* **2009**, *9*, 7234–7249. [CrossRef] [PubMed]

162. Amal, H.; Leja, M.; Funka, K.; Lasina, I.; Skapars, R.; Sivins, A.; Ancans, G.; Kikuste, I.; Vanags, A.; Tolmanis, I. Breath testing as potential colorectal cancer screening tool. *Int. J. Cancer* **2016**, *138*, 229–236. [CrossRef] [PubMed]

163. Amal, H.; Shi, D.Y.; Ionescu, R.; Zhang, W.; Hua, Q.L.; Pan, Y.Y.; Tao, L.; Liu, H.; Haick, H. Assessment of ovarian cancer conditions from exhaled breath. *Int. J. Cancer* **2015**, *136*, E614–E622. [CrossRef]

164. Nakhleh, M.K.; Amal, H.; Jeries, R.; Broza, Y.Y.; Aboud, M.; Gharra, A.; Ivgi, H.; Khatib, S.; Badarneh, S.; Har-Shai, L. Diagnosis and classification of 17 diseases from 1404 subjects via pattern analysis of exhaled molecules. *ACS Nano* **2017**, *11*, 112–125. [CrossRef]

165. Jombart, T.; Devillard, S.; Balloux, F. Discriminant analysis of principal components: A new method for the analysis of genetically structured populations. *BMC Genet.* **2010**, *11*, 94. [CrossRef]

166. Lee, D.; Ahn, C.; Kim, B.; Pyo, H.; Kim, J.; Huh, C.; Kim, S. Intelligent Olfactory Sensor. *Electron. Telecommun. Trends* **2019**, *34*, 76–88.

167. Turner, A.P.; Magan, N. Electronic noses and disease diagnostics. *Nat. Rev. Microbiol.* **2004**, *2*, 161–166. [CrossRef]

168. Srivastava, N.; Hinton, G.; Krizhevsky, A.; Sutskever, I.; Salakhutdinov, R. Dropout: A simple way to prevent neural networks from overfitting. *J. Mach. Learn. Res.* **2014**, *15*, 1929–1958.

169. Chung, H.; Lee, S.J.; Park, J.G. Deep neural network using trainable activation functions. In Proceedings of the 2016 International Joint Conference on Neural Networks (IJCNN), Vancouver, BC, Canada, 24–29 July 2016.

170. Gutiérrez, J.; Horrillo, M.C. Advances in artificial olfaction: Sensors and applications. *Talanta* **2014**, *124*, 95–105. [CrossRef] [PubMed]

171. Lee, D.-S.; Jung, J.-K.; Lim, J.-W.; Huh, J.-S.; Lee, D.-D. Recognition of volatile organic compounds using SnO_2 sensor array and pattern recognition analysis. *Sens. Actuators B Chem.* **2001**, *77*, 228–236. [CrossRef]

172. Lee, Y.-S.; Moon, P.-J. A comparison and Analysis of deep learning framework. *J. Korea Inst. Electron. Commun. Sci.* **2017**, *12*, 115–122.

173. Peng, P.; Zhao, X.; Pan, X.; Ye, W. Gas classification using deep convolutional neural networks. *Sensors* **2018**, *18*, 157. [CrossRef] [PubMed]

Article

Fluorescent Alloyed CdZnSeS/ZnS Nanosensor for Doxorubicin Detection

Svetlana A. Mescheryakova, Ivan S. Matlakhov, Pavel D. Strokin, Daniil D. Drozd, Irina Yu. Goryacheva and Olga A. Goryacheva *

Department of General and Inorganic Chemistry, Chemistry Institute, Saratov State University Named after N.G. Chernyshevsky, Astrakhanskaya 83, 410012 Saratov, Russia; meshcheryakova.s.a@gmail.com (S.A.M.); goryachevaiy@mail.ru (I.Y.G.)
* Correspondence: olga.goryacheva.93@mail.ru

Abstract: Doxorubicin (DOX) is widely used in chemotherapy as an anti-tumor drug. However, DOX is highly cardio-, neuro- and cytotoxic. For this reason, the continuous monitoring of DOX concentrations in biofluids and tissues is important. Most methods for the determination of DOX concentrations are complex and costly, and are designed to determine pure DOX. The purpose of this work is to demonstrate the capabilities of analytical nanosensors based on the quenching of the fluorescence of alloyed CdZnSeS/ZnS quantum dots (QDs) for operative DOX detection. To maximize the nanosensor quenching efficiency, the spectral features of QDs and DOX were carefully studied, and the complex nature of QD fluorescence quenching in the presence of DOX was shown. Using optimized conditions, turn-off fluorescence nanosensors for direct DOX determination in undiluted human plasma were developed. A DOX concentration of 0.5 μM in plasma was reflected in a decrease in the fluorescence intensity of QDs, stabilized with thioglycolic and 3-mercaptopropionic acids, for 5.8 and 4.4%, respectively. The calculated Limit of Detection values were 0.08 and 0.03 μg/mL using QDs, stabilized with thioglycolic and 3-mercaptopropionic acids, respectively.

Keywords: quantum dots; doxorubicin; nanosensor; fluorescence quenching; Stern–Volmer constants; anthracycline antibiotics; human plasma

Citation: Mescheryakova, S.A.; Matlakhov, I.S.; Strokin, P.D.; Drozd, D.D.; Goryacheva, I.Y.; Goryacheva, O.A. Fluorescent Alloyed CdZnSeS/ZnS Nanosensor for Doxorubicin Detection. *Biosensors* **2023**, *13*, 596. https://doi.org/10.3390/bios13060596

Received: 10 April 2023
Revised: 24 May 2023
Accepted: 26 May 2023
Published: 31 May 2023

1. Introduction

Doxorubicin (DOX) is an effective chemotherapeutic agent from the group of anthracycline antibiotics (Figure 1), and has high activity against many types of cancer. It is one of the commonly used agents in the treatment of different cancers types, including pediatric cancer, leukemia, breast cancer, etc. [1]. However, the use of DOX has significant disadvantages because it affects both cancer and healthy cells. DOX is highly cardiotoxic, cytotoxic and neurotoxic. It causes cell death through multiple intracellular targets: DNA-adduct formation, topoisomerase II enzyme inhibition, reactive oxygen species generation, histone eviction, Ca^{2+} and iron hemostasis regulation and ceramide overproduction [2]. There are several mechanisms of DOX-induced cardiotoxicity, among which oxidative stress, free radical generation and apoptosis are the most widely reported [1]. The side effects are often observed when the concentration of DOX exceeds a certain level in blood, so the therapeutic range of DOX concentration in blood is narrow. For this reason, the continuous monitoring of DOX concentrations in body fluids is important [3].

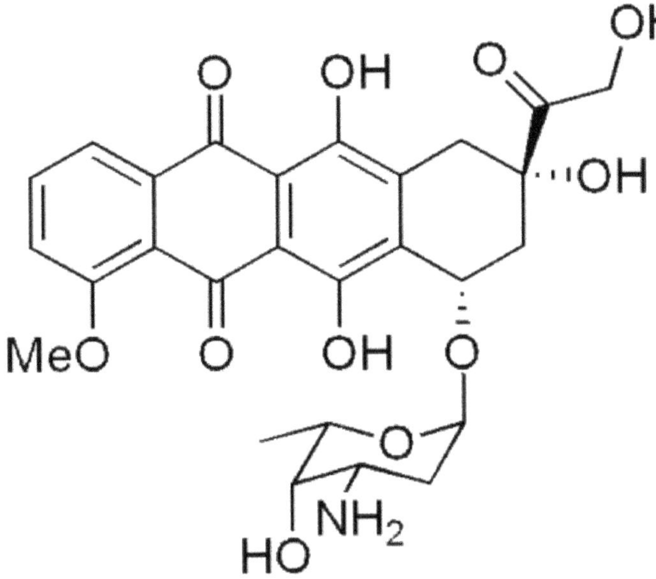

Figure 1. Doxorubicin structure.

Number methods are known for quantitative DOX determination in body fluids and tissues. One way is to measure the DOX concentration by registering its own absorption in the visible spectral range. In this case, a thorough purification of the sample or extraction is necessary. Such sample preparation takes up to one hour and allows for DOX determination with a Limit of Detection (LOD) of ~1.2 µg/mL DOX in plasma [3]. Spectrophotometric detectors are also used in tandem with high performance liquid chromatography and provide low DOX LOD values of 5–5000 ng/mL in rat serum and tissues. Here, performing a sample extraction using a methanol-chloroform mixture is also necessary. The method is able to determine the DOX concentration in both blood and tissues [4,5], as well as in gall and lymph [6]. The optical properties of DOX also allow its concentration to be determined on the basis of its emission. Thus, the DOX passage across the blood–brain barrier in the form of a low-density lipoprotein receptor-targeted liposomal drug was studied by Pinzón-Daza et al. DOX concentrations can be determined based on its fluorescence, at excitation and emission wavelengths of 475 and 553 nm, respectively [7]. DOX concentrations as low as 18 ng/mL in rabbit plasma were detected by capillary electrophoresis with in-column double optical-fiber LED-induced fluorescence detection using rhodamine B as an internal standard. Before analysis, rabbit serum samples were subsequently diluted twice with acetonitrile to precipitate the proteins. For electrophoretic separation, a borate buffer (15 mM, pH 9.0) containing 50% acetonitrile (*v/v*) was used [8]. Electrochemical methods are traditionally used for operative high-throughput detection. Differential pulse cathodic stripping voltammetry on a polished silver solid amalgam electrode in a specially designed micro-volume voltametric cell was used for the DOX determination with a LOD of 0.44 µM. The applicability of this method was verified through an analysis of spiked tap water samples and human urine [9]. The same authors developed differential pulse voltammetry on a polarized liquid/liquid interface impregnated with a ionic liquid polyvinylidenfluoride microporous filter with a DOX LOD of 0.84 µM. Due to the affection of some body fluids interfering compounds, this method must be preceded by a separation step. The ability of DOX to bind with DNA was used in a voltammetric DNA sensor using a glassy carbon electrode covered with electropolymerized Azure B film and physically adsorbed native DNA. In optimal conditions, the DNA sensor provided a DOX LOD of 0.07 nM for commercial DOX formulations and on artificial samples mimicking the electrolyte content of human serum [10]. The modification of the glassy carbon electrode with a vertically-ordered mesoporous silica-nanochannel film with electrochemically reduced graphene oxide by a one-step electrochemically assisted self-assembly method allowed us to reach a DOX LOD of 0.77 nM in human whole blood [11]. The application of rolling circle amplification allowed us to develop an ultrasensitive electrochemical DNA sensor with DOX@tetrahedron-Au as the electrochemical indicator, and reach a DOX LOD of 0.29 fM [12]. The application of surface enhanced raman spectroscopy for DOX detection using silica nanoparticles covered with 10 nm think gold film allowed us to archive a LOD of 20 nM in undiluted serum [13]. These methods for determining DOX concentrations are complex and costly and/or need exhaustive sample preparation.

DOX has a fairly intense native fluorescence in the range of 540–660 nm (quantum yield, QY is 4.39%) [14]. However, the optical properties of DOX (absorption and emission) are sensitive to the form of DOX molecules in the solution, which in turn, depends on plenty of factors, including DOX concentration, ionic strength, pH, additives and so on [15]. DOX, like other anthracyclines, has the propensity to form dimers and associates; DOX fluorescence dramatically drops upon dimerization (QY~10^{-5}) [16]. This makes trouble for direct fluorescence-based DOX detection. There is a demand for the development of simple nanosensors that enable the fast determination of analytes in solutions based on changes in the properties of nanoparticles as they interact with analytes [17,18]. One example of such a system is localized surface plasmon resonance technology (LSPR). This LSPR-based biosensing system utilizes the sensitivity of the plasmonic frequency to changes in the local index of refraction at the nanoparticle surface. Optimizing the nanoparticle material and geometry alters the plasmonic properties towards sensitivity improvement [19,20].

An important property of DOX is the effective quenching of the emission of a large number of luminophores, which allows for the use of quenching-based methods for DOX detection. Different luminophores were used as emission turn-off probes, such as fluorescent polyethyleneimine-functionalized carbon dots [21] and other carbon nanostructures [22,23], gold nanoclusters [24], and phosphorescent Mn-doped ZnS quantum dots (QDs) [25], fluorescent Mn-doped ZnSe D-dots [26] and CdSe/ZnS QDs [27]. The DOX quenching of nanoparticles fluorescence can be so efficacious that the DOX quenching of QDs emission has been used for the detection of DNA [26] and analysis of telomerase activity [27]. While characterizing the DOX—luminophore interaction, the authors mostly focus on the quenching in model solutions, demonstrating perspectives for its application in biofluids.

In this work, we use fluorescent alloyed semiconductor QDs, which are of particular interest because they are photostable, homogeneous in size and properties and can form a stable aqueous colloid. In addition, the possibility to select surface ligands to achieve optimal interactions and improve the sensitivity of turn-off nanosensors is important and useful [28,29]. The analytical application of the quenching of QD fluorescence is based on changes in the emission intensity during the interaction between the nanosensors and modulating molecules (quenchers). The quenching of QD probe fluorescence is the simplest analytical method for DOX detection compared to other known protocols. Compared with the carbon-based nanostructures mentioned above, semiconductor QDs have excellent uniformity, high synthesis reproducibility, photo- and chemical stability, as well as well-studied synthetic routs and emission mechanism. As QDs, we used alloyed CdZnSeS/ZnS core/shell nanocrystals, covered with 3-mercaptopropionic acid (MPA) and thioglycolic acid (TGA) (Figure S1 in Supplementary Material). Compared to traditional core/shell QDs, alloyed QDs are characterized by a simple synthesis route and smaller size, while maintaining narrow emission peaks and high emission QY, and the possibility to use various surface ligands. The comparison of two QD samples with the same architecture of a semiconductor core, but stabilized with different surface ligands, makes it possible to determine the effect of the ligand on the sensitivity of QDs to emission quenching.

2. Materials and Methods

2.1. Materials

Cadmium (II) acetate (99.995%), zinc acetate (99.99%), zinc stearate (technical grade), elemental selenium (powder), elemental sulfur (powder), trioctylphosphine (90%), 1-octadecene (90%), oleic acid (90%), MPA and TGA were purchased from Merck. All other chemicals and solvents were of analytical grade and used without additional purification. In this work, we used Sindroxocin (50 mg, Actavis, Iceland), an antitumor antibiotic whose active ingredient is doxorubicin hydrochloride (50 mg). It also contains the following excipients: methyl parahydroxybenzoate (5 mg) and lactose monohydrate (263.15 mg). Before analysis, 3 mg (corresponding to 0.5 mg of DOX) of the lyophilizate was dissolved in 1 mL of Milli-Q water. From the resulting solution (500 µg/mL), solutions with concentrations of 50, 5 and 1 µg/mL were obtained.

2.2. QD Synthesis and Hydrophilization

The alloyed QD synthesis was adapted from [30]. For hydrophilization, 25 µmol of TGA or MPA were added to 50 µL of QDs aliquot in 1 mL of toluene. The following steps were performed according to [30].

2.3. QD Characterization and Spectral Measurements

Absorption spectra were obtained with a Shimadzu UV-1800 spectrophotometer (Shimadzu, Kyoto, Japan). Emission spectra were recorded with a Cary Eclipse fluorescence spectrometer (Agilent Technologies, Santa Clara, CA, USA) and multimode microplate reader Synergy H1 (BioTek Instruments, Charlotte, NC, USA). The ζ-potential of the QDs samples were determined via dynamic light scattering measurements using a Zetasizer Ultra instrument (Malvern Instruments, Malvern, UK). A Libra 120 transmission electron microscope (TEM) (Carl Zeiss, Jena, Germany) was used to take photomicrographs and determine the size of QDs.

2.4. Quantum Yield Calculation

The relative QY of QD emission was calculated as described in [31], relative to fluorescent dye Coumarin 153 (ethanol solution, peak optical density 0.1, QY 53%). The excitation and emission wavelengths of the QY measurement were 360 and 514 nm, respectively.

$$\Phi_x = \Phi_{st} \frac{I_x}{I_{st}} \frac{A_{st}}{A_x} \frac{n_x^2}{n_{st}^2}$$

Φ_x—relative QY of the QDs sample;
Φ_{st}—the relative QY of the reference (coumarin-153);
I_x—is the integral fluorescence intensity of the QDs sample;
I_{st}—integral intensity of the reference (coumarin-153);
A_x—optical density of the QDs sample;
A_{st}—optical density of the reference (coumarin-153);
n_x—the refractive index of the sample (water);
n_{st}—the refractive index of the reference solvent (ethyl alcohol).

2.5. Optical Measurements

In order to avoid the internal filter effect and maintain the same QD concentration in the samples, all QD samples were brought to the same optical density equal to 0.1 at excitation wavelength (λ_{ex} = 360 nm). All experiments were performed in three repetitions to evaluate the accuracy of the measurements.

2.6. Analysis Performance in Plasma Samples

Syndroxocin solution in Milli-Q (1000 mg/mL) water was used to determine the DOX concentration in plasma. It was added to plasma to obtain DOX concentrations in the range of 0.5, 1, 5, 10, 50, 100 and 500 μg/mL. Plasma was incubated for 30 min at 37 °C. The test was performed in a 96-well plate, and 50 μL of QDs colloid with an optical density of 0.1 was added to each well. Subsequently, 50 μL of spiked plasma was added to the QDs colloid and measured immediately. To evaluate the stability of the quenching magnitude, the incubation of DOX with plasma samples for 24 h was also performed.

3. Results and Discussion

3.1. Optical Properties of QDs

The nanosensor for DOX detection is based on QD fluorescence quenching. The most probable quenching mechanisms are the Förster resonance energy transfer (FRET) and photoinduced electron transfer (PIET). Therefore, to enhance quenching, it was necessary to obtain QDs with a fluorescence peak overlapping with the absorption band of DOX (Figure S2) [32,33].

CdZnSeS/ZnS QDs (λ_{em} = 540 nm) (Figure 2) were synthesized and transferred into water by the ligand exchange process according to the procedure described by Drozd et al. [30]. TEM images show a size of about 10 nm (Figure S3). As hydrophilizing agents, two mercaptoacids with a different carbon chain length—TGA (two carbon atoms) and MPA (three carbon atoms)—were used (Figure S1). After TGA coverage, the zeta potential of QD is—76 mV, and for MPA it was— 46 mV. The emission QY for QD@TGA and QD@MPA was calculated as 0.55 and 0.60, respectively.

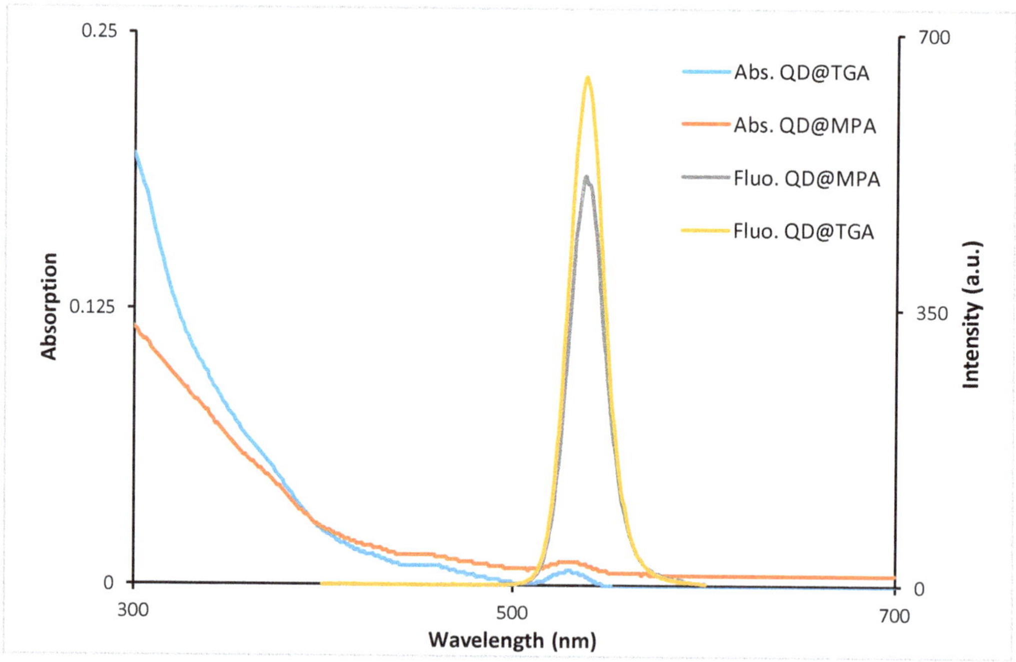

Figure 2. Absorbance and fluorescence (λ_{ex} = 360 nm) spectra of QDs stabilized with 3-mercaptopropionic acid (QD@MPA) and thioglycolic acid (QD@TGA).

It is important to note that in most cases, studies were performed with pure DOX hydrochloride, while chemotherapy often uses DOX formulations with excipients, which can affect the efficacy and accuracy of assays. In this work, we used the pharmaceutical formulation Syndroxocin, which contains Doxorubicin hydrochloride (50 mg), methyl p-hydroxybenzoate (1 mg) and lactose monohydrate (50 mg). To control the possible effect of lactose on QDs emission, we added lactose (5 mg/mL) to the QD solutions and compared the optical properties. The results show that no changes in the QD fluorescence intensity and spectra shape were observed, so we can conclude that lactose has no effect on QDs fluorescence (Figure S4).

3.2. Optical Properties of DOX

DOX has a characteristic wide absorption peak at 450–550 nm (λ_{max} = 480 nm) and fluorescence spectra with three distinct peaks at 560, 594 and 638 nm. DOX molecules in solutions can be presented in different forms, which are reflected in their spectral properties.

As can be seen from the 3D fluorescence spectra (Figure 3), there is no shift in the DOX emission maxima when excited with the light of different wavelengths and when the DOX concentration is changed. At the same time, there are notable changes in the ratio of peaks when the DOX concentration increases. These effects are related to the dimerization of DOX as well as the inner filter effect. Thus, DOX undergoes dimerization with increasing concentration. An increase in the absorption intensity at 415–540 nm indicates the formation of aggregates [16].

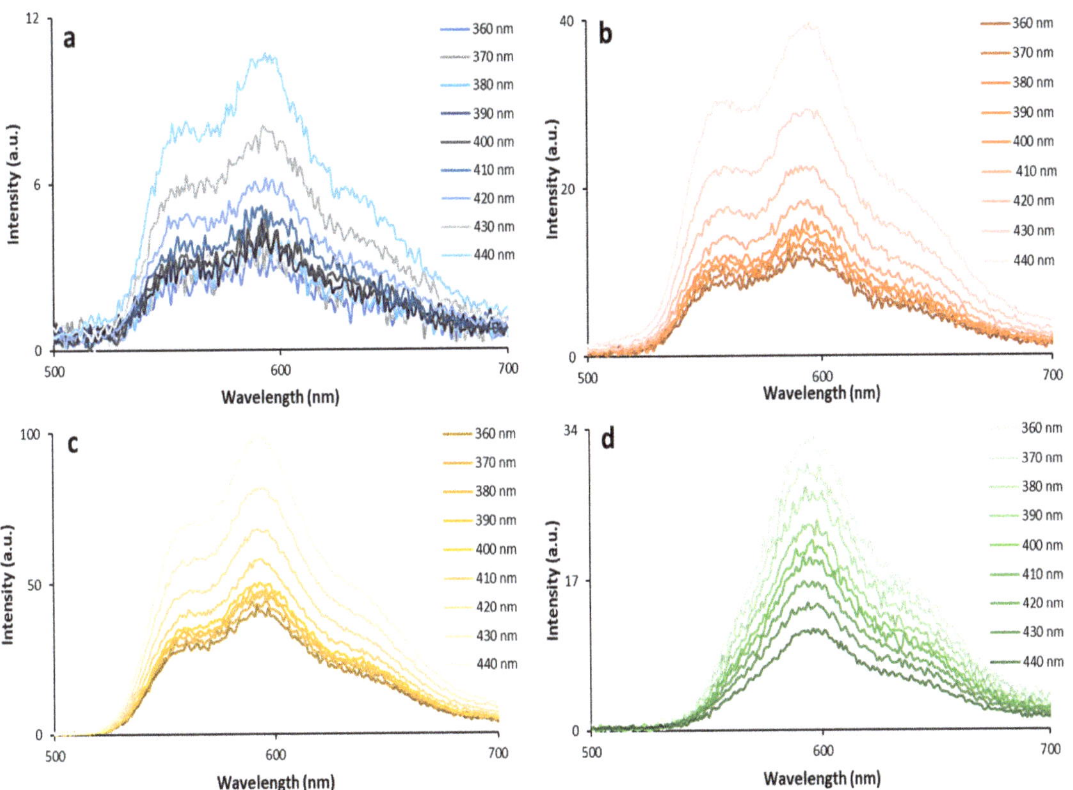

Figure 3. 3D fluorescence spectra of DOX aqua solutions with concentrations of 1 µg/mL (**a**), 5 µg/mL (**b**), 50 µg/mL (**c**) and 500 µg/mL (**d**).

3.3. Influence of QDs on the DOX Fluorescence

Various possible mechanisms of the interaction between QDs in the excited state and DOX molecules have been suggested, including FRET and PIET. Since the emission spectra of QDs and the absorption spectra of DOX overlap (Figure S2), we assume that FRET is the main reason for QD emission quenching. Because both QDs and DOX are emissive, the mutual influence of QDs and DOX molecules on each other's fluorescence should be checked.

To examine the possible effect of QDs on DOX, fluorescence spectra were recorded for a series of DOX and QD-DOX solutions with an increasing concentration of DOX and a fixed concentration of QDs (Figure 4). It was shown that the fluorescence intensity of DOX concentrations < 50 μM is lower than the fluorescence intensity of QDs at a peak near 540 nm. This suggests that at DOX concentrations around or above 50 μM, the contribution of DOX to the total fluorescence intensity of QD-DOX solutions will be very significant. A DOX concentration < 18 μM practically does not contribute to the total QD-DOX fluorescence.

At the same time, when QDs are added to the DOX solution, the fluorescence spectra of DOX are almost unchanged. However, at high concentrations (>50 μM), a slight quenching of the fluorescence in the presence of the QDs occurs, which may be due to the inner filter effect. This effect is clearly seen from the comparison of the intensities of the fluorescence of the DOX, QD@MPA + DOX and QD@TGA + DOX solutions at increasing DOX concentrations (Figure 4c). The shape of the DOX spectrum has not visibly changed, indicating that there are no changes in the molecule.

3.4. QDs Fluorescence Quenching by DOX

Since the DOX concentration used for chemotherapy is large (244 mg per square meter) [34], we studied a wide range of DOX concentrations. To determine the magnitude of QDs fluorescence quenching with DOX, 0–920 μM (0 to 500 μg/mL) DOX concentrations (Table S1) at fixed QDs concentration were used. The comparison was made for QDs coated with TGA and MPA to find the more sensitive QDs (Figure S1). Since quenching is based on the adsorption of DOX on the QD surface, the modifying surface layer can play a key role. The obtained dependences of QDs emission intensity from DOX concentrations are presented in Figure 5a,b. It is possible to see two different ranges; at low DOX concentrations (<50 μM), there is the quenching of QDs emission, and at the range > 100 μM of DOX, an approximately stable emission intensity is observed due to the increasing contribution of DOX fluorescence. For the evaluation of the quenching of QDs fluorescence, Stern–Volmer plots were built and Stern–Volmer constants were evaluated (Formula (1)).

$$\frac{I_0}{I} = 1 + k_{SV}[DOX] \tag{1}$$

I_0 and I—fluorescence intensity in the absence and presence of DOX, [DOX]—concentration of the DOX [mol/L], k_{SV}—Stern–Volmer constant [L/mol]. The deviation of the Stern–Volmer plot from liner dependency suggests that several processes accompany quenching and that the nature of quenching may combine both static and dynamic interactions (Figure 5b) [35].

The plots are linear for a DOX concentration of 0–50 μM, which corresponds to the range in which there is no significant DOX fluorescence effect on the total emission of QD-DOX solutions. The value of the Stern–Volmer constants was calculated at 0.074 M^{-1} for QD@TGA and 0.039 M^{-1} for QD@MPA. The quenching for QD@TGA in the presence of DOX is about two times more effective than that for QD@MPA. Two reasons, both related to the surface ligand size, can cause such a difference. The smaller size of the TGA molecules provides a better interaction of the quenching agent on the QD surface with the emissive semiconductor core. On the other hand, the smaller size of the TGA molecule makes the hydrophilic ligand layer on the QD surface more dynamic, opening up more opportunities for DOX interactions with the QD surface.

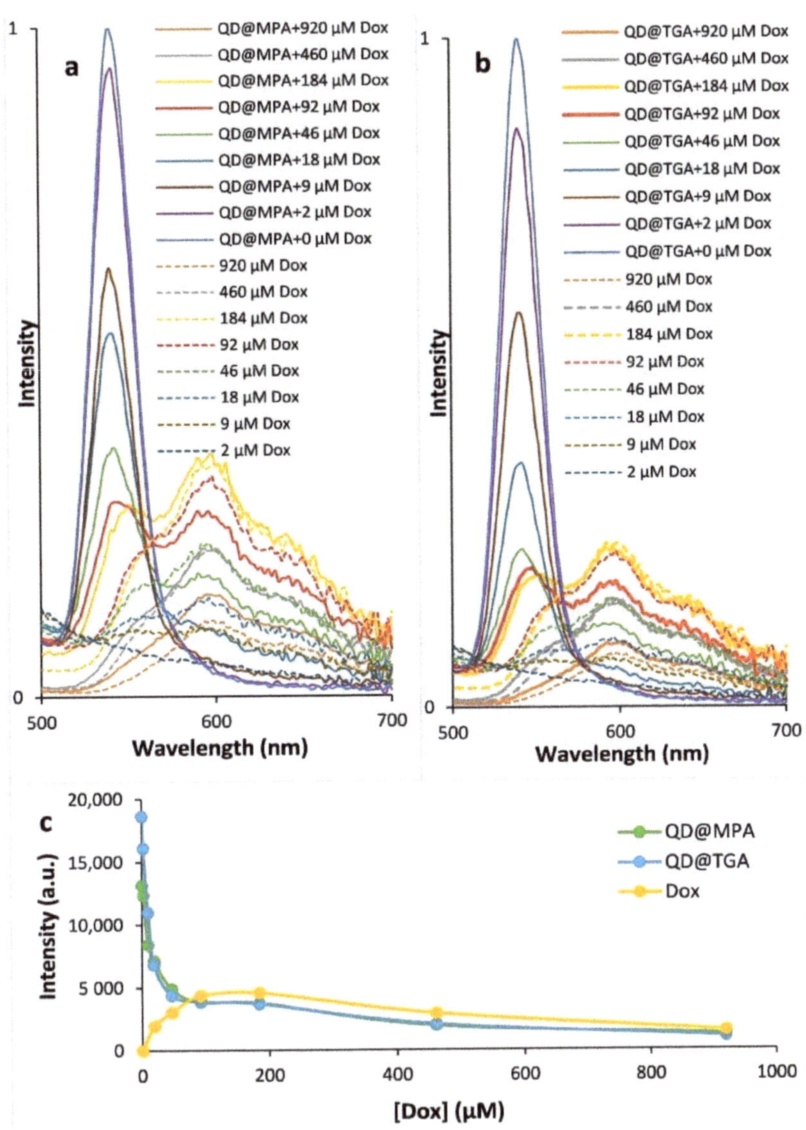

Figure 4. Normalized fluorescence spectra (λ_{ex} = 360 nm) of QDs stabilized by MPA (**a**) and TGA (**b**) in the presence of different DOX concentrations (solid lines), for comparison DOX fluorescence spectra are presented (dashed lines); (**c**) intensity of fluorescence of DOX, QD@MPA + DOX and QD@TGA + DOX solutions (λ_{ex} = 360 nm, λ_{em} = 540 nm), depending on DOX concentration.

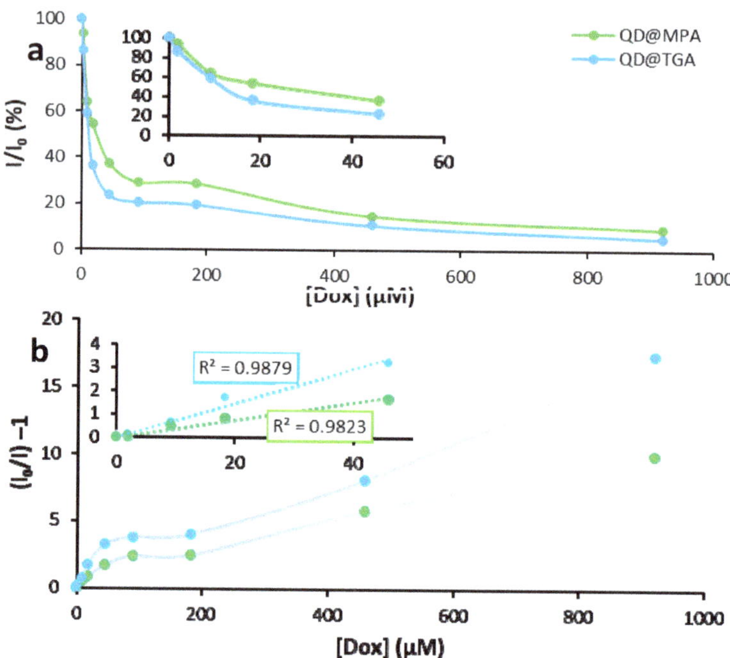

Figure 5. Dependence of normalized QDs emission on DOX concentration (**a**) and Stern–Volmer plots (**b**); insert (**b**): calculation of Stern–Volmer constants for 0–50 µM DOX.

3.5. Determination of DOX in Human Blood Plasma

DOX is usually determined in patients' blood, plasma or urine. The intrinsic feature of DOX as an anthracycline antibiotic is its ability to bind with DNA and other compounds. The determination of DOX concentration can be inaccurate due to its elimination from the sample with biomaterials. Therefore, for DOX determination, it is important to use biofluid samples that are as unaltered as possible. To determine the feasibility of analysis in biological fluids, the matrix effect of plasma on QDs fluorescence was tested. For this purpose, 50 µL of plasma was added to 50 µL of the QDs colloid in the well of a microtiter plate. As a result, a slight change in the fluorescence intensity can be seen due to the inner filter effect of the serum components and the change in the pH in comparison with the previous experiment, which was performed in Milli-Q water (Figure 6) [36].

Freshly prepared Syndroxocin solutions were used for the spiking plasma experiments. Incubation at body temperature and stirring were carried out in order to uniformly distribute DOX over the plasma components and obtain an equilibrium of possible interactions. The Corning Costar 96-well plate was chosen as the plate with the least adsorbing effect for bio-objects. The 96-well plate also allows you to build a calibration curve and use the test for multiple samples at once. The fluorescence spectra show that the concentration of DOX in the plasma affects the QDs fluorescence. Moreover, the fluorescence intensity of TGA-stabilized QDs decreases more than that of MPA-stabilized QDs as well as DOX in water solutions (Figure 7). Consequently, TGA-stabilized QDs are more sensitive to DOX in plasma, but their dependence on concentration is not sufficiently uniform throughout the DOX concentration range. Nevertheless, QD@MPA shows a more uniform dependence of fluorescence quenching on DOX concentration, which may be due to the increased stability of QD@MPA relative to QD@TGA over a wide range of pH [37].

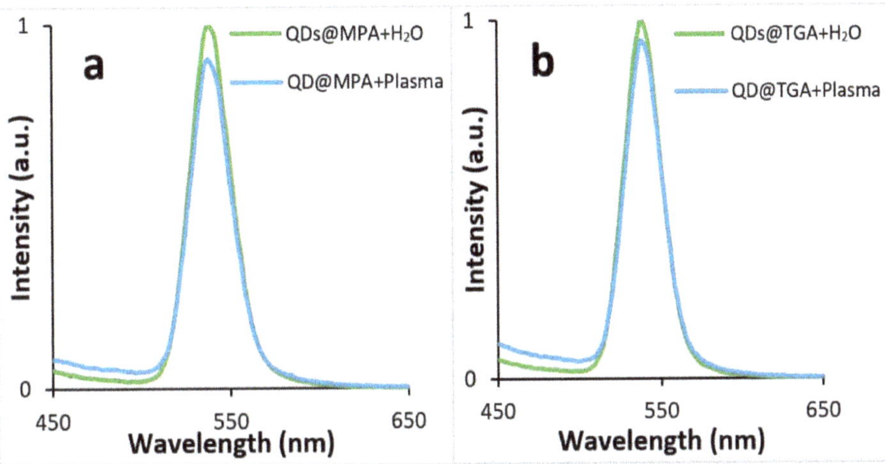

Figure 6. Fluorescence of QDs in water and plasma: TGA stabilized QDs (QD@TGA) (**a**) MPA stabilized QDs (QD@MPA) (**b**).

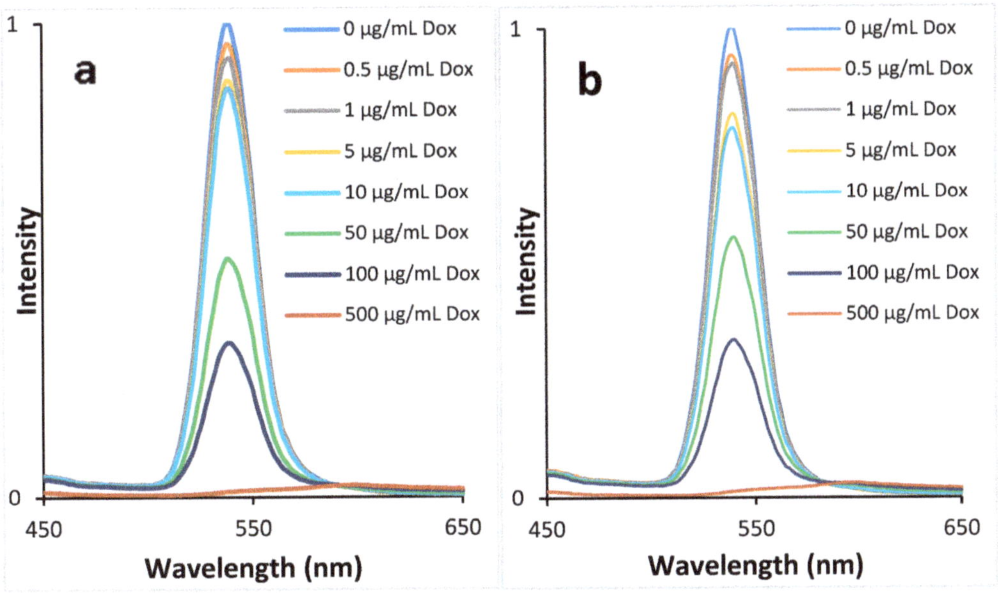

Figure 7. Normalized fluorescence spectra of QDs stabilized with MPA (**a**) and TGA (**b**) in the presence of plasma spiked with different DOX concentrations.

To reveal the dependence of the QDs fluorescence intensity decrease on the DOX concentration, we plotted the fluorescence quenching profiles and Stern–Volmer plots (Figure 8). The DOX concentration range was chosen taking into account DOX dosing recommendations. The linear range of DOX concentrations was 0–184 μM (corresponding to 0–100 μg/mL), with a Stern–Volmer constant of 0.011 M^{-1} for QD@TGA and 0.010 M^{-1} for QD@MPA, so that the fluorescence reduction values could be determined accurately. The better linearity of Stern–Volmer plots in plasma (R^2 = 0.977 and 0.9809 for 0–184 μM of DOX) compared to Stern–Volmer plots in aqua solutions (R^2 = 0.9634 and 0.9765 for a more narrow DOX concentration range of 0–50 μM) may result from the increased stability of

QDs coatings with TGA and MPA in more alkaline media, and thus the uniform interaction of DOX with the QD surface [38]. A DOX concentration of 0.5 µM in plasma was reflected in a decrease in the fluorescence intensity of QDs, stabilized with thioglycolic and 3-mercaptopropionic acids, for 5.8 and 4.4%, respectively. To calculate the Limit of Detection (LOD) values, two approaches were used. The calculated instrumental LOD value by Formula (S1) was 4.0 ng/mL for QDs@TGA and 1.2 ng/mL for QDs@MPA, because of the low deviation in the QDs emission intensity measurements. So, we used an approach based on the standard deviation of response in the presence of DOX (Formula (S2)), and obtained more realistic DOX LOD values of 0.06 µg/mL for QDs@TGA and 0.02 µg/mL for QDs@MPA.

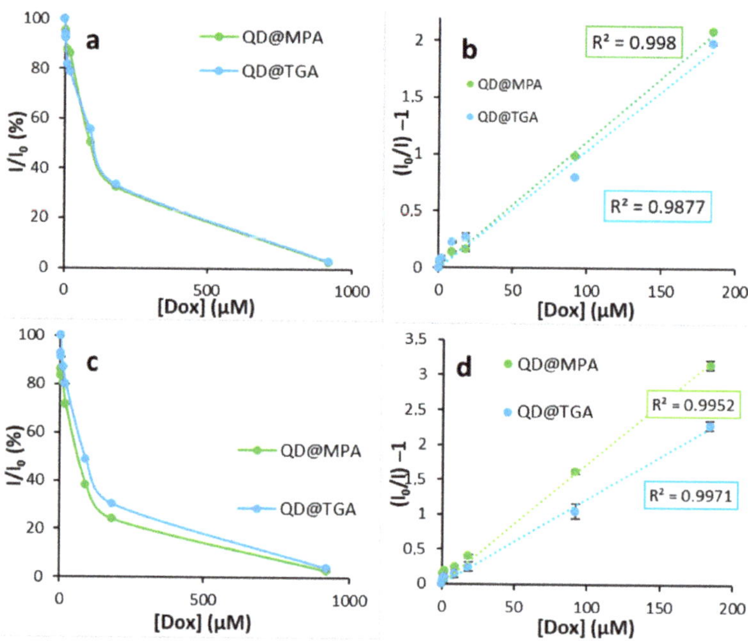

Figure 8. Dependence of QDs fluorescence intensity of spiked plasma vs. DOX concentration (**a,c**) and Stern–Vomer plots (**b,d**) after 30 min (**a,b**) and 24 hours (**c,d**) incubation.

Since plasma has been found to affect QDs fluorescence quenching, we decided to try diluting the plasma by a factor of 10 when preparing the samples. The results showed that the dilution resulted in a significant change in QDs sensitivity due to a dramatic decrease in the DOX concentration. Therefore, undiluted plasma was used for further studies.

Because the half-life of DOX in the human body varies between 30 and 150 h [39], DOX is prescribed in several cycles. Each cycle consists of several days with an injection of DOX until therapeutic blood concentrations are achieved [40]. Thus, there is a period of DOX accumulation until the next injection in one cycle. Therefore, in order to simulate a round in the intrinsic cycle, we analyzed the spiked plasma with different concentrations of DOX plasma samples after incubation for 24 h. QDs fluorescence quenching in the presence of DOX was preserved even after this long incubation period (Figure 8), indicating the high affinity of DOX with the QDs surface and stability of the obtained equilibrium between the DOX and plasma components. The Stern–Volmer plot has liner dependence in the area of 0–184 µM. The Stern–Volmer constants were 0.017 M^{-1} for QDs@TGA and 0.012 M^{-1} for QDs@MPA.

Due to the fact that after 24 h, the Stern–Volmer constant values experienced minor changes, but the plots' linearity was better, we assume that both static and dynamic interactions may have played a contribution. The LOD values for DOX detection after 24 hours incubation in plasma were calculated by Formula (S2), and it was 0.08 μg/mL for QDs@TGA and 0.03 μg/mL for QDs@MPA.

4. Conclusions

The ability of DOX to quench the fluorescence of alloyed CdZnSeS/ZnS QDs was used to develop turn-off fluorescence nanosensors for DOX determination in human blood plasma. The quenching depends on the concentration of DOX in the solution. By comparing the magnitude of quenching for two samples of QD covered with mercaptoacids (TGA and MPA), the effect of surface hydrofilic ligands was shown:

The QDs covered with shorter carbon chain TGA are more sensitive to DOX presence both in aqua solution and in plasma. The smaller ligand predetermines slightly less reproducibility of QD fluorescence intensity values. With longer carbon chain MPA, the dependence of fluorescence on the DOX concentration has smaller coefficients of variation. The quenching system was also implemented in blood plasma. In this case, the blood sample preparation only consists of obtaining plasma and does not need any additional purification. It is interesting to mention our unexpected discovery that the magnitude of QD fluorescence quenching is higher in plasma than in aqua, resulting in about a double increase in Stern–Volmer constants, as well as additionally increasing the plasma long incubation with DOX. The quenching of QDs fluorescence in the DOX-spiked plasma has a linear dependence in the range of 0–200 μM.

DOX calculated LOD values for fluorescence turn-off nanosensors were 0.08 μg/mL for QDs@TGA and 0.03 μg/mL for QDs@MPA, respectively (we used 24 h DOX incubation with plasma before assay). The suggested format using undiluted plasma without any sample preparation and lengthy complex analysis methods opens the way for routine studies of DOX transformation in the patient's body and the selection of individual chemotherapy protocols. An easy-to-use nanosensor has been developed and could become a meaningful tool for medical applications, as it can be used as an indicator for chemotherapy protocol correction.

The developed nanosensor can be used to monitor the concentration of DOX in patients through cycles of chemotherapy in order to optimize the cytostatic dose. Potentially, the developed nanosensor (after appropriate studies) can be used for the detection of other anthracycline cytostatics, as they are able to bind to the surface of the nanosensor and quench its emission. The issue of selectivity is not a limiting factor for nanosensors in biofluids, since the chemotherapy regimen and the particular drugs are known for each patient. The lack of selectivity for specific anthracycline antibiotics can be a problem, for example, in the analysis of hospital wastewater. In this case, the properties of the nanosensor must be optimized for each specific task.

Supplementary Materials: The following supporting information can be downloaded at: https://www.mdpi.com/article/10.3390/bios13060596/s1, Figure S1: Formulas of (a) Thioglycolic acid and (b) 3-mercaptopropionic acid; Figure S2: An overlap between the QDs fluorescence spectrum (red) and DOX absorption spectrum (blue); Figure S3: Fluorescence spectra of QDs stabilized with 3-mercaptopropionic acid (A) and thioglycolic acid (B) with and without lactose; Figure S4: Fluorescence spectra of QDs stabilized with 3-mercaptopropionic acid (a) and thioglycolic acid (b) in X10 diluted plasma, spiked with DOX; Table S1: Correspondence of DOX concentration in μg/mL and μM (DOX Molecular Weight 543.5).

Author Contributions: Conceptualization, O.A.G. and I.Y.G.; methodology, O.A.G. and S.A.M.; validation, O.A.G.; investigation, D.D.D., P.D.S. and I.S.M.; resources, O.A.G.; data curation, S.A.M.; writing—original draft preparation, review and editing, O.A.G. and I.Y.G.; visualization, S.A.M.; supervision, O.A.G.; project administration, O.A.G. All authors have read and agreed to the published version of the manuscript.

Funding: The reported study was funded by the Russian Science Foundation: project number 21-73-10046.

Institutional Review Board Statement: The research volunteer blood plasma was used for the experiments of DOX determination. Protocol № 8 from 02.03.2021, Saratov State Medical University named after V.I. Razumovsky.

Informed Consent Statement: Informed consent was obtained from all subjects.

Acknowledgments: Light scattering measurements were performed using Zetasizer Ultra (Resource Sharing Center of Saratov State University). TEM measurements were performed at the Simbioz Center for the Collective Use of Research Equipment in the Field of Physical–Chemical Biology and Nanobiotechnology at the Russian Academy of Sciences' Institute of Biochemistry and Physiology of Plants and Microorganisms.

Conflicts of Interest: The authors declare no conflict of interest.

References

1. Rawat, P.S.; Jaiswal, A.; Khurana, A.; Bhatti, J.S.; Navik, U. Doxorubicin-Induced Cardiotoxicity: An Update on the Molecular Mechanism and Novel Therapeutic Strategies for Effective Management. *Biomed. Pharmacother.* **2021**, *139*, 111708. [CrossRef] [PubMed]
2. Sritharan, S.; Sivalingam, N. A Comprehensive Review on Time-Tested Anticancer Drug Doxorubicin. *Life Sci.* **2021**, *278*, 119527. [CrossRef] [PubMed]
3. Sikora, T.; Morawska, K.; Lisowski, W.; Rytel, P.; Dylong, A. Application of Optical Methods for Determination of Concentration of Doxorubicin in Blood and Plasma. *Pharmaceuticals* **2022**, *15*, 112. [CrossRef] [PubMed]
4. Gulyaev, A.E.; Gelperina, S.E.; Skidan, I.N.; Antropov, A.S.; Kivman, G.Y.; Kreuter, J. Significant Transport of Doxorubicin into the Brain with Polysorbate 80- Coated Nanoparticles. *Pharm. Res.* **1999**, *16*, 1564–1569. [CrossRef]
5. Álvarez-Cedrón, L.; Sayalero, M.L.; Lanao, J.M. High-Performance Liquid Chromatographic Validated Assay of Doxorubicin in Rat Plasma and Tissues. *J. Chromatogr. B Biomed. Sci. Appl.* **1999**, *721*, 271–278. [CrossRef] [PubMed]
6. Shinozawa, S.; Oda, T. Determination of Adriamycin (Doxorubicin) and Related Fluorescent Compounds in Rat Lymph and Gall by High-Performance Liquid Chromatography. *J. Chromatogr. A* **1981**, *212*, 323–330. [CrossRef] [PubMed]
7. Pinzón-Daza, M.L.; Garzón, R.; Couraud, P.O.; Romero, I.; Weksler, B.; Ghigo, D.; Bosia, A.; Riganti, C. The Association of Statins plus LDL Receptor-Targeted Liposome-Encapsulated Doxorubicin Increases in Vitro Drug Delivery across Blood–brain Barrier Cells. *Br. J. Pharmacol.* **2012**, *167*, 1431–1447. [CrossRef] [PubMed]
8. Yang, X.; Gao, H.; Qian, F.; Zhao, C.; Liao, X. Internal Standard Method for the Measurement of Doxorubicin and Daunorubicin by Capillary Electrophoresis with in-Column Double Optical-Fiber LED-Induced Fluorescence Detection. *J. Pharm. Biomed. Anal.* **2016**, *117*, 118–124. [CrossRef] [PubMed]
9. Skalová, Š.; Langmaier, J.; Barek, J.; Vyskočil, V.; Navrátil, T. Doxorubicin Determination Using Two Novel Voltammetric Approaches: A Comparative Study. *Electrochim. Acta* **2020**, *330*, 135180. [CrossRef]
10. Porfireva, A.; Vorobev, V.; Babkina, S.; Evtugyn, G. Electrochemical Sensor Based on Poly(Azure B)-DNA Composite for Doxorubicin Determination. *Sensors* **2019**, *19*, 2085. [CrossRef]
11. Yan, F.; Chen, J.; Jin, Q.; Zhou, H.; Sailjoi, A.; Liu, J.; Tang, W. Fast One-Step Fabrication of a Vertically-Ordered Mesoporous Silica-Nanochannel Film on Graphene for Direct and Sensitive Detection of Doxorubicin in Human Whole Blood. *J. Mater. Chem. C* **2020**, *8*, 7113–7119. [CrossRef]
12. Li, D.; Xu, Y.; Fan, L.; Shen, B.; Ding, X.; Yuan, R.; Li, X.; Chen, W. Target-Driven Rolling Walker Based Electrochemical Biosensor for Ultrasensitive Detection of Circulating Tumor DNA Using Doxorubicin@tetrahedron-Au Tags. *Biosens. Bioelectron.* **2020**, *148*, 111826. [CrossRef] [PubMed]
13. Panikar, S.S.; Banu, N.; Escobar, E.R.; García, G.R.; Cervantes-Martínez, J.; Villegas, T.C.; Salas, P.; De la Rosa, E. Stealth Modified Bottom up SERS Substrates for Label-Free Therapeutic Drug Monitoring of Doxorubicin in Blood Serum. *Talanta* **2020**, *218*, 121138. [CrossRef] [PubMed]
14. Shah, S.; Chandra, A.; Kaur, A.; Sabnis, N.; Lacko, A.; Gryczynski, Z.; Fudala, R.; Gryczynski, I. Fluorescence Properties of Doxorubicin in PBS Buffer and PVA Films. *J. Photochem. Photobiol. B Biol.* **2017**, *170*, 65–69. [CrossRef] [PubMed]
15. Sturgeon, R.J.; Schulman, S.G. Electronic Absorption Spectra and Protolytic Equilibria of Doxorubicin: Direct Spectrophotometric Determination of Microconstants. *J. Pharm. Sci.* **1977**, *66*, 958–961. [CrossRef] [PubMed]
16. Changenet-Barret, P.; Gustavsson, T.; Markovitsi, D.; Manet, I.; Monti, S. Unravelling Molecular Mechanisms in the Fluorescence Spectra of Doxorubicin in Aqueous Solution by Femtosecond Fluorescence Spectroscopy. *Phys. Chem. Chem. Phys.* **2013**, *15*, 2937–2944. [CrossRef] [PubMed]
17. Lu, Z.; Chen, X.; Wang, Y.; Zheng, X.; Li, C.M. Aptamer Based Fluorescence Recovery Assay for Aflatoxin B1 Using a Quencher System Composed of Quantum Dots and Graphene Oxide. *Microchim. Acta* **2014**, *182*, 571–578. [CrossRef]
18. Khan, S.; Carneiro, L.S.A.; Vianna, M.S.; Romani, E.C.; Aucelio, R.Q. Determination of Histamine in Tuna Fish by Photoluminescence Sensing Using Thioglycolic Acid Modified CdTe Quantum Dots and Cationic Solid Phase Extraction. *J. Lumin.* **2017**, *182*, 71–78. [CrossRef]

19. Unser, S.; Bruzas, I.; He, J.; Sagle, L. Localized Surface Plasmon Resonance Biosensing: Current Challenges and Approaches. *Sensors* **2015**, *15*, 15684–15716. [CrossRef]

20. Wang, J.; Zhang, H.Z.; Li, R.S.; Huang, C.Z. Localized Surface Plasmon Resonance of Gold Nanorods and Assemblies in the View of Biomedical Analysis. *TrAC Trends Anal. Chem.* **2016**, *80*, 429–443. [CrossRef]

21. Yang, M.; Yan, Y.; Liu, E.; Hu, X.; Hao, H.; Fan, J. Polyethyleneimine-Functionalized Carbon Dots as a Fluorescent Probe for Doxorubicin Hydrochloride by an Inner Filter Effect. *Opt. Mater. (Amst).* **2021**, *112*, 110743. [CrossRef]

22. Zhu, J.; Chu, H.; Shen, J.; Wang, C.; Wei, Y. Green Preparation of Carbon Dots from Plum as a Ratiometric Fluorescent Probe for Detection of Doxorubicin. *Opt. Mater.* **2021**, *114*, 110941. [CrossRef]

23. Zhang, W.; Ma, R.; Gu, S.; Zhang, L.; Li, N.; Qiao, J. Nitrogen and Phosphorus Co-Doped Carbon Dots as an Effective Fluorescence Probe for the Detection of Doxorubicin and Cell Imaging. *Opt. Mater.* **2022**, *128*, 112323. [CrossRef]

24. Huang, K.Y.; He, H.X.; He, S.B.; Zhang, X.P.; Peng, H.P.; Lin, Z.; Deng, H.H.; Xia, X.H.; Chen, W. Gold Nanocluster-Based Fluorescence Turn-off Probe for Sensing of Doxorubicin by Photoinduced Electron Transfer. *Sens. Actuators B Chem.* **2019**, *296*, 126656. [CrossRef]

25. Miao, Y.; Zhang, Z.; Gong, Y.; Yan, G. Phosphorescent Quantum Dots/doxorubicin Nanohybrids Based on Photoinduced Electron Transfer for Detection of DNA. *Biosens. Bioelectron.* **2014**, *59*, 300–306. [CrossRef] [PubMed]

26. Gao, X.; Niu, L.; Su, X. Detection of DNA via the Fluorescence Quenching of Mn-Doped ZnSe D-dots/doxorubicin/DNA Ternary Complexes System. *J. Fluoresc.* **2012**, *22*, 103–109. [CrossRef] [PubMed]

27. Raichlin, S.; Sharon, E.; Freeman, R.; Tzfati, Y.; Willner, I. Electron-Transfer Quenching of Nucleic Acid-Functionalized CdSe/ZnS Quantum Dots by Doxorubicin: A Versatile System for the Optical Detection of DNA, Aptamer–substrate Complexes and Telomerase Activity. *Biosens. Bioelectron.* **2011**, *26*, 4681–4689. [CrossRef] [PubMed]

28. Goryacheva, O.A.; Guhrenz, C.; Schneider, K.; Beloglazova, N.V.; Beloglazova, N.V.; Goryacheva, I.Y.; De Saeger, S.; Gaponik, N. Silanized Luminescent Quantum Dots for the Simultaneous Multicolor Lateral Flow Immunoassay of Two Mycotoxins. *ACS Appl. Mater. Interfaces* **2020**, *12*, 24575–24584. [CrossRef]

29. Goryacheva, O.A.; Wegner, K.D.; Sobolev, A.M.; Häusler, I.; Gaponik, N.; Goryacheva, I.Y.; Resch-Genger, U. Influence of Particle Architecture on the Photoluminescence Properties of Silica-Coated CdSe Core/shell Quantum Dots. *Anal. Bioanal. Chem.* **2022**, *414*, 4427–4439. [CrossRef] [PubMed]

30. Drozd, D.D.; Byzova, N.A.; Pidenko, P.S.; Tsyupka, D.V.; Strokin, P.D.; Goryacheva, O.A.; Zherdev, A.V.; Goryacheva, I.Y.; Dzantiev, B.B. Luminescent Alloyed Quantum Dots for Turn-off Enzyme-Based Assay. *Anal. Bioanal. Chem.* **2022**, *414*, 4471–4480. [CrossRef]

31. Podkolodnaya, Y.A.; Kokorina, A.A.; Goryacheva, I.Y. A Facile Approach to the Hydrothermal Synthesis of Silica Nanoparticle/Carbon Nanostructure Luminescent Composites. *Materials* **2022**, *15*, 8469. [CrossRef]

32. Savla, R.; Taratula, O.; Garbuzenko, O.; Minko, T. Tumor Targeted Quantum Dot-Mucin 1 Aptamer-Doxorubicin Conjugate for Imaging and Treatment of Cancer. *J. Control. Release* **2011**, *153*, 16–22. [CrossRef] [PubMed]

33. Bagalkot, V.; Zhang, L.; Levy-Nissenbaum, E.; Jon, S.; Kantoff, P.W.; Langery, R.; Farokhzad, O.C. Quantum Dot-Aptamer Conjugates for Synchronous Cancer Imaging, Therapy, and Sensing of Drug Delivery Based on Bi-Fluorescence Resonance Energy Transfer. *Nano Lett.* **2007**, *7*, 3065–3070. [CrossRef]

34. Lipshultz, S.E.; Lipsitz, S.R.; Mone, S.M.; Goorin, A.M.; Sallan, S.E.; Sanders, S.P.; Orav, E.J.; Gelber, R.D.; Colan, S.D. Female Sex and Higher Drug Dose as Risk Factors for Late Cardiotoxic Effects of Doxorubicin Therapy for Childhood Cancer. *N. Engl. J. Med.* **1995**, *332*, 1738–1744. [CrossRef]

35. Liu, Y.; Ji, F.; Liu, R. The Interaction of Bovine Serum Albumin with Doxorubicin-Loaded Superparamagnetic Iron Oxide Nanoparticles: Spectroscope and Molecular Modelling Identification. *Nanotoxicology* **2013**, *7*, 97–104. [CrossRef]

36. Kumar Panigrahi, S.; Kumar Mishra, A. Inner Filter Effect in Fluorescence Spectroscopy: As a Problem and as a Solution. *J. Photochem. Photobiol. C Photochem. Rev.* **2019**, *41*, 100318. [CrossRef]

37. Vale, B.R.C.; Mourão, R.S.; Bettini, J.; Sousa, J.C.L.; Ferrari, J.L.; Reiss, P.; Aldakov, D.; Schiavon, M.A. Ligand Induced Switching of the Band Alignment in Aqueous Synthesized CdTe/CdS Core/shell Nanocrystals. *Sci. Rep.* **2019**, *9*, 8332. [CrossRef]

38. Daramola, O.A.; Siwe Noundou, X.; Nkanga, C.I.; Tseki, P.F.; Krause, R.W.M. Synthesis of pH Sensitive Dual Capped CdTe QDs: Their Optical Properties and Structural Morphology. *J. Fluoresc.* **2020**, *30*, 557–564. [CrossRef] [PubMed]

39. Gołuński, G.; Borowik, A.; Derewońko, N.; Kawiak, A.; Rychłowski, M.; Woziwodzka, A.; Piosik, J. Pentoxifylline as a Modulator of Anticancer Drug Doxorubicin. Part II: Reduction of Doxorubicin DNA Binding and Alleviation of Its Biological Effects. *Biochimie* **2016**, *123*, 95–102. [CrossRef]

40. Hines, A.; Flin, H. Doxorubicin and Ifosfamide Sarcoma PROTOCOL REF: MPHADOXIFO (Version No: 1.0). 2016. Available online: https://www.clatterbridgecc.nhs.uk/application/files/9814/9787/1938/Doxorubicin_and_Ifosfamide_Sarcoma_Protocol_V1.0.pdf (accessed on 1 April 2023).

Article

Developing a Fluorescent Inducible System for Free Fucose Quantification in *Escherichia coli*

Samantha Nuñez, Maria Barra and Daniel Garrido *

Department of Chemical and Bioprocess Engineering, School of Engineering, Pontificia Universidad Católica de Chile, Vicuña Mackenna, 4860, Santiago 8331150, Chile
* Correspondence: dgarridoc@ing.puc.cl

Abstract: L-Fucose is a monosaccharide abundant in mammalian glycoconjugates. In humans, fucose can be found in human milk oligosaccharides (HMOs), mucins, and glycoproteins in the intestinal epithelium. The bacterial consumption of fucose and fucosylated HMOs is critical in the gut microbiome assembly of infants, dominated by *Bifidobacterium*. Fucose metabolism is important for the production of short-chain fatty acids and is involved in cross-feeding microbial interactions. Methods for assessing fucose concentrations in complex media are lacking. Here we designed and developed a molecular quantification method of free fucose using fluorescent *Escherichia coli*. For this, low- and high-copy plasmids were evaluated with and without the transcription factor *fucR* and its respective fucose-inducible promoter controlling the reporter gene sfGFP. *E. coli* BL21 transformed with a high copy plasmid containing *pFuc* and *fucR* displayed a high resolution across increasing fucose concentrations and high fluorescence/OD values after 18 h. The molecular circuit was specific against other monosaccharides and showed a linear response in the 0–45 mM range. Adjusting data to the Hill equation suggested non-cooperative, simple regulation of FucR to its promoter. Finally, the biosensor was tested on different concentrations of free fucose and the supernatant of *Bifidobacterium bifidum* JCM 1254 supplemented with 2-fucosyl lactose, indicating the applicability of the method in detecting free fucose. In conclusion, a bacterial biosensor of fucose was validated with good sensitivity and precision. A biological method for quantifying fucose could be useful for nutraceutical and microbiological applications, as well as molecular diagnostics.

Keywords: biosensor; *Escherichia coli*; fucose; GFP; gene regulation

Citation: Nuñez, S.; Barra, M.; Garrido, D. Developing a Fluorescent Inducible System for Free Fucose Quantification in *Escherichia coli*. *Biosensors* **2023**, *13*, 388. https://doi.org/10.3390/bios13030388

Received: 12 February 2023
Revised: 10 March 2023
Accepted: 13 March 2023
Published: 15 March 2023

1. Introduction

L-Fucose is an important monosaccharide that exerts functional roles in multiple biological processes [1]. L-Fucose is a deoxy hexose sugar characterized by missing a hydroxyl group at C-6 [1]. It is commonly present in mammalian mucins, human milk oligosaccharides (HMOs), and glycoconjugates of the intestinal epithelium [2,3], such as glycolipids, N-glycans, and O-glycans. In these glycoconjugates, fucose is usually found at terminal positions in α- linkages (such as α1-2, α1-3, α1-4, and α1-6; [4]). The main enzyme responsible for these fucosylations is α-1,2-fucosyltransferase (Fut2), which is expressed in epithelial cells and links fucose to the terminal β-D-galactose of mucosal glycans [5].

Fucose is abundant in the gastrointestinal tract (GT) and influences the complex microbial ecosystem that inhabits there. Several microorganisms are equipped with α-fucosidases targeting all existing fucose linkages [6–8]. Therefore, gut microbes can release fucose from dietary glycans, which is used as a microbial carbon source [9]. In addition, fucose may promote the growth of beneficial bacteria in the gut, such as *Bifidobacterium* and *Bacteroides* [10]. Fucose is finally a common exchange molecule involved in multiple microbial cross-feeding interactions [11–13].

In addition to serving as an energy source for some microbes, fucose is involved in diverse metabolic pathways, including the regulation of quorum sensing and suppression of virulence genes in pathogens [14,15]. Detecting low levels of free fucose in biological samples could be a valuable indicator of infection or inflammation [16]. Some pathogens can use fucose as a signaling molecule regulating pathogenesis [17]. Fucose and fucose-containing oligosaccharides usually act as a decoy, preventing the binding of some viral and bacterial pathogens [18]. High levels of fucose in urine have been associated with cirrhosis and certain types of cancer [19,20]. Finally, loss of function mutations of fucosyl-transferase Fut2 have been associated with Crohn's disease [21].

Free fucose is usually quantified by HPLC [22–25] and enzymatic assays with L-fucose dehydrogenase [19,26]. Recently, lectin-based microfluidic detection assays have been developed [27,28], as well as fluorescence-based assays with probes and electrochemical sensors [20,29]. Shin et al. [30] developed a molecular biosensor for quantifying 2-fucosyllactose (2FL) in breast milk samples. Their circuit contained a constitutive α-fucosidase expressed in an *E. coli* strain mutant for lactose consumption. Therefore, the detection of 2FL and emission of fluorescence were coupled to cell growth and 2FL degradation [30].

Bacterial biosensors are genetically modified organisms that detect an input, usually a substance or the changes in the concentrations of a specific molecule, which are sensed and internally translated into a genetic output that emits a quantifiable signal [31,32]. Bacterial biosensors are usually constructed of transcription factors and their corresponding promoters. The most common outputs are fluorescent proteins such as Green Fluorescent Protein (GFP). An excellent candidate to develop a bacterial biosensor is *Escherichia coli*, a model widely used in biotechnological research and development since its genome and metabolic pathways are fully known [30,32,33]

Escherichia coli K12 can use multiple sugars as a carbon source for its growth, including fucose [34,35]. Fucose can induce the expression of genes allowing its transport and metabolism, a genetic system known as the fucose regulon. It consists of six genes: L-fucose permease (*fucP*), L-fucose isomerase (*fucI*), L-fuculose kinase (*fucK*), L-fuculose phosphate aldolase (*fucA*), L-1,2-propanediol oxidoreductase (*fucO*) and the transcription regulator of the regulon (*fucR*) [36,37]. These genes are clustered into three operons, fucPIK, fucA (which is transcribed in a clockwise direction), and fucO (which is transcribed counterclockwise). FucR is an activator [37], which is induced by fuculose 1-phosphate, an intermediate molecule from fucose metabolism. FucR also shows positive autoregulation [38].

In this study, we used the *fuc* molecular system for developing a method of quantifying free L-fucose, using FucR and the fuc promoter triggering the induction of sfGFP in *E. coli*. We first compared the detection of fucose in high- and low-copy plasmids with or without *fucR*. The best system was evaluated for specificity, and calibration curves were obtained at low (0–3 mM) and high concentrations (0–50 mM) with good resolution. The biosensor was successfully applied to quantify fucose in bacterial supernatants. This molecular biosensor could be further studied to quantify free fucose in complex biological samples with good resolution and specificity.

2. Materials and Methods

Mediums, reagents, and sugars. Miller Luria-Bertani (LB) liquid and agar medium was obtained from Merck (Boston, MA, USA) and autoclaved at 121 °C for 15 min. Minimum medium MM9 was prepared with KH_2PO_4 (15% *w/v*), NaCl (2.5% *w/v*), Na_2HPO_4 (33.9% *w/v*), and NH_4Cl (5% *w/v*). These reagents were obtained from Sigma Aldrich (St. Louis, MO, USA). The liquid medium ZMB was prepared according to Medina et al. [39]. Solid media contained 1.5% *w/v* agar. Carbohydrates used were L-fucose, 2-O-fucosyllactose, 3-O-fucosyllactose, and sialic acid (Neu5ac), which were kindly donated by Glycom (Hørshol, Denmark). Mannose, glucose, galactose, and lactose were obtained from Sigma Aldrich (St. Louis, MO, USA). Carbohydrate solutions were prepared with Milli-Q water and then filtered with Millex-gv filters (0.22 μm).

Plasmid construction. The in-silico plasmid construction was carried out in the SnapGene program, obtaining all the constructs and the primers (Table 1). Fucose biosensors were created in a high copy plasmid backbone (pTAC_sfGFP ColE1) and a low copy plasmid backbone (pTAC_sfGFP SC101). The high copy plasmid contains an ampicillin resistance gene as a selection marker and superfolder GFP (sfGFP) as a reporter molecule [40], while the low copy plasmid contains a chloramphenicol resistance and sfGFP controlled by *pTac*. These plasmids were a kind donation from Dr. Tal Danino (Columbia University). The fucose-induced promoter (*pFuc*) was synthesized as a gBlock from Integrated DNA Technologies, Inc. (IDT), including the PstI and EcoRI restriction sites at the 5′ and 3′ ends, respectively. The sequence was obtained from the *E. coli* K12 MG1655 genome, specifically from the fucose *fucPIK* operon [38]. Both DNA fragments were digested with the PstI-HF and EcoRI-HF restriction enzymes for 1 h at 37 °C, gel purified, and ligated with T4 DNA ligase at room temperature for 1 h (New England Biolabs, Inc., Ipswich, MA, USA). The resulting plasmids were named pFUC_sfGFP_colE1 and pFUC_sfGFP_SC101.

Table 1. Plasmids used in this study. These plasmids were used to transform bacteria that did not metabolize fucose as a carbon source. BL21 and DH5α colonies were obtained for these plasmids.

	Plasmids	Antibiotic	Description
1	Pfuc *sfGFP*_ColE1	Ampicillin	High-copy plasmid colE1 with fucose promoter and sfGFP with an ampicillin resistance gene.
2	Pfuc *sfGFP*_SC101	Chloramphenicol	Low-copy plasmid SC101 with fucose promoter and sfGFP with a chloramphenicol resistance gene.
3	Pfuc+FucR1_colE1	Ampicillin	High-copy plasmid colE1 with transcription factor FucR, fucose promoter and sfGFP with an ampicillin resistance gene

Cloning of FucR. Later, the transcription factor gene (*fucR*) was obtained from the genome of *E. coli* K12 MG1655 by PCR with the primers 5′-tctcatACCGGTacgcccgcc-3′ and 5′-ctatCCCGGGtcaggctgttaccaaagaag-3′. These primers contain restriction sites for the enzymes *AgeI* and *XmaI*. PCR reactions were performed with Q5 high-fidelity polymerase (New England BioLabs, Ipswich, MA, USA) using manufacturer instructions. Exceptions were an annealing temperature of 70 °C and an extension time of 20 s, using 0.5 µM of the primers and 1 U of polymerase Q5. PCR products were recovered from a 1% agarose gel with the Zymoclean Gel DNA Recovery Zymo research kit (Irvine, CA, USA). The pFUC_sfGFP_colE1 plasmid containing the fuc promoter was digested with the same enzymes and amplified with the primers 5′-TGAcgctagaactagtggatcc-3′ and 5′-tcagACCGGtagaccgagatagggttgag-3. PCR products were recovered (FucR and the fucose-induced promoter in the high-copy plasmid (Table 1)). Digestions were carried out for 16 h at 37 °C with *AgeI-HF* and *XmaI* enzymes following manufacturer instructions (New England Biolabs; Ipswich, MA, USA). Digested plasmids and fragments were ligated with a T4 DNA ligase (New England Biolabs; Ipswich, MA, USA) at room temperature for 16 h.

Bacterial transformations. All plasmids were transformed into chemically competent *E. coli* strains. Plasmids were stored in DH5α, and biosensors were produced in the BL21 strain. Two microliters of ligation mixture or Gibson Assembly Master Mix were added to 50 µL of cells and incubated on ice for 30 min. Heat shock was performed at 42 °C for 50 s, followed by 2 min on ice. One ml of SOC media was added, and the bacteria were incubated at 37 °C with shaking at 200 rpm for 1 h. The transformation volume was plated onto LB agar plates with the corresponding antibiotic. Single colonies were picked and cultured in LB media with the antibiotic for stock preparation and miniprep. Carbenicillin was used at 100 µg/mL, and chloramphenicol was prepared in ethanol at 25 µg/mL. Correct insertion of genes of interest was verified through plasmid sequencing at Macrogen Inc (Seoul, Republic of Korea).

Fluorescence kinetics. Four colonies were selected per plate and cultured in 2 mL of LB-antibiotic broth with 200-rpm agitation for 16 h at 37 °C. Filtered fucose (100 mM) was used to prepare 200 µL triplicate reactions with decreasing monosaccharide concentrations, inoculated at 1% *w/v* with fresh LB-antibiotic medium. All kinetics were performed on black with transparent bottom Nunc™ F96 MicroWell™ 96-well polystyrene plates (Thermo Scientific, Waltham, MA, USA)), in a Synergy H1 Biotek multi-plate reader (Winooski, VT, USA). Growth curves were monitored for 24 h with agitation, measuring OD_{600} and fluorescence every 30 min with excitation at 485 nm and emission at 510 nm. The Gen5 3.09 software was used for absorbance and fluorescence measurements and data analysis.

B. bifidum culture and fucose quantification. *B. bifidum* JCM 1254 was inoculated in de Mann Rogose Sharp (MRS) broth supplemented with cysteine 0.05% for 48 h in an anaerobic jar at 37 °C with an anaerobic GasPak EZ patch (Becton Dickson, Franklin Lakes, NJ, USA). Cells were centrifuged at $12{,}000 \times g$ for 1 min after 48 h and resuspended in reduced mZMB broth [39] with no carbon source. *B. bifidum* was then cultured at 4% *w/v* in 5 mL of mZMB supplemented with 2FL (81 mM) or with 3FL (20.4 mM) for 40 h under anaerobic conditions as above. Supernatants were recovered at 0, 12, 16, 20, 24, and 40 h. All supernatants were filtered with Millex-gv 0.22 µm filters (Sigma Aldrich, St. Louis, MO, USA), and pH was adjusted to 7 using NaOH 1 M. Supernatants were later analyzed using thin layer chromatography (TLC) in parallel to biosensor detection. Standards of fucose, 2FL, 3FL, lactose, galactose (all at 1% *w/v*), and *B. bifidum* supernatants JCM1254 were used. TLC DC-Fertigfolien ALUGRAM Xtra Silica Gel 0.20 mm plates were used (Macherey-Nagel, Allentown, PA, USA), with 1 µL of each sample. A run solution was prepared with 50% *v/v* of n-butanol and 25% *v/v* of acetic acid in distilled water. Two ascents were performed to improve resolution. A staining solution was prepared with 0.5% *w/v* naphthol and 5% *v/v* sulfuric acid in ethanol.

Statistical analysis. All curves represent the average of triplicates, and the standard deviation is shown. Statistical analyses, including linear regressions, were performed in GraphPad Prism 9. To determine the sensitivity of the regulation and potential cooperativity, the Hill equation for an activator was fitted to fluorescence/OD values [41]. Hill equation parameters were minimized to experimental data using Solver in Excel. β is the maximum expression rate, K represents the dissociation constant, [S] is the substrate concentration and n is the Hill cooperativity coefficient [42].

$$\frac{d\,(F/OD)}{d\,t} = \frac{\beta K^n}{K^n + [s]^n} \tag{1}$$

3. Results

3.1. Biosensor Properties and Functions

The three plasmids constructed in this study are shown in Table 1 and depicted in Supplementary Figure S1. The reporter gene sfGFP is controlled by *pFuc* and is a marker of selection that gives resistance to antibiotics. Low- and high-copy plasmids were evaluated. The transcription factor gene *fucR* was also cloned upstream of *pFuc*, with constitutive expression. The plasmids were used for chemical transformations of the BL21 and DH5α *E. coli* strains and tested with increasing fucose concentrations.

E. coli BL21 transformed only with *pFuc* displayed good behavior with increased F/OD ratio to higher fucose concentrations (Figure 1A). This molecular system did not fluoresce with 0 mM fucose, suggesting a tight control (Figure 1A). Interestingly, the cloning of *fucR* and *pFuc* into the high copy plasmid increased fucose detection values by nearly 50%, with good resolution and increasing F/OD ratios in response to higher fucose concentrations in BL21 (Figure 1C). The low-copy plasmid biosensor (SC101) emitted smaller fluorescence values than high-copy plasmids (Figure 1B; $p < 0.0001$ at 50 mM fucose). The three plasmids transformed in *E. coli* DH5α generated F/OD curves that did not correlate well with fucose concentrations (Figure 1D–F).

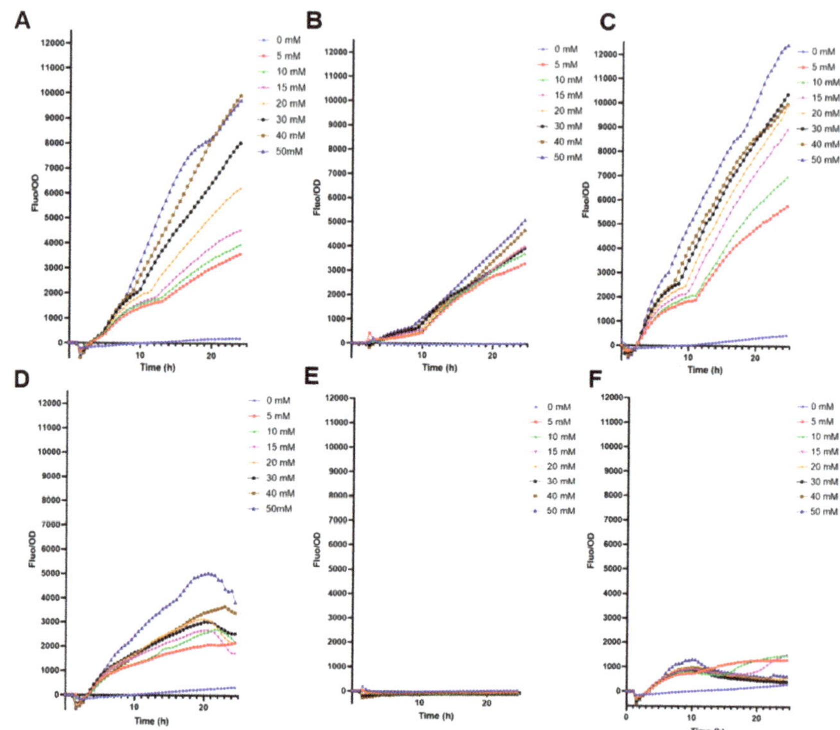

Figure 1. Fluorescence kinetics of different *E. coli* biosensors in response to increasing fucose concentrations. (**A**) *E. coli* BL21 containing pFuc_sfGFP_colE1; (**B**) *E. coli* BL21 containing pFuc_sfGFP_SC101; (**C**) *E. coli* BL21 containing pFuc_FucR1_colE1; (**D**) *E. coli* DH5α containing pFuc_sfGFP_colE1; (**E**) *E. coli* DH5α containing pFuc_sfGFP_SC101; (**F**) *E. coli* DH5α containing pFuc_FucR1_colE1. Kinetic curves were performed in triplicates.

3.2. Comparison of Biosensor Specificities

High-copy biosensors with *pFuc* and pFuc+FucR in BL21 were preliminary evaluated for non-specific cross-detection of other monosaccharides. The system with only *pFuc* showed a crossed response with galactose, irrespective of its concentration (Figure 2A). This result is in part explained by the vigorous growth on galactose (Figure 2E). Low concentrations of glucose and mannose (5 and 10 mM), but not higher, also triggered GFP production (Figure 2B,D). Sialic acid appeared not to induce GFP expression (Figure 2C). These results indicate that the sole inclusion of the *pFuc* promoter is insufficient to provide a specific response to fucose.

Interestingly, the inclusion of the transcription factor increased biosensor specificity (Figure 3). No positive F/OD values were obtained in the presence of galactose, glucose, sialic acid, or mannose (Figure 3). Specificity to these carbohydrates was highlighted by the good growth the biosensor showed in these sugars, with no fluorescence emitted (Figure 3). Finally, rhamnose is another 6-deoxyhexose sugar that could interfere with fucose sensing. A small crossed response was observed for rhamnose, indicating the molecular system needs further improvements in its specificity (Supplementary Figure S2).

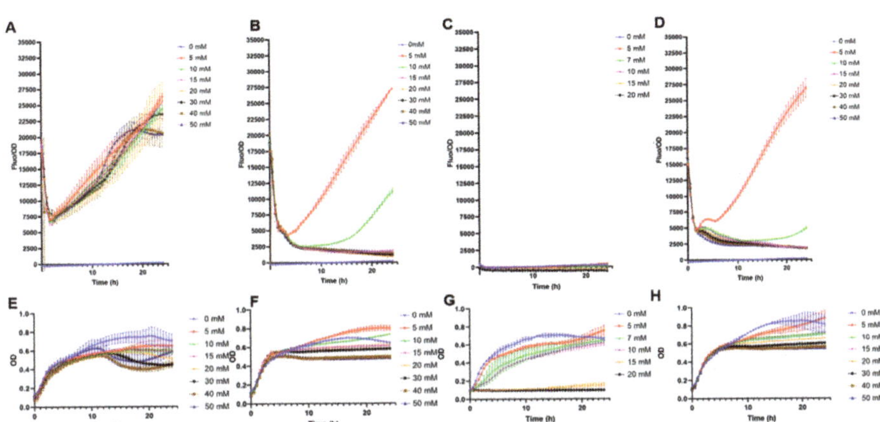

Figure 2. Specificity tests of *E. coli* BL21 containing pFuc_sfGFP_colE1. (**A**) F/OD rations in the presence of increasing concentrations of galactose; (**B**) glucose; (**C**) sialic acid; (**D**) mannose; (**E**) growth curves (OD values) of this strain in the presence of increasing concentrations of galactose; (**F**) glucose; (**G**) sialic acid; (**H**) mannose. Kinetics and growth curves were performed in triplicates and are presented as average ± SD.

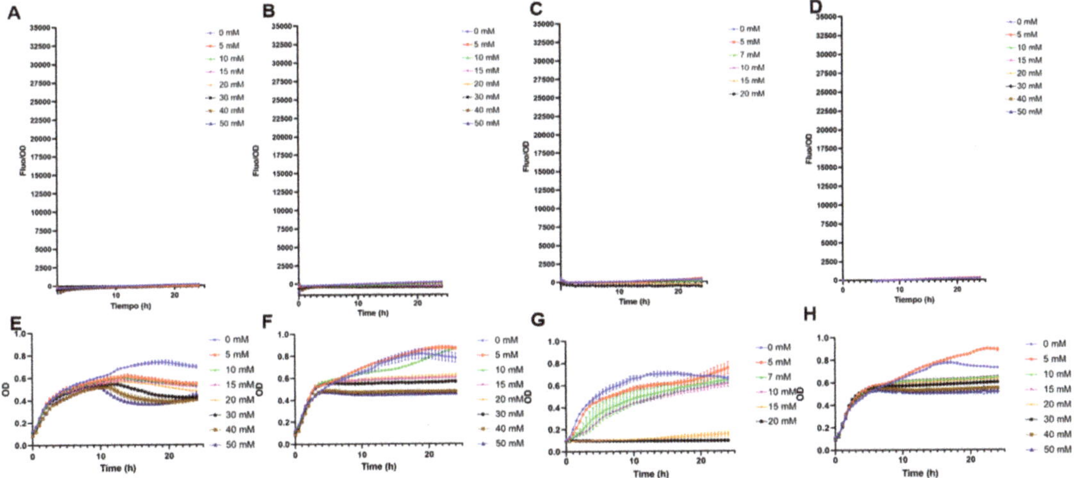

Figure 3. Specificity tests of *E. coli* BL21 containing pFuc_FucR_colE1, using sfGFP as a reporter. (**A**) F/OD values in the presence of increasing concentrations of galactose; (**B**) glucose; (**C**) sialic acid; (**D**) mannose; (**E**) growth curves (OD values) of this strain in the presence of increasing concentrations of galactose; (**F**) glucose; (**G**) sialic acid; (**H**) mannose. Kinetics and growth curves were performed in triplicates and are presented as average ± SD.

3.3. Calibration Curves

The biosensor *E. coli* BL21 pFuc + FucR high copy (colE1) was evaluated in a range of 0 to 3 mM of fucose to assess its performance under low concentrations (Figure 4). Even at 0.4 mM fucose, the system generated a measurable output (Figure 4A). It can be observed that from 15 h and after, fucose concentrations were well differentiated, with a positive linear correlation between fucose amounts and F/OD values (Figure 4B). Applying a linear regression to these parameters (Supplementary Table S1), the best correlation (higher R squared value) was obtained at 15.5 h (Figure 4B, Supplementary Table S1).

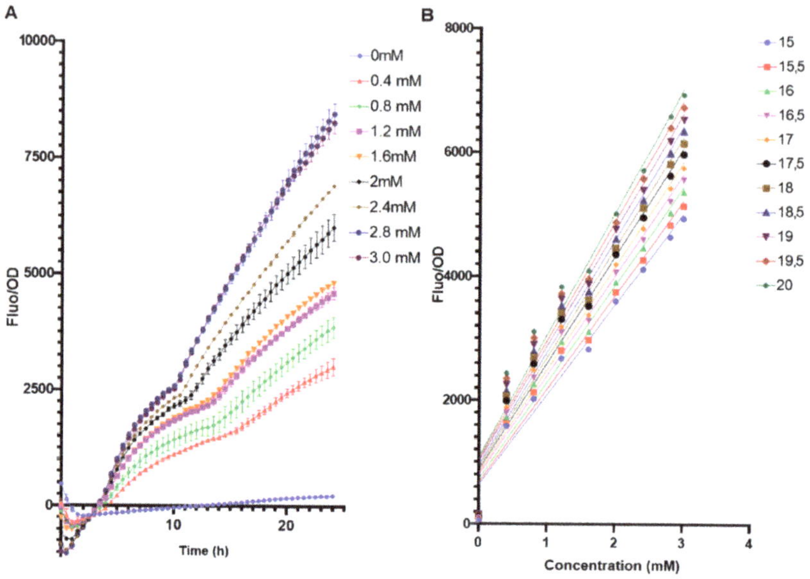

Figure 4. (**A**) Standardized F/OD values of *E. coli* BL21 pFuc + FucR at different concentrations of fucose for 24 h, in the range of 0–3 mM fucose. All values are presented as average ±SD; n = 3. (**B**) Calibration curves of F/OD vs. fucose concentration for the *E. coli* BL21 pFuc+ FucR biosensor from 0 mM to 3 mM for 20 h. Numbers on the right indicate time points (h).

3.4. Sensitivity of the E. coli BL21 pFuc+FucR colE1 Biosensor

The behavior of the biosensor was later evaluated in a broader range to be used for measurements of free fucose, from 0 mM to 45 mM (Figure 5A). As expected, increasing F/OD values were obtained. These data were used to determine the regulatory parameters of the Hill equation ([42]; see methods). F/OD values at 15.5 h were used to fit experimental data to the equation. Modeling results indicate a Hill coefficient value n of 1, which suggests that FucR regulates its promoter via simple non-cooperative regulation. This suggests that FucR binds only one fucose molecule upon binding its DNA. K is a dissociation constant, and a small value was obtained (5 mM). K indicates the affinity of FucR for its promoter, representing the concentration of fucose required to activate 50% of the maximal response.

3.5. Measuring Fucose in the Supernatant of B. bifidum JCM1254

Finally, the biosensor was used to measure fucose concentrations from a bacterial supernatant (Figure 6). *B. bifidum* can ferment HMOs, especially 2FL and 3FL, as carbon sources. This microorganism displays extracellular α1-2 and α1-3 fucosidase activities, releasing free fucose in the medium and allowing the bacterium to use lactose [29]. The bacterium was cultured anaerobically for 40 h in a medium supplemented with either 2FL or 3FL. Samples were taken regularly (Figure 6) and incubated with the *E. coli* biosensor. OD and fluorescence measurements were taken for 24 h at an interval of 30 min.

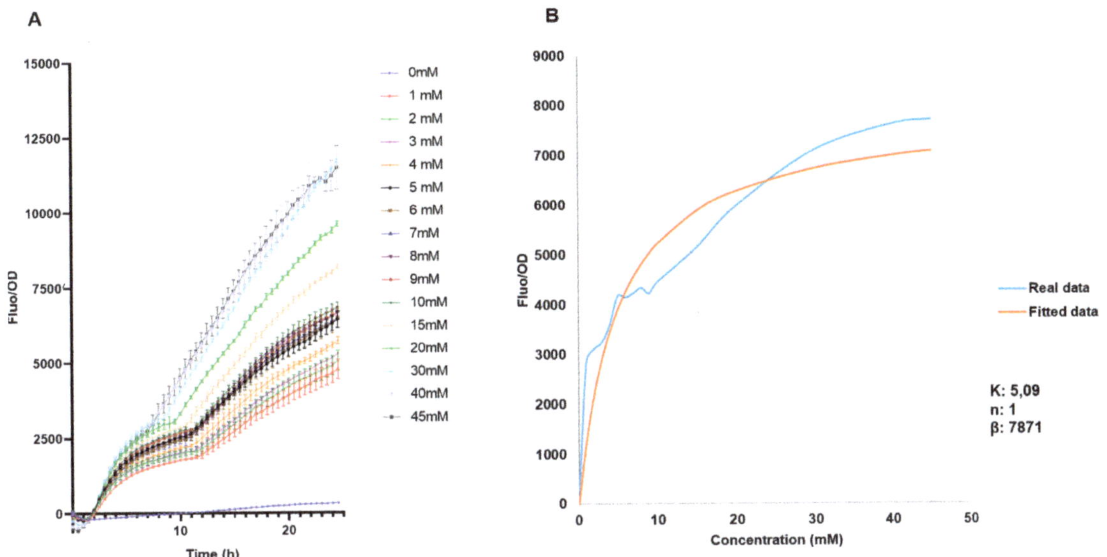

Figure 5. (A) Standardized F/OD values of *E. coli* BL21 pFuc + FucR at different concentrations of fucose for 24 h, in the range of 0–45 mM fucose. **(B)** Biosensor response and fitted response to Hill equation parameters for an activator. Concentrations of K are mM, n has no units, and β has units of h^{-1}.

Figure 6A shows normalized F/OD values of the supernatants obtained from *B. bifidum* growing on 2FL. Supernatants from time points at 12–24 h generated low but increasing F/OD values in time (Figure 6A). A strong fucose signal was detected at 40 h. These results correlated well with a visual assessment of carbohydrates in TLC (Figure 6E), where a strong band with the same migration as fucose was observed. Finally, no major differences in growth were observed for the biosensors using the supernatants from multiple time points, suggesting they were not inhibitory.

In the case of *B. bifidum* supernatants with 3FL, a similar result was observed compared to 2FL (Figure 6B). The released fucose concentration at 40 h was lower than in 2FL (Figure 6B), and the supernatant sample at 24 h also showed a significant fluorescent output. Similar to 2FL supernatants, only the 40 h sample showed a strong fucose band in the TLC, which correlated with fluorescence data.

At 15.5 h of incubation, the biosensor incubated with the 2FL supernatant at 40 h presented a normalized F/OD value average of 8206.12, while for 3FL at 40 h was 3717.66 after 15.5 h. These values were used in a calibration curve obtained from the linear regression analysis (Figure 5A). The extrapolation of fucose concentrations in these supernatants was 42.4 mM for 2FL and 6.47 mM for 3FL. These values correlated well with TLC band intensity and appeared in the correct range compared with fucose standards of 1 and 10 mM (Figure 6E).

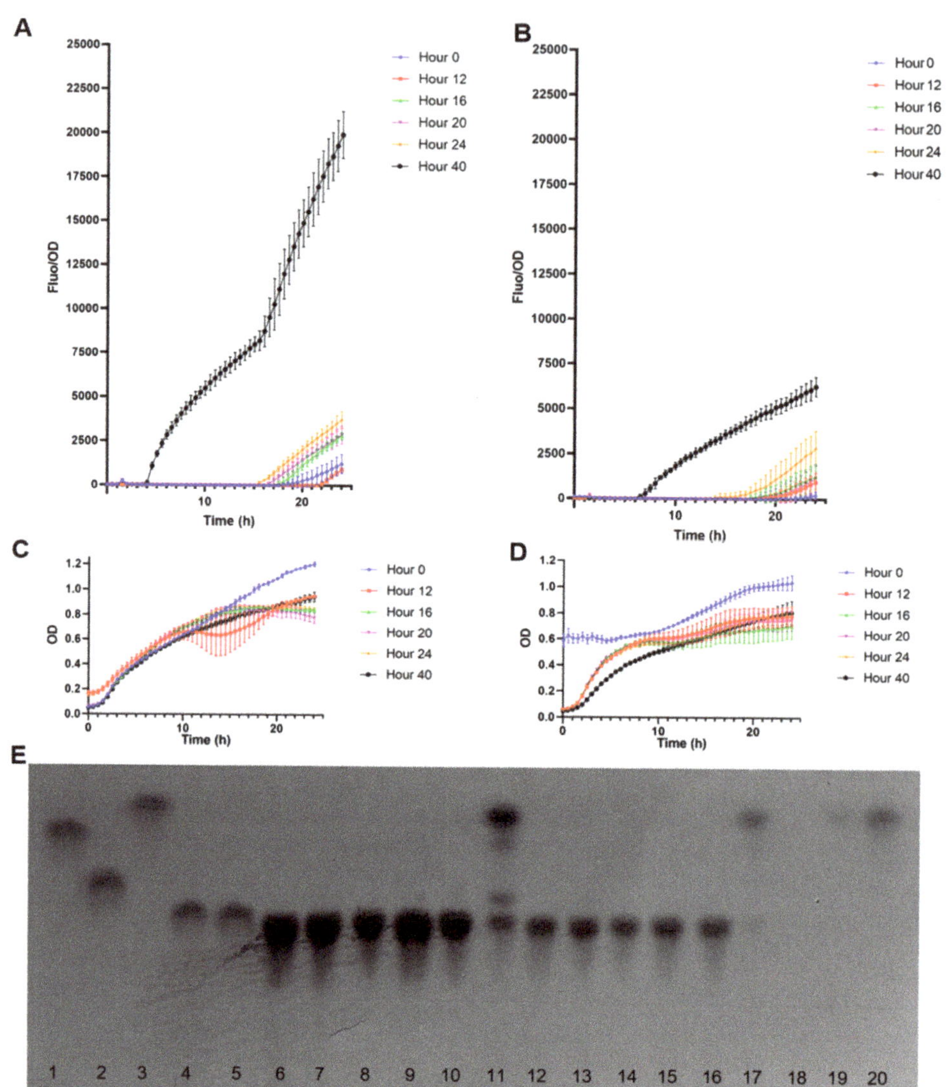

Figure 6. Quantification of fucose in *B. bifidum* supernatants using the biosensor *E. coli* BL21 pFuc + FucR colE1. (**A**) Normalized F/OD values for supernatants after growth in 2FL; (**B**) Normalized F/OD values for supernatants after growth in 3FL; (**C**) growth curves of the biosensors in the presence of supernatants *from B. bifidum* grown in 2FL; (**D**) growth curves of the biosensors in the presence of supernatants *from B. bifidum* grown in 3FL; (**E**) TLC analysis of *B. bifidum* supernatants. Standards used were 1: galactose, 2: lactose, 3: fucose, 4: 2FL, 5: 3FL, at 1 mg/mL. 6–11: supernatants from 2FL growth at 0 (6), 12 (7), 16 (8), 20 (9), 24 (10), and 48 h (11). 12–17: supernatants from 3FL growth at 0 (12), 12 (13), 16 (14), 20 (15), 24 (16), and 48 h (17). Lane 19 corresponds to 1 mM fucose, and lane 20 to 10 mM fucose. Kinetics and growth values shown correspond to average ± SD; n = 3.

4. Discussion

In this study, we constructed a biosensor for quantifying fucose in biological samples, using a molecular promoter and transcription factor naturally occurring in *E. coli* and using sfGFP as output. *E. coli* is well characterized by its L-fucose utilization mechanism [36]. A permease allows fucose entrance, and a feeder pathway allows L-fucose conversion into lactate and 1,2-propanediol, generating NADH and FADH [35]. An intermediate in this pathway, fuculose-1-phosphate, is the ligand recognized by the system regulator, FucR [37]. Therefore, our biosensor is expected to sense fuculose-1-phosphate and not directly fucose. The system requires that any external fucose sensed be first metabolized to generate an output.

Fucose is not among the most preferred carbon sources for *E. coli*, compared to glucose, galactose, or arabinose [43,44]. It shows a slow growth in this substrate in minimal media [35]. Catabolic repression exerted by CRP on the fucose promoter is also complemented with small RNA regulation via Spot42 [45]. The biosensor developed here is based on the activation role of FucR, which binds its promoter in the presence of fuculose-1-phosphate and allows the expression of sfGFP. We were able to determine in this study that FucR displays simple regulation, showing no cooperativity and suggesting it acts as a monomer. It is known that rhamnose appears to induce the operon [46], and fucose can also activate the galactose *galETK* system in *E. coli* [43]. These findings indicate that crossed regulatory responses of fucose and FucR are common in *E. coli* and might alter biosensor specificity.

Results in this study showed a high increase in specificity attributed to the presence of FucR. *pFuc* alone showed little specificity, indicating that other molecules can still induce leaky expression. The cloning of additional copies of FucR dramatically reduced crossed regulatory responses, probably increasing the threshold of fucose activation and resulting in a much tighter response. Finally, a much better resolution for strain BL21 compared to DH5α could be explained by the mutation in the *lon* protease in BL21, which allows a smaller reporter protein degradation and increased half-life [47].

The biosensor characterized showed a good linear response in the low concentration range (0–3 mM) or higher (0–45 mM). Some applications of the biosensor are as a diagnostic tool. Fucose is metabolized in the liver, and excess fucose is secreted in the urine [2,48]. A rare genetic disorder is fucosidosis, where fucose found in glycoconjugates cannot be removed and accumulates in the body resulting in severe consequences [49]. Therefore, quantifying fucose in urine could be of interest, especially since there is an increase in fucose concentrations in certain liver diseases [19,50].

Another field of application of biosensors is in GIT and gut microbiome research [21]. The rapid quantification of HMOs in breast milk samples is desirable, especially fucosylated molecules. Similarly, there is great interest in the enzymatic biosynthesis of these molecules, which requires quantifying fucose [51]. Finally, several gut microbes display α-fucosidase activities and use fucose as a carbon and energy source [6,9,52]. *Bifidobacterium* and *Bacteroides* species are well known for their extracellular activities, which release fucose from HMO, mucins, other glycoproteins, or glycolipids [3,10]. Therefore, free fucose can be expected to be detected in GIT contents in mammals. Free fucose is also known to participate in cross-feeding interactions, where the fucose released by one microorganism is imported and used by another. This has been observed during the consumption of mucin glycans and HMO, for example, between *B. bifidum* and *Bifidobacterium breve* [12,13,53]. Therefore, an accurate and inexpensive method for quantifying fucose could show how this monosaccharide is shared between species. In this study, the developed biosensor displayed a good performance in quantifying fucose derived from 2FL utilization by *B. bifidum* and could be used in studying cross-feeding interactions.

5. Conclusions

A fluorescent quantification method of fucose was developed in this study in *E. coli* with a high copy plasmid containing a reporter sfGFP, a fucose promoter, and FucR. The biosensor showed good sensitivity and specificity, showing a linear response to increasing fucose concentrations from 0 to 45 mM, a range within physiological concentrations. A validation to quantify fucose in a bacterial supernatant during HMO utilization was achieved. This method could be coupled to other enzymes (fucosidases, endoglycosidases, peptidases) to determine the concentration of fucosylated glycoconjugates. This assay could be an important tool for clinical research (serum, urine), foods, bioprocessing in the production of fucosylated glycoproteins, or detecting fucose in bacterial supernatants and quantifying fucose in fecal samples or bioreactors.

Supplementary Materials: The following supporting information can be downloaded at: https://www.mdpi.com/article/10.3390/bios13030388/s1, Table S1. Linear regression of the calibration curves obtained for 0 mM to 3 mM with a resolution of 0.4 mM at different incubation times; Figure S1. Representation of the plasmids used in this study. A: pFUC_sfGFP_colE1 is a high copy plasmid containing the fucose promoter controlling GFP expression and an ampicillin resistance gene; B: pFUC_sfGFP_SC101 is a low copy plasmid containing the fucose promoter controlling GFP expression and a chloramphenicol resistance gene: C: pFUC+FucR1_colE1 is a high copy plasmid containing the fucose promoter controlling GFP expression, in addition to a cloned FucR encoding gene and an ampicillin resistance gene. Internal arrows correspond to transcription units; Figure S2. Specificity tests of E. coli BL21 containing pFuc_FucR_colE1, using sfGFP as a reporter, for rhamnose. A: F/OD values in the presence of increasing concentrations of rhamnose; B: growth curves (OD values) of this strain in the presence of increasing concentrations of rhamnose. Kinetics and growth curves were performed in triplicates and are presented as average ± SD.

Author Contributions: Conceptualization: S.N., M.B. and D.G.; methodology: S.N. and M.B.; investigation: S.N. and M.B.; formal analysis: S.N.; writing—original draft preparation: D.G.; visualization: S.N.; writing—review and editing: D.G.; supervision: D.G.; resources: D.G.; funding acquisition: D.G. All authors have read and agreed to the published version of the manuscript.

Funding: This research was funded by ANID Fondecyt Regular 1190074, ANID Fondequip EQM190070, and School of Engineering Seed Funds 2018, 2020, and 2022.

Institutional Review Board Statement: Not applicable.

Informed Consent Statement: Not applicable.

Data Availability Statement: The data presented in this study are available upon request.

Acknowledgments: The authors thank Tal Danino for providing the initial plasmids used in this work.

Conflicts of Interest: The authors declare no conflict of interest.

References

1. Schneider, M.; Al-Shareffi, E.; Haltiwanger, R.S. Biological functions of fucose in mammals. *Glycobiology* **2017**, *27*, 601–618. [CrossRef] [PubMed]
2. Becker, D.J.; Lowe, J.B. Fucose: Biosynthesis and biological function in mammals. *Glycobiology* **2003**, *13*, 41R–53R. [CrossRef]
3. González-Morelo, K.J.; Vega-Sagardía, M.; Garrido, D. Molecular Insights Into O-Linked Glycan Utilization by Gut Microbes. *Front. Microbiol.* **2020**, *11*, 591568. [CrossRef]
4. Varki, A.; Lowe, J.B. Biological roles of glycans. In *Essentials of Glycobiology*; Varki, A., Cummings, R., Esko, J., Eds.; Cold Spring Harbor Laboratory Press: Cold Spring Harbor, NY, USA, 2009.
5. Goto, Y.; Uematsu, S.; Kiyono, H. Epithelial glycosylation in gut homeostasis and inflammation. *Nat. Immunol.* **2016**, *17*, 1244–1251. [CrossRef] [PubMed]
6. Sela, D.A.; Garrido, D.; Lerno, L.; Wu, S.; Tan, K.; Eom, H.-J.; Joachimiak, A.; Lebrilla, C.B.; Mills, D.A. Bifidobacterium longum subsp. infantis ATCC 15697 α-fucosidases are active on fucosylated human milk oligosaccharides. *Appl. Environ. Microbiol.* **2012**, *78*, 795–803. [CrossRef] [PubMed]
7. Ojima, M.N.; Asao, Y.; Nakajima, A.; Katoh, T.; Kitaoka, M.; Gotoh, A.; Hirose, J.; Urashima, T.; Fukiya, S.; Yokota, A.; et al. Diversification of a Fucosyllactose Transporter within the Genus Bifidobacterium. *Appl. Environ. Microbiol.* **2022**, *88*, e01437-21. [CrossRef]
8. Marcobal, A.; Barboza, M.; Sonnenburg, E.D.; Pudlo, N.; Martens, E.C.; Desai, P.; Lebrilla, C.B.; Weimer, B.C.; Mills, D.A.; German, J.B.; et al. Bacteroides in the Infant Gut Consume Milk Oligosaccharides via Mucus-Utilization Pathways. *Cell Host Microbe* **2011**, *10*, 507–514. [CrossRef]
9. Ruiz-Moyano, S.; Totten, S.M.; Garrido, D.A.; Smilowitz, J.T.; Bruce German, J.; Lebrilla, C.B.; Mills, D.A. Variation in consumption of human milk oligosaccharides by infant gut-associated strains of bifidobacterium breve. *Appl. Environ. Microbiol.* **2013**, *79*, 6040–6049. [CrossRef]

10. Hooper, L.V.; Xu, J.; Falk, P.G.; Midtvedt, T.; Gordon, J.I. A molecular sensor that allows a gut commensal to control its nutrient foundation in a competitive ecosystem. *Proc. Natl. Acad. Sci. USA* **1999**, *96*, 9833–9838. [CrossRef]

11. Gotoh, A.; Katoh, T.; Sakanaka, M.; Ling, Y.; Yamada, C.; Asakuma, S.; Urashima, T.; Tomabechi, Y.; Katayama-Ikegami, A.; Kurihara, S.; et al. Sharing of human milk oligosaccharides degradants within bifidobacterial communities in faecal cultures supplemented with Bifidobacterium bifidum. *Sci. Rep.* **2018**, *8*, 13958. [CrossRef]

12. Tsukuda, N.; Yahagi, K.; Hara, T.; Watanabe, Y.; Matsumoto, H.; Mori, H.; Higashi, K.; Tsuji, H.; Matsumoto, S.; Kurokawa, K.; et al. Key bacterial taxa and metabolic pathways affecting gut short-chain fatty acid profiles in early life. *ISME J.* **2021**, *15*, 2574–2590. [CrossRef] [PubMed]

13. Bunesova, V.; Lacroix, C.; Schwab, C. Mucin Cross-Feeding of Infant Bifidobacteria and Eubacterium hallii. *Microb. Ecol.* **2018**, *75*, 228–238. [CrossRef] [PubMed]

14. Pacheco, A.R.; Munera, D.; Waldor, M.K.; Sperandio, V.; Ritchie, J.M. Fucose sensing regulates bacterial intestinal colonization. *Nature* **2012**, *492*, 113–117. [CrossRef] [PubMed]

15. Pickard, J.M.; Chervonsky, A.V. Intestinal Fucose as a Mediator of Host–Microbe Symbiosis. *J. Immunol.* **2015**, *194*, 5588–5593. [CrossRef]

16. Pickard, J.M.; Maurice, C.F.; Kinnebrew, M.A.; Abt, M.C.; Schenten, D.; Golovkina, T.V.; Bogatyrev, S.R.; Ismagilov, R.F.; Pamer, E.G.; Turnbaugh, P.J.; et al. Rapid fucosylation of intestinal epithelium sustains host-commensal symbiosis in sickness. *Nature* **2014**, *514*, 638–641. [CrossRef]

17. Furniss, R.C.D.; Clements, A. Regulation of the locus of enterocyte effacement in attaching and effacing pathogens. *J. Bacteriol.* **2018**, *200*, e00336-17. [CrossRef]

18. Gibson, G.R.; McCartney, A.L.; Rastall, R.A. Prebiotics and resistance to gastrointestinal infections. *Br. J. Nutr.* **2005**, *93*, S31–S34. [CrossRef]

19. Yamauchi, M.; Nishikawa, F.; Kimura, K.; Maezawa, Y.; Ohata, M.; Toda, G. Urinaryl-fucose: A new marker for the diagnosis of chronic liver disease. *Int. Hepatol. Commun.* **1993**, *1*, 222–227. [CrossRef]

20. Tsiafoulis, C.G.; Prodromidis, M.I.; Karayannis, M.I. Development of an amperometric biosensing method for the determination of L-fucose in pretreated urine. *Biosens. Bioelectron.* **2004**, *20*, 620–627. [CrossRef]

21. McGovern, D.P.B.; Jones, M.R.; Taylor, K.D.; Marciante, K.; Yan, X.; Dubinsky, M.; Ippoliti, A.; Vasiliauskas, E.; Berel, D.; Derkowski, C.; et al. Fucosyltransferase 2 (FUT2) non-secretor status is associated with Crohn's disease. *Hum. Mol. Genet.* **2010**, *19*, 3468–3476. [CrossRef]

22. Wang, H.; Walaszczyk, E.J.; Li, K.; Chung-Davidson, Y.-W.; Li, W. High-performance liquid chromatography with fluorescence detection and ultra-performance liquid chromatography with electrospray tandem mass spectrometry method for the determination of indoleamine neurotransmitters and their metabolites in sea lamprey plasma. *Anal. Chim. Acta* **2012**, *721*, 147–153. [CrossRef]

23. Chen, Y.; Yao, Q.; Zeng, X.; Hao, C.; Li, X.; Zhang, L.; Zeng, P. Determination of monosaccharide composition in human serum by an improved HPLC method and its application as candidate biomarkers for endometrial cancer. *Front. Oncol.* **2022**, *12*, 1014159. [CrossRef] [PubMed]

24. Suzuki, J.; Kondo, A.; Kato, I.; Hase, S.; Ikenaka, T. Assay of urinary free fucose by fluorescence labeling and high-performance liquid chromatography. *Clin. Chem.* **1992**, *38*, 752–755. [CrossRef] [PubMed]

25. Chaturvedi, P.; Warren, C.D.; Altaye, M.; Morrow, A.L.; Ruiz-Palacios, G.; Pickering, L.K.; Newburg, D.S. Fucosylated human milk oligosaccharides vary between individuals and over the course of lactation. *Glycobiology* **2001**, *11*, 365–372. [CrossRef] [PubMed]

26. Sakai, T.; Yamamoto, K.; Yokota, H.; Hakozaki-Usui, K.; Hino, F.; Kato, I. Rapid, simple enzymatic assay of free L-fucose in serum and urine, and its use as a marker for cancer, cirrhosis, and gastric ulcers. *Clin. Chem.* **1990**, *36*, 474–476. [CrossRef]

27. Erlandsson, P.G.; Åström, E.; Påhlsson, P.; Robinson, N.D. Determination of Fucose Concentration in a Lectin-Based Displacement Microfluidic Assay. *Appl. Biochem. Biotechnol.* **2019**, *188*, 868–877. [CrossRef]

28. Gao, Z.; Chen, S.; Du, J.; Wu, Z.; Ge, W.; Gao, S.; Zhou, Z.; Yang, X.; Xing, Y.; Shi, M.; et al. Quantitative analysis of fucosylated glycoproteins by immobilized lectin-affinity fluorescent labeling. *RSC Adv.* **2023**, *13*, 6676–6687. [CrossRef]

29. Thurl, S.; Müller-Werner, B.; Sawatzki, G. Quantifbation of individual oligosaccharide compounds from human milk using high-pH anion-exchange chromatography. *Anal. Biochem.* **1996**, *235*, 202–206. [CrossRef]

30. Shin, J.; Park, M.; Kim, C.; Kim, H.; Park, Y.; Ban, C.; Yoon, J.W.; Shin, C.S.; Lee, J.W.; Jin, Y.S.; et al. Development of fluorescent Escherichia coli for a whole-cell sensor of 2′-fucosyllactose. *Sci. Rep.* **2020**, *10*, 10514. [CrossRef]

31. Della Corte, D.; van Beek, H.L.; Syberg, F.; Schallmey, M.; Tobola, F.; Cormann, K.U.; Schlicker, C.; Baumann, P.T.; Krumbach, K.; Sokolowsky, S.; et al. Engineering and application of a biosensor with focused ligand specificity. *Nat. Commun.* **2020**, *11*, 4851. [CrossRef]

32. Barra, M.; Danino, T.; Garrido, D. Engineered Probiotics for Detection and Treatment of Inflammatory Intestinal Diseases. *Front. Bioeng. Biotechnol.* **2020**, *8*, 265. [CrossRef]

33. Sharma, S.; Kumar, R.; Jain, A.; Kumar, M.; Gauttam, R.; Banerjee, R.; Mukhopadhyay, J.; Tyagi, J.S. Functional insights into Mycobacterium tuberculosis DevR-dependent transcriptional machinery utilizing Escherichia coli. *Biochem. J.* **2021**, *478*, 3079–3098. [CrossRef] [PubMed]

34. Pierre, M.; Julien, P.; Fabien, L. Functional Analysis of Deoxyhexose Sugar Utilization in Escherichia coli Reveals Fermentative Metabolism under Aerobic Conditions. *Appl. Environ. Microbiol.* **2021**, *87*, e00719-21. [CrossRef]

35. Kim, J.; Cheong, Y.E.; Jung, I.; Kim, K.H. Metabolomic and transcriptomic analyses of Escherichia coli for efficient fermentation of L-Fucose. *Mar. Drugs* **2019**, *17*, 82. [CrossRef] [PubMed]
36. Bartkus, J.M.; Mortlock, R.P. Isolation of a mutation resulting in constitutive synthesis of L-fucose catabolic enzymes. *J. Bacteriol.* **1986**, *165*, 710–714. [CrossRef] [PubMed]
37. Chen, Y.-M.; Zhu, Y.; Lin, E.C.C. The organization of the fuc regulon specifying l-fucose dissimilation in Escherichia coli K12 as determined by gene cloning. *Mol. Gen. Genet. MGG* **1987**, *210*, 331–337. [CrossRef] [PubMed]
38. Zhu, Y.; Lin, E.C. A mutant crp allele that differentially activates the operons of the fuc regulon in Escherichia coli. *J. Bacteriol.* **1988**, *170*, 2352–2358. [CrossRef] [PubMed]
39. Medina, D.A.; Pinto, F.; Ovalle, A.; Thomson, P.; Garrido, D. Prebiotics mediate microbial interactions in a consortium of the infant gut microbiome. *Int. J. Mol. Sci.* **2017**, *18*, 2095. [CrossRef]
40. Pédelacq, J.-D.; Cabantous, S.; Tran, T.; Terwilliger, T.C.; Waldo, G.S. Engineering and characterization of a superfolder green fluorescent protein. *Nat. Biotechnol.* **2006**, *24*, 79–88. [CrossRef]
41. Madar, D.; Dekel, E.; Bren, A.; Alon, U. Negative auto-regulation increases the input dynamic-range of the arabinose system of Escherichia coli. *BMC Syst. Biol.* **2011**, *5*, e111. [CrossRef]
42. Alon, U. *An Introduction to Systems Biology: Design Principles of Biological Circuits*, 1st ed.; Chapman and Hall: London, UK; CRC: Boca Raton, FL, USA, 2006.
43. Mangan, S.; Itzkovitz, S.; Zaslaver, A.; Alon, U. The Incoherent Feed-forward Loop Accelerates the Response-time of the gal System of Escherichia coli. *J. Mol. Biol.* **2006**, *356*, 1073–1081. [CrossRef] [PubMed]
44. Aidelberg, G.; Towbin, B.D.; Rothschild, D.; Dekel, E.; Bren, A.; Alon, U. Hierarchy of non-glucose sugars in Escherichia coli. *BMC Syst. Biol.* **2014**, *8*, e133. [CrossRef]
45. Beisel, C.L.; Storz, G. The Base-Pairing RNA Spot 42 Participates in a Multioutput Feedforward Loop to Help Enact Catabolite Repression in Escherichia coli. *Mol. Cell* **2011**, *41*, 286–297. [CrossRef]
46. Chen, Y.M.; Tobin, J.F.; Zhu, Y.; Schleif, R.F.; Lin, E.C. Cross-induction of the L-fucose system by L-rhamnose in Escherichia coli. *J. Bacteriol.* **1987**, *169*, 3712–3719. [CrossRef] [PubMed]
47. Gottesman, S. Proteases and their targets in Escherichia coli. *Annu. Rev. Genet.* **1996**, *30*, 465–506. [CrossRef] [PubMed]
48. Coffey, J.W.; Miller, O.N.; Sellinger, O.Z. The Metabolism of l-Fucose in the Rat. *J. Biol. Chem.* **1964**, *239*, 4011–4017. [CrossRef]
49. Stepien, K.M.; Ciara, E.; Jezela-Stanek, A. Fucosidosis—Clinical Manifestation, Long-Term Outcomes, and Genetic Profile—Review and Case Series. *Genes* **2020**, *11*, 1383. [CrossRef]
50. Wang, H.; Zhang, X.; Peng, Y.; Pan, B.; Wang, B.; Peng, D.H.; Guo, W. A LC-MS/MS method to simultaneously profile 14 free monosaccharides in biofluids. *J. Chromatogr. B Anal. Technol. Biomed. Life Sci.* **2022**, *1192*, 123086. [CrossRef]
51. Li, M.; Luo, Y.; Hu, M.; Li, C.; Liu, Z.; Zhang, T. Module-Guided Metabolic Rewiring for Fucosyllactose Biosynthesis in Engineered Escherichia coli with Lactose De Novo Pathway. *J. Agric. Food Chem.* **2022**, *70*, 14761–14770. [CrossRef]
52. Crost, E.H.; Tailford, L.E.; Le Gall, G.; Fons, M.; Henrissat, B.; Juge, N. Utilisation of Mucin Glycans by the Human Gut Symbiont Ruminococcus gnavus Is Strain-Dependent. *PLoS ONE* **2013**, *8*, e76341. [CrossRef]
53. Everard, A.; Belzer, C.; Geurts, L.; Ouwerkerk, J.P.; Druart, C.; Bindels, L.B.; Guiot, Y.; Derrien, M.; Muccioli, G.G.; Delzenne, N.M.; et al. Cross-talk between Akkermansia muciniphila and intestinal epithelium controls diet-induced obesity. *Proc. Natl. Acad. Sci. USA* **2013**, *110*, 9066–9071. [CrossRef] [PubMed]

MDPI

St. Alban-Anlage 66

4052 Basel

Switzerland

www.mdpi.com

Biosensors Editorial Office

E-mail: biosensors@mdpi.com

www.mdpi.com/journal/biosensors